# CHEMISTRY IN THE LABORATORY

## Second Edition

**John J. Alexander**
University of Cincinnati

**Margaret J. Steffel**
The Ohio State University Marion Campus

BURGESS INTERNATIONAL GROUP INC
Bellwether Press Division

ISBN 0-8087-0026-X



Printed in the United States of America.
J I H G F E D C B A

Address orders to:

Burgess International Group, Inc.
7110 Ohms Lane
Edina, MN 55435
Telephone 612/831-1344
Telex 29-0458

Bellwether Press
A Division of Burgess International Group, Inc.

# Preface

*Chemistry in the Laboratory* was written for use in a one-year general chemistry course for students majoring in natural sciences, engineering, and medicine-related areas. There are thirty-one experiments in the manual, which would require forty three-hour laboratory periods if each experiment were allotted the suggested time; consequently, the instructor has flexibility in choosing experiments. Nearly every experiment has distinct sections and can be shortened for a two-hour period; in the instructor's manual we suggest such variations.

We encourage students to work as independently as possible, without a partner (except for one experiment, #8, that requires a two-person team for precise use of equipment) and without excessive consultation with an instructor. Therefore, we have tried to make the Procedure sections as complete and unambiguous as possible. However, we have avoided procedures that a young friend calls "witnessing chemistry": "If you do this, what you will see happen is ... Now do this and record your observations." Virtually every experiment requires the student to make observations and comparisons, measure properties, or draw conclusions that cannot be known in advance without doing the laboratory work. Of course this is accomplished partly by having the students work with unknowns; in addition to the four qualitative analysis experiments (#28, 29, 30, 31), five unknowns must be identified (#2, 9, 27) and some particular physical or chemical property of eight others (#11, 12, 13, 15, 16, 23, 24) must be determined.

Chemistry, like other natural sciences, was *originally* only qualitative, descriptive, and the ability to make discriminating qualitative judgments is still an essential skill. To assist students in developing this skill, eight experiments (#6, 9, 14, 17, 18, 21, 24, 26) emphasize comparing and contrasting materials or reactions that have some characteristics in common.

But *modern* chemistry, like other modern physical sciences, is also ubiquitously quantitative. It is essential that students learn to make quantitative measurements and to use measured values in calculations with proper attention to precision and accuracy. Therefore, even the experiments that emphasize qualitative observations and comparisons have some quantitative aspects; in addition to the qualitative analysis experiments, there is only one other experiment (#9) that does not require some sort of calculations. Eight of the extensively quantitative experiments require students to graph their raw data or some quantity calculated from their data, and then to draw some conclusion or make some final calculation from their graphs (#4, 8, 10, 13, 14, 15, 22, 27); in three of these experiments (#4, 8, 15) students make graphs while still in the laboratory and use the graphs to determine what further laboratory work is necessary.

We have tried to make every experiment as safe as possible. To draw attention to materials or procedures that can be dangerous, especially if used incorrectly, we present each of these in bold-face type. We have avoided the use of materials known to present health hazards. The only chlorinated hydrocarbon that is required is dichloromethane in one experiment (#6). In the qualitative analysis, ammonium sulfide is used, rather than carcinogenic thioacetamide, to precipitate sulfides. To avoid the paperwork and security measures that use of ethanol entails, it is not required for any experiment. Every experiment includes brief instructions about proper disposal of waste materials to avoid risk to other students and maintenance personnel, and to minimize avoidable pollution.

We thank our former students for their suggestions and mistakes that helped us make continual improvements in these experiments. We welcome comments and suggestions from future students and from our colleagues.

J.J.A and M.J.S.

# TABLE OF CONTENTS


| | | |
|---|---|---|
| Introduction | | 1 |
| Laboratory Safety and Protocol | | 3 |
| Laboratory Equipment and Techniques | | 5 |
| Experiment 1 | Measurement of Volume and Mass | 21 |
| Experiment 2 | Physical Properties of Substances | 29 |
| Experiment 3 | Compound Stoichiometry | 41 |
| Experiment 4 | Reaction Stoichiometry | 47 |
| Experiment 5 | Synthesis of an Alum, $KAl(SO_4)_2 \cdot 12H_2O$ | 53 |
| Experiment 6 | A Simple Covalent Compound: Synthesis and Properties | 59 |
| Experiment 7 | Synthesis of $[Co(NH_3)_6]Cl_3$ | 67 |
| Experiment 8 | Line Spectra of Elements | 75 |
| Experiment 9 | Salt Identification Using Visible and Infrared Spectra | 85 |
| Experiment 10 | Enthalpy of Reaction | 103 |
| Experiment 11 | Gravimetric Determination of Iron | 113 |
| Experiment 12 | Molar Weight by Vapor Density | 121 |
| Experiment 13 | Vapor Pressure and Heat of Vaporization | 129 |
| Experiment 14 | Molecular Interactions in Solution | 135 |
| Experiment 15 | Molar Weight by Freezing Point Depression | 147 |
| Experiment 16 | Acid–Base Titration | 157 |
| Experiment 17 | Redox Reactions | 165 |
| Experiment 18 | Electrochemistry: Voltaic Cells | 169 |
| Experiment 19 | Electrochemistry: Electrolytic Cells | 179 |
| Experiment 20 | A Complete Rate Law | 185 |
| Experiment 21 | Chemical Equilibrium | 193 |
| Experiment 22 | Spectrophotometric Determination of an Equilibrium Constant | 205 |

| | | |
|---|---|---|
| Experiment 23 | Antacids | 213 |
| Experiment 24 | Acidic and Basic Salts | 219 |
| Experiment 25 | Solubility of a Slightly Soluble Electrolyte | 231 |
| Experiment 26 | Metal Complexes With Monodentate and Chelating Ligands | 237 |
| Experiment 27 | Synthesis and Identification of Coordination Compounds Using Spectrophotometry | 247 |
| Introduction II: | Inorganic Qualitative Analysis | 261 |
| Experiment 28 | Qualitative Analysis of Cation Group I | 267 |
| Experiment 29 | Qualitative Analysis of Cation Group II | 273 |
| Experiment 30 | Qualitative Analysis of Cation Group III | 287 |
| Experiment 31 | Qualitative Analysis of Anions | 297 |
| Appendix I | Sample Laboratory Report | 303 |
| Appendix II | Scientific Measurement | 305 |
| Appendix III | Graphing Experimental Data | 309 |
| Tables | | |
| | SI (Systeme Internationale) Units | 316 |
| | Conversion Factors for Commonly Used Metric and English Units | 317 |
| | Periodic Table of the Elements with Element Names | 318 |

# INTRODUCTION

## Why Have a Chemistry Laboratory?

In the chemistry course you are taking you will spend time each week working in a chemistry laboratory. If you do not intend to become a chemist you might wonder why you must do chemistry laboratory work. After all, you don't have to write a novel when you take a literature course. Why isn't just learning about chemistry sufficient? Why must you actually carry out experiments in a laboratory? There are at least three important reasons why students who are preparing for careers based on natural sciences, engineering, and medicine must have personal experience with scientific experimentation.

First, chemistry, like other natural sciences, is based on experimentation. Scientific conclusions are reached by studying the results of experimental observations, not by making plausible conjectures. To illustrate this point consider a simple example. Someone might wonder whether iron(II) chloride will react with iodine. A balanced equation for a possible reaction can easily be written:

$$FeCl_2 + I_2 \rightarrow FeI_2 + Cl_2 \qquad \text{(i-l)}$$

But does this equation represent the way these materials actually behave? The way to find out is to try the reaction to see whether it occurs; it should be obvious if it does occur because iodine is a purple solid and chlorine is a yellow gas. (It doesn't occur). Information in your textbook resulted from experiments by many scientists. What you do in the laboratory should help you realize that scientific knowledge is based on experimentation rather than just on reasoning.

Second, you should develop manipulative skills in doing experiments. Good laboratory technique becomes increasingly important as you progress to more advanced science courses, and it will be essential when you practice your profession.

Third, you need to learn to keep complete and accurate records of your experimental observations and measurements, and to communicate, both orally and in writing, with others about your results and the conclusions you draw from them. These skills are developed by your laboratory notebook and by the reports you hand in to your instructor.

In keeping with these educational goals each of the experiments in this manual was developed with at least one of the following criteria in mind:

1. To show the relationship between experimental results and chemical theory. Sometimes this involves confirming or applying results that you read about in your textbook; at other times you design simple experiments to test theoretical possibilities.

2. To teach basic laboratory techniques.

3. To help you learn to interpret and work with numerical data.

## Your Laboratory Responsibilities

1. It is impossible to overemphasize the importance of thorough preparation prior to the laboratory period. If you are to do your work correctly and are to complete it within the allotted time, *you must be prepared* when you come to class. This means that you must *study* the experiment in advance; try to visualize the materials you will use and each of the actions you will perform. The

laboratory instructor will answer questions that you have, but this cannot help you if you are not familiar enough with the assigned work to know what questions to ask.

2. You will be expected to keep a laboratory notebook. Reserve a notebook exclusively for chemistry laboratory; take it to each laboratory session and to any other class meeting in which laboratory work will be discussed.

Several kinds of entries are appropriate in your notebook. First, you might need to prepare a work outline or set up a data table before you go to the laboratory, or you might have to calculate the weight or volume that you will need of a particular reagent. Any preparatory information of this kind should be recorded in your notebook.

Second, you should record your qualitative observations describing what happens and what materials look like as you carry out an experiment. For example, the colors of starting materials and of products, the texture of a precipitate, and whether a reaction evolves heat should be recorded. Also, you should record any mistakes or difficulties that might cause you to distrust your results in some part of an experiment. For example, using the wrong concentration of a solution, spilling some of a carefully measured substance, or using the wrong amount of a material might affect your results. However, if you know that you make an error and you record it so that you will remember it, you might be able to account for inaccurate results.

Third, you should record all numerical data in your notebook; label all numbers with the correct units. The purpose of keeping a laboratory notebook is not to provide a work of art for your instructor to admire; do not record data on scraps of paper and then copy them, perhaps more neatly, into your notebook at a later time. Have your notebook with you at all times in the laboratory and completely record data as you obtain them. If, for any reason, you decide you are not going to use a particular piece of data you can simply draw a line through it. But don't run the risk of losing necessary data just to keep your records neat. Many types of practicing technicians and researchers record data and observations this way; you should learn now to do this properly.

Fourth, most quantitative experiments require calculations using measurements made in the laboratory. These should be shown clearly and completely in your notebook, from initial equation through step-by-step solutions to final answer.

3. In all your laboratory work you should try to develop the ability to observe what *actually* happens in an experiment, not what you think *should* happen. Most scientific experiments are designed with some notion of what will occur. However, if expectations are not realized it does not necessarily mean the experiment has failed. The task then is to try to explain what happened without being disappointed that it wasn't what was expected. You should do the same: observe what happens and try to understand and explain it.

4. The content of lecture and laboratory sessions will probably not be completely correlated. Sometimes ideas already discussed in lectures will be applied to a laboratory problem. However, some experiments might involve material that has not yet been covered in a lecture. Perhaps this seems like "putting the cart before the horse", but you should be able to learn new chemistry in the laboratory just as you do from a lecture or book.

5. The experiments in this manual require you to submit written reports. Most experiments have a Report Sheet to guide you in preparing your report. The Results and Discussion section at the end of an experiment gives information on how to treat your data when filling out the Report Sheet. Questions that should stimulate your thinking about the experiment are also included. An example of a report is presented in Appendix I.

# LABORATORY SAFETY AND PROTOCOL

## Safety

One way in which chemistry laboratory differs from, say, a philosophy or calculus course is that in the laboratory you can be physically injured if you are careless. To avoid accidents that might cause injury to yourself or others, you must constantly pay attention to safety precautions while you are in the laboratory. The following list indicates areas where caution is essential.

1. **Wear suitable clothing**. Shorts leave large areas of skin exposed to potential chemical or thermal burns. Long full sleeves might drag in chemicals or catch on equipment; both the sleeves and the experiment could be ruined. Never go barefooted in a chemistry laboratory; there might be broken glass or spilled chemicals on the floor.

2. **Wear safety goggles** all the time you are in the laboratory. Even if you are careful not to splash chemicals into your eyes, someone near you might accidentally do so. Goggles pushed up onto your forehead or left in your desk offer your eyes no protection. It is not advisable to wear contact lenses in a chemistry laboratory unless special safety goggles are provided.

3. **Know the locations and proper use of safety equipment**. This might include eye shower, overhead shower, fire extinguisher, and fire blanket.

4. **Do not eat, drink, or smoke in the chemistry laboratory**. Do not bring any food or beverages into the laboratory. Never taste any chemicals.

5. **Work under a hood when so instructed**. Some experiments involve malodorous, irritating, or potentially harmful substances; a hood draws their vapors out of the laboratory.

6. **Always pour acid into water, not water into acid** when diluting concentrated acid solutions or mixing them with other materials.

7. **Avoid cutting yourself**. Discard cracked or chipped glassware. Fire-polish rough ends of glass rods and tubing. Use a lubricant and a towel when inserting tubing into a stopper.

8. **Report any injury to your instructor**, no matter how minor it might seem to you. Even a small scratch can be serious if chemicals get into it.

9. **Dispose of waste chemicals as directed**. Usually your instructor will tell you to place waste chemicals in specific waste containers or to wash them down a sink drain with plenty of running water. When in doubt, ask.

10. **Never attempt an unauthorized experiment**. The procedures in this manual have been designed with your safety in mind. Don't risk injury to yourself or others by mixing chemicals randomly or by setting up unauthorized equipment or reactions.

## Protocol

Following a few general rules of procedure in the laboratory can make your work and the work of others around you much more orderly and pleasant.

1. **Be neat**. This is especially important if your laboratory is crowded. Keep your materials within the work space allotted to you. Put away or dispose of materials you no longer need so that they do not take up space or get in the way unnecessarily.

2. **Don't take shared reagent bottles to your desk**. Another student's work might be needlessly delayed by a search for a missing reagent. The proper way to obtain a reagent from the reagent shelf is to take a beaker or other rigid, nonpermeable container to the reagent shelf and pour the required amount of material into it. If you take a shared reagent bottle to a balance to weigh out the required amount of material, return the bottle to the reagent shelf as soon as you complete the weighing.

3. **Don't put anything into shared reagent bottles**. This is to eliminate the possibility of contamination. Always pour liquids and solids from bottles; do not use medicine droppers, pipets, spatulas, or any other objects to transfer chemicals from shared reagent bottles.

4. **Don't return leftover chemicals to shared reagent bottles**. A chemical might be returned to the wrong bottle or a contaminated chemical might be returned, thereby contaminating the entire contents of a reagent bottle. This can be avoided if you dispose of any excess material as your instructor directs.

5. **Don't waste chemicals**. Large quantities of material and energy resources are used to make and purify the chemicals you use. Take what you need, but do not waste resources or unnecessarily increase your school's expenses by taking more of a chemical than you need. Also, remember that any excess chemical that must be discarded is a potential air, soil, or water pollutant. Plan ahead to avoid waste.

6. **Clean up**. You are responsible for cleaning up anything you spill in the reagent and balance areas. Thoroughly clean your own work area before you leave the laboratory. Plan your work so that you have enough time to clean up.

# LABORATORY EQUIPMENT AND TECHNIQUES

Many operations are performed so often in the laboratory that a special section describing them is provided here rather than in individual experiments. Some common laboratory equipment is illustrated and discussed on the following pages. Read this section now and study the relevant parts again before you carry out each experiment.

## A. Laboratory Burner

A mixture of natural gas and oxygen (from air) is burned in Bunsen-type burners. Figure 1 shows burners; if yours is not shown, its adjustments will be similar to those described here.

**Figure 1**

Bunsen-type burners: (left) a Bunsen burner; (right) a Tirril burner.

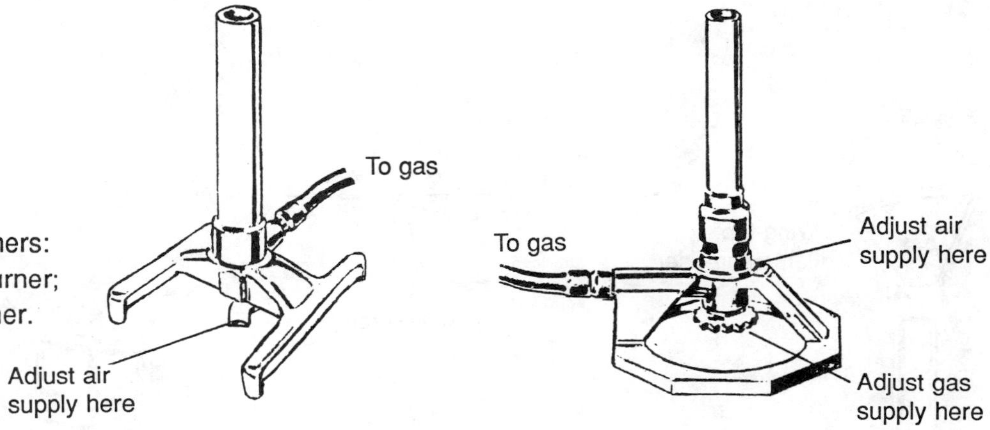


A Tirril burner has a needle valve under the barrel to control gas flow. To admit air, partially unscrew the barrel of the burner or change the position of the slots around the lower portion of the barrel. When the air intake is properly adjusted the flame should look like the one depicted in Figure 2; it will have a bright blue inner cone and a darker, less distinct outer cone. If insufficient air is mixed with the gas the flame will be long and yellow; if this happens, open the air intake until you have the proper blue flame. Too much air can cause the flame to "flash back" (burn inside the barrel) and it might even flare out the bottom of the barrel; if this happens, turn off the gas, decrease the air supply, and relight the burner.

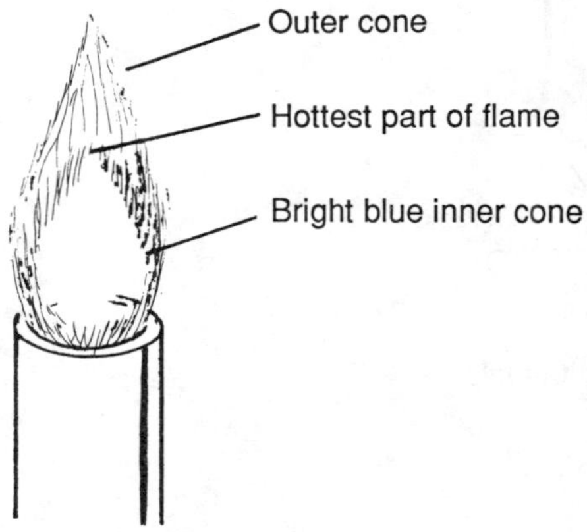


**Figure 2**

Flame of a properly adjusted burner.

# I
# Laboratory Equipment

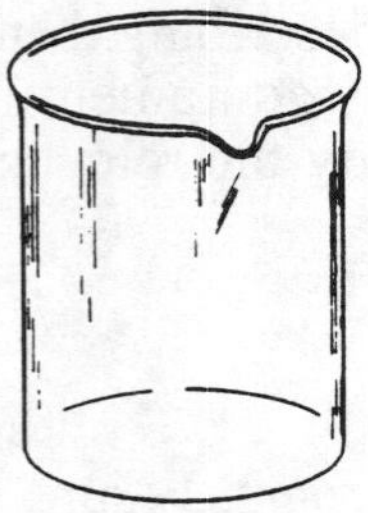

Beaker

Plastic wash bottle

Büchner funnel

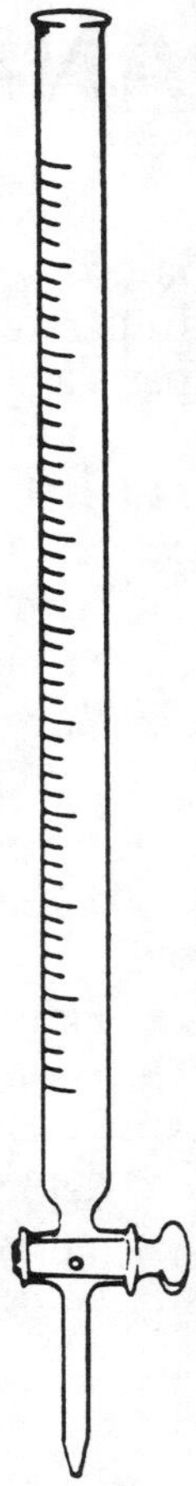

Buret

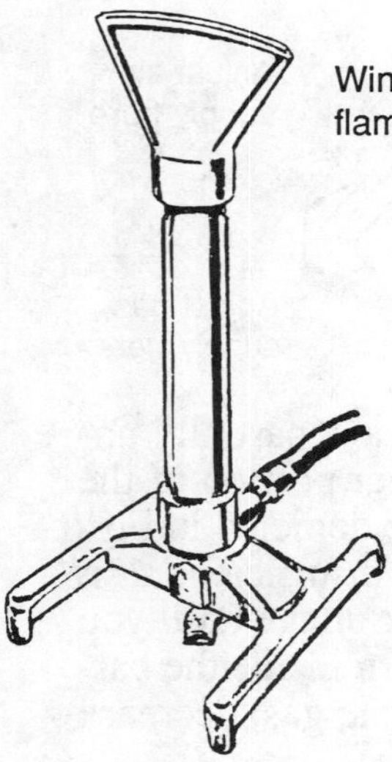

Wing-top flame spreader

Bunsen burner

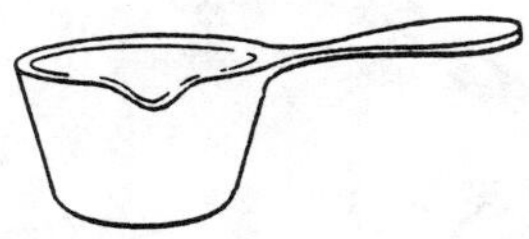

Casserole

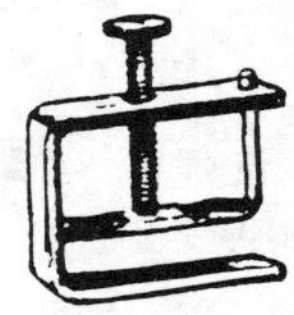

Screw clamp

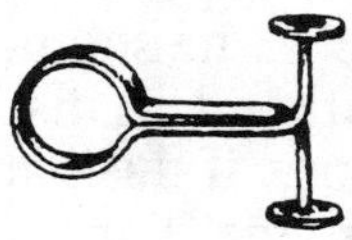

Pinchcock clamp

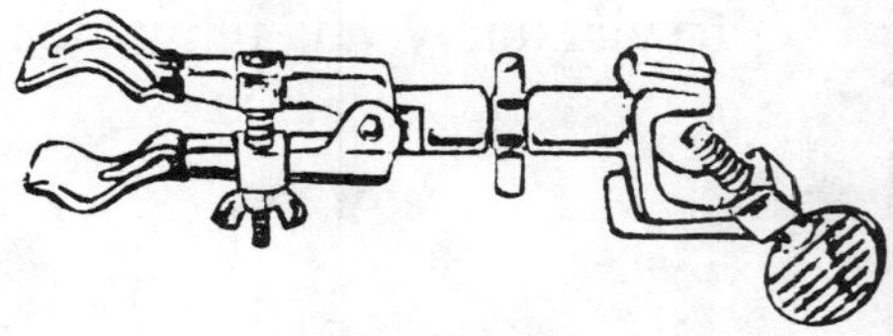

Universal clamp

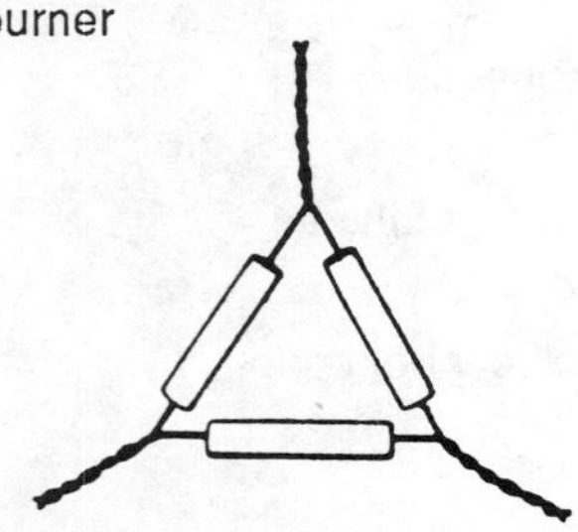

Clay triangle

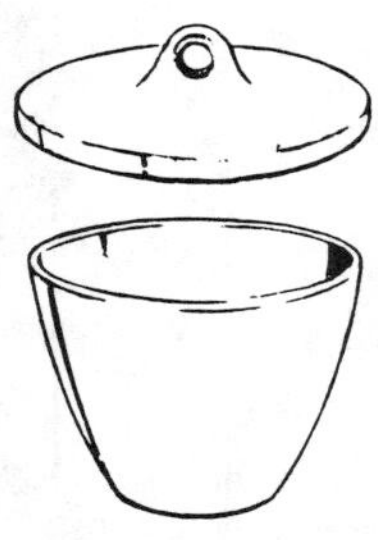

Crucible and lid

## Laboratory Equipment

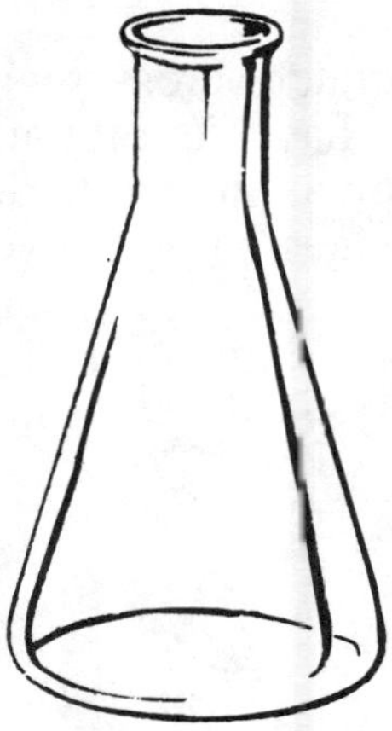

Erlenmeyer flask

Filter flask

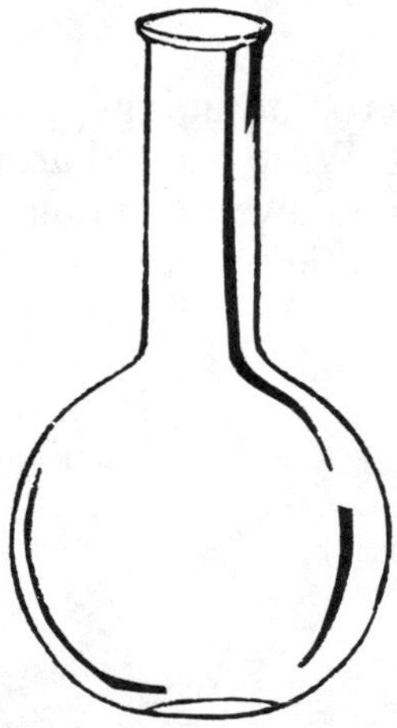

Florence flask

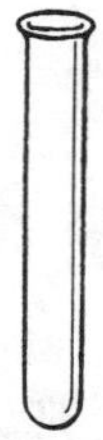

Test tube

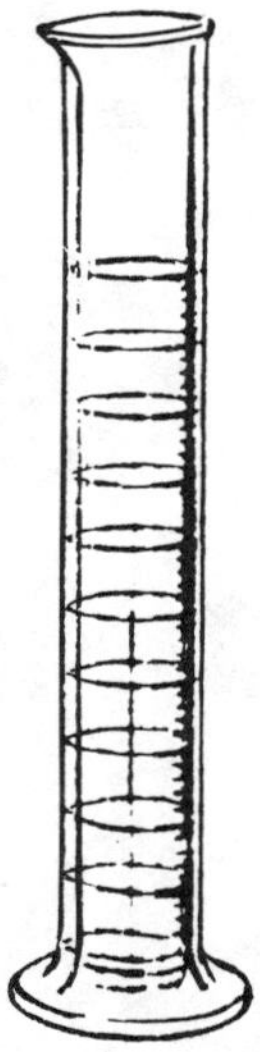

Graduated cylinder

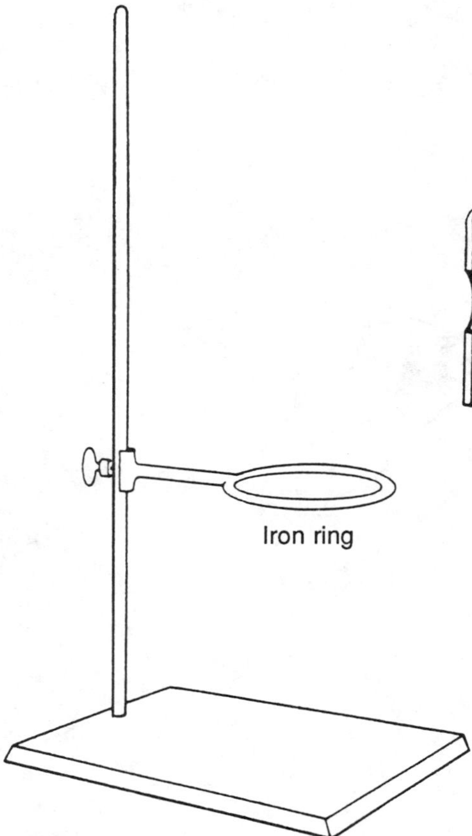

Iron ring

Ring stand

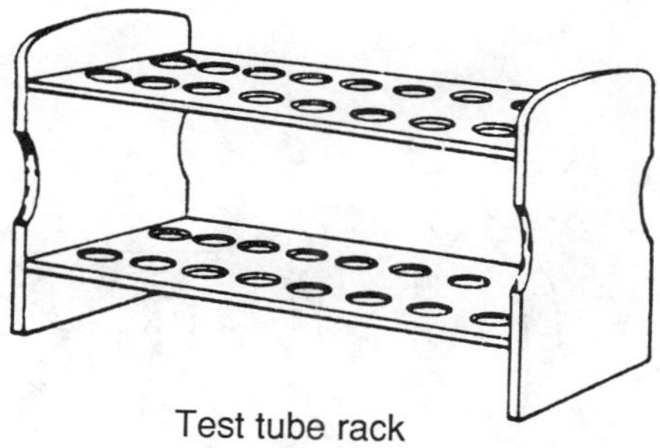

Test tube rack

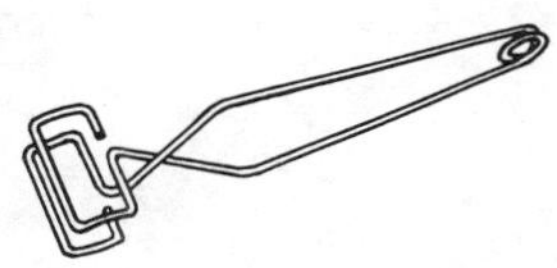

Test tube holder

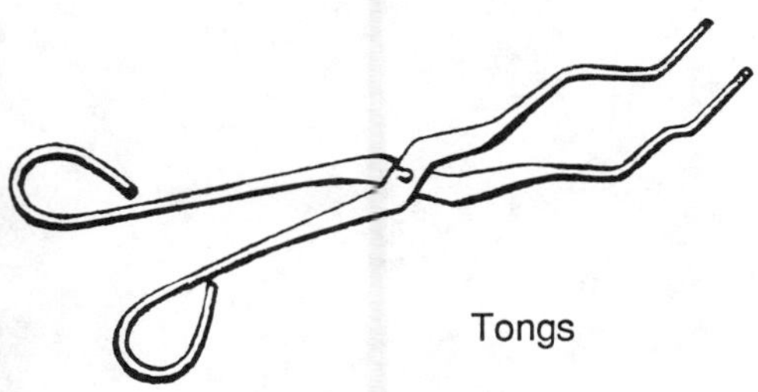

Tongs

Watch glass

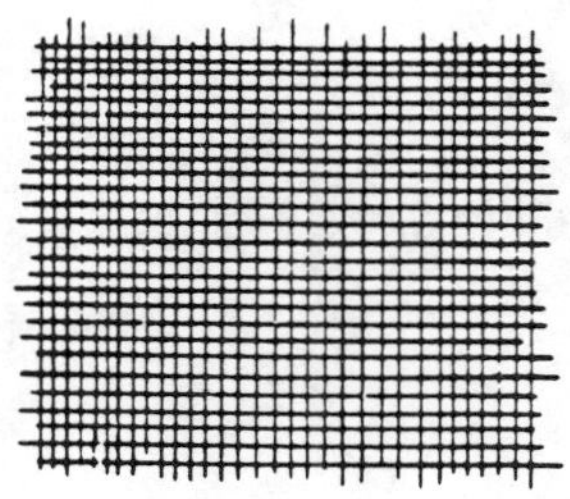

Wire gauze

## B. Cutting Glass

To cut soft glass rod or tubing, place it on the desk top and make a single scratch across it with a file or glass scorer. Wrap a towel around the glass and grasp the glass firmly; have your thumbs on the side of the glass away from the scratch. Break the tubing away from you as shown in Figure 3. Glass that is broken in this way will have sharp, and perhaps rough, edges on which you

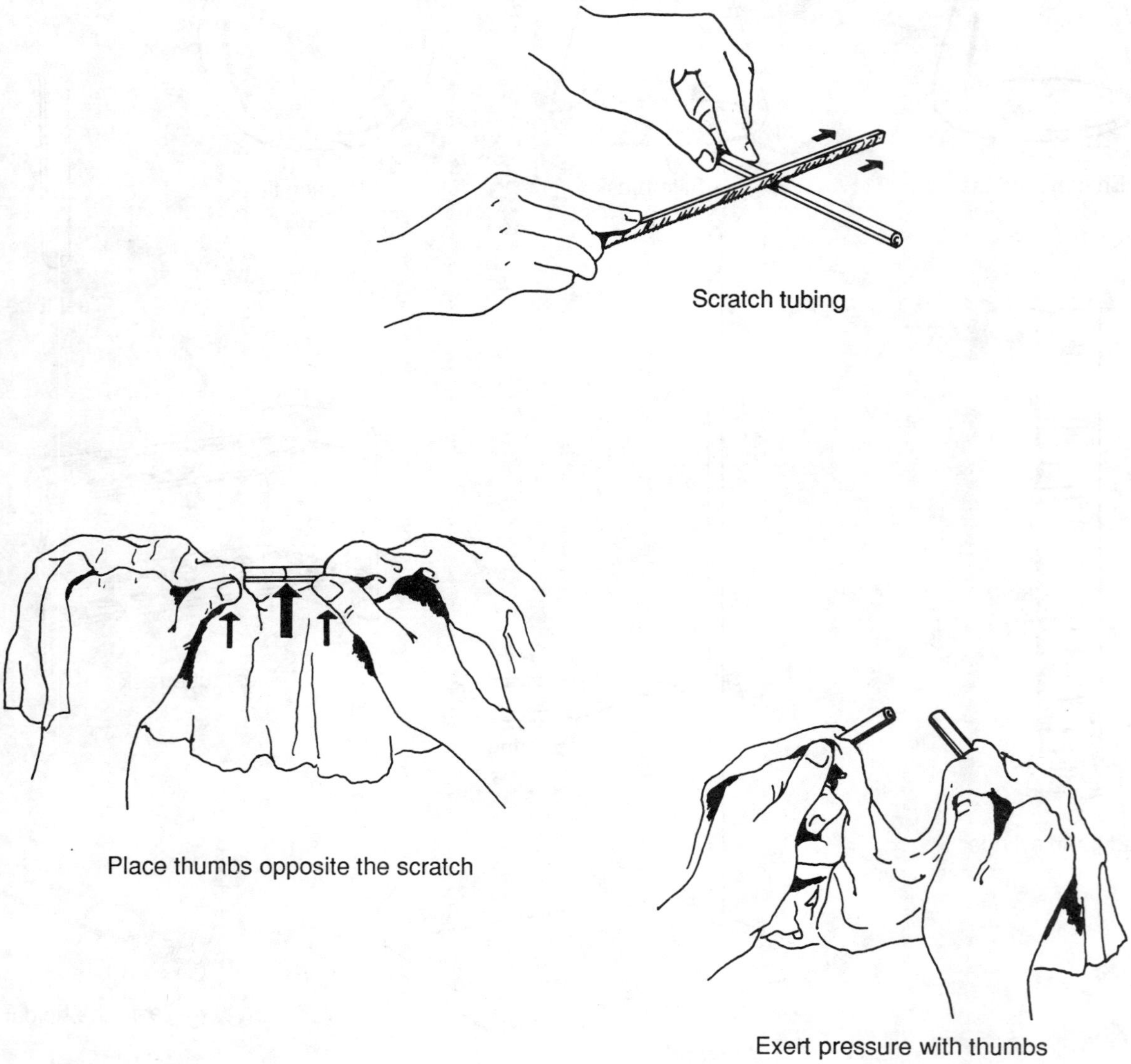


**Figure 3**

Breaking glass tubing.

could cut yourself; always take the few minutes that are required to fire-polish the cut ends. First, remove jagged edges by rubbing the end of the glass on wire gauze held over a waste can. Then hold the end of the glass in the tip of a burner flame as shown in Figure 4 and turn the glass so that it is heated evenly. After the glass is partially melted and the end is smooth, place the glass on wire gauze or a ceramic mat to cool.

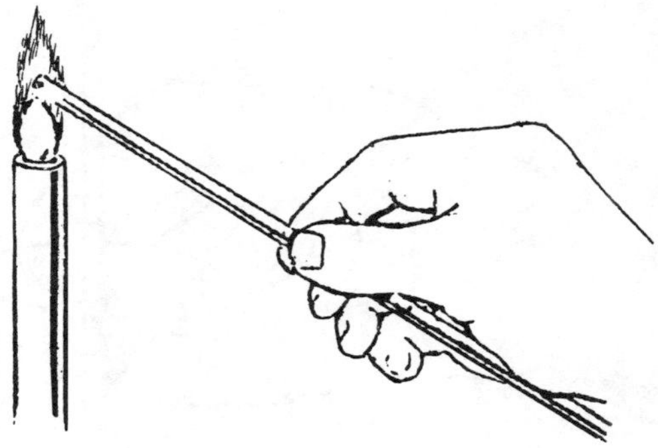

**Figure 4**

Fire-polishing glass tubing

## C. Inserting Glass Tubing into a Stopper

Some of the worst injuries in a first-year chemistry laboratory result from trying to force a thermometer, the stem of a glass funnel, or a piece of glass tubing into a stopper. If the glass breaks, it can be shoved forcefully into your hand causing a long, deep gash. To insert glass safely, wet it first with glycerine or soapy water. Then grasp the glass just above the area to be inserted into the stopper. Hold *both* the glass and the stopper with a towel and *twist* the stopper onto the glass (see Figure 5). Don't be reluctant to use a lot of glycerine or soap as a lubricant; it can always be washed off with water afterwards.

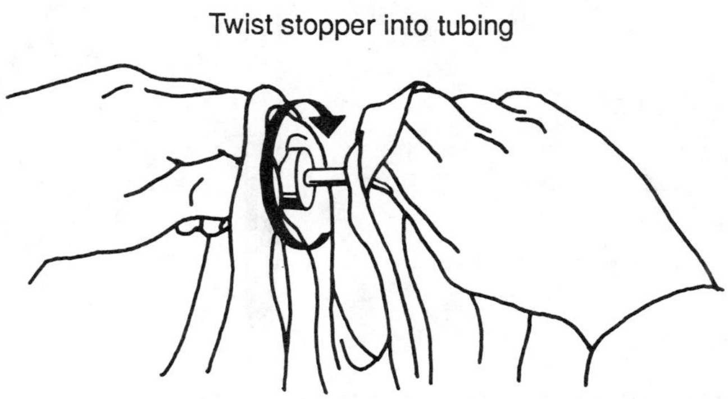


**Figure 5**

Inserting tubing into a stopper.

Use this same procedure, including the lubricant and towel, to insert plastic tubing into a stopper. Smooth plastic can't cut you, but it can cause a painful bruise if it is shoved forcefully against your bare hand.

In some experiments you must use a thermometer that is inserted into a rubber stopper and then the thermometer is held in place by a clamp around the rubber stopper. In these experiments you will use a rubber stopper that has been cut vertically from the outside of the stopper to a hold.

Hold the thermometer in one hand, force the cut open with your other hand, and pass the thermometer through the cut into the stopper. Be sure to hold the thermometer just barely above or below the part that goes into the stopper. (see Figure 6).

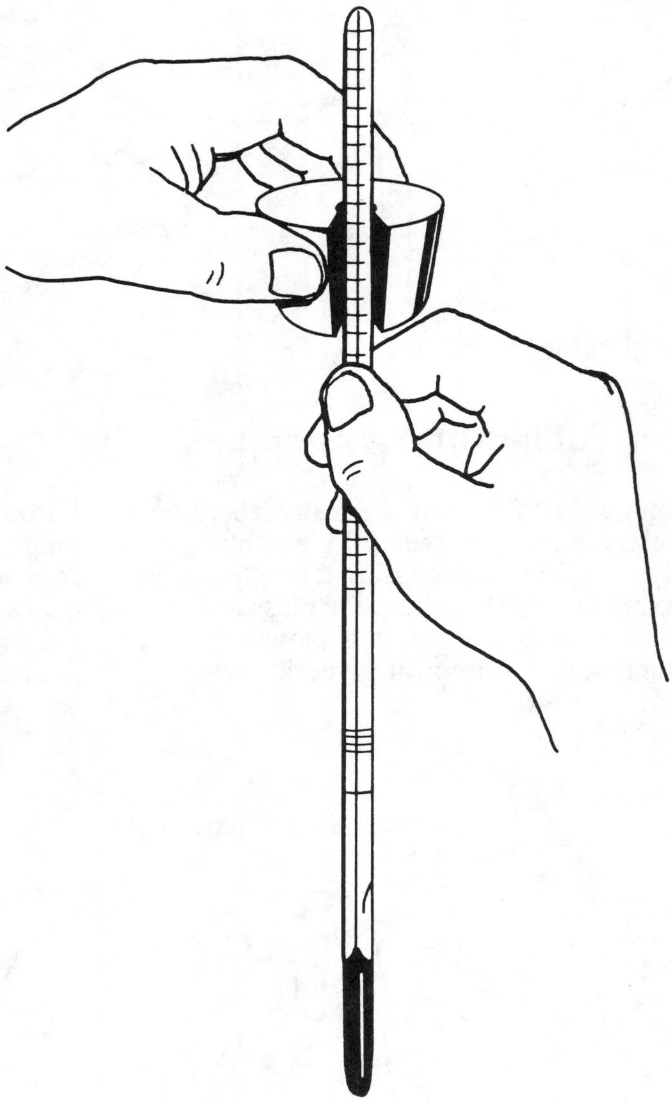

**Figure 6**

Inserting a thermometer into a slit rubber stopper.

## D. Pouring Chemicals

Many liquid reagents will be provided in glass-stoppered bottles. To pour from the bottle, first remove the stopper by grasping the top of it between your second and third fingers with your hand turned palm up. Then pick up the bottle with the same hand (see Figure 7).

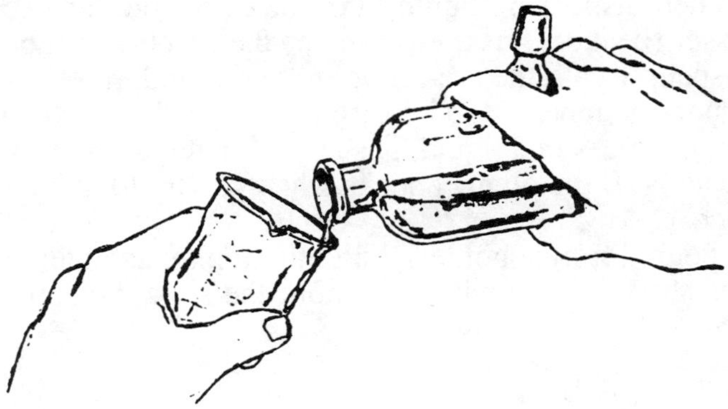

**Figure 7**

Pouring a liquid.

Advantages of this method are that the stopper is never placed where it can become contaminated and that your other hand is free to hold the container into which you are pouring the liquid. If a bottle has a stopper or lid that cannot be handled in this manner, take it off with one hand and pick up the bottle with the other hand.

Some containers of liquid will have bottle-top dispensers that can be pre-set to deliver a specific volume with reasonable accuracy (see Figure 8). If you are to set the dispenser volume control,

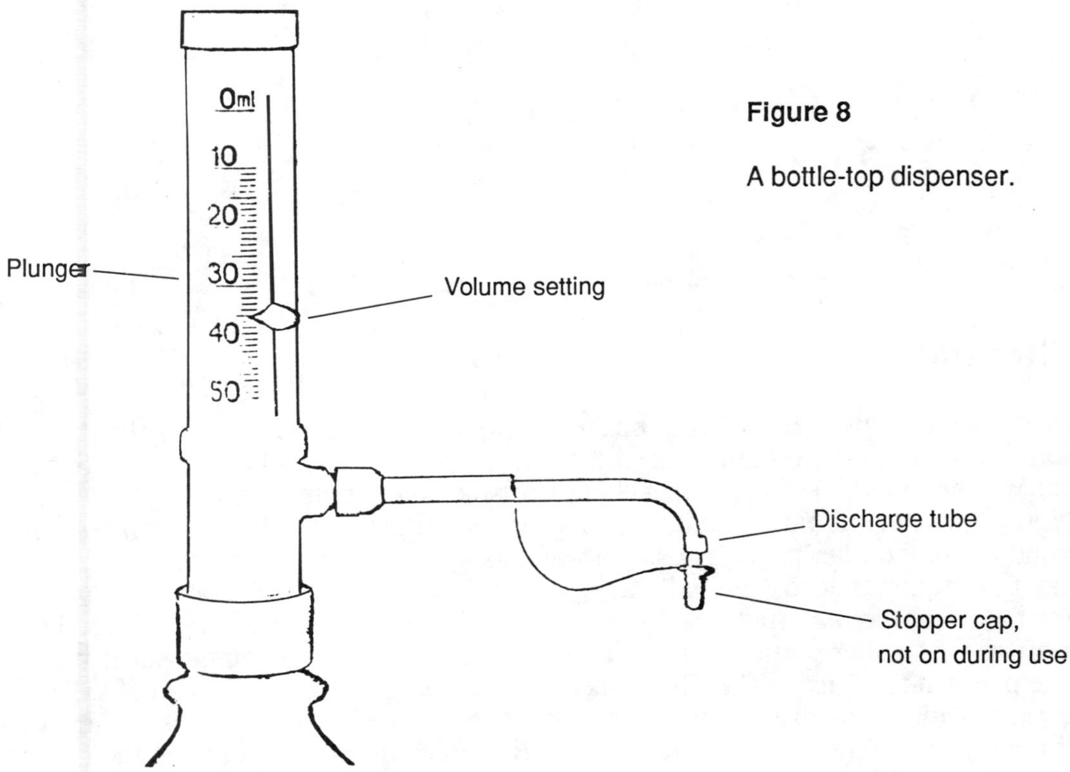


**Figure 8**

A bottle-top dispenser.

your instructor will provide instructions; if the dispenser has been pre-set for you, do not change the volume setting. When dispensing liquid, turn the discharge tube so that it is not pointing toward you; hold a vessel (beaker, flask, etc.) under the discharge tube (with its stopper cap removed). To fill the dispenser with the pre-set volume of liquid, raise the plunger (the external cylinder of the dispenser) to its upper "stop" position; do this slowly to avoid getting air bubbles in the liquid. To dispense the measured liquid, push the plunger all the way down; do this slowly and evenly to prevent splashing. If the plunger sticks when you try to move it, do not force it; ask your instructor for assistance.

A solid should be poured from a bottle by slightly tipping and then rotating the bottle (see Figure 9). If you must put the lid down while you pour the solid, be sure to place the *top* of the lid down on the desk.

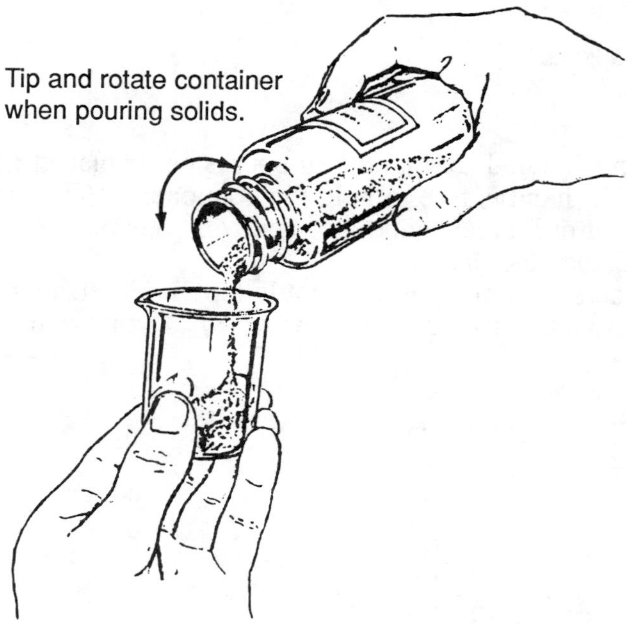


**Figure 9**

Pouring a solid.

## E. Filtration

A solid can be separated by filtration from a liquid in which it is not dissolved. Either gravity filtration or vacuum filtration can be used. Both methods leave the solid on a piece of filter paper in a funnel and the liquid, the filtrate, in a container beneath the funnel. In *gravity filtration*, gravitational force draws the liquid through the funnel and the filter paper. In *vacuum filtration*, the liquid is drawn through the filter paper by using a water aspirator to create suction.

Gravity filtration is accomplished with a conical funnel. Fold a piece of filter paper in half; then fold one half over the other half leaving an angle of about 10° between the edges. Tear the outer corner off of the smaller portion of the filter paper (see Figure 10), open the paper into a cone, and place the paper in the funnel. The filter paper should be slightly below the rim of the funnel. Moisten the paper with a little distilled water (or other appropriate liquid) and press it firmly against the side of the funnel; having torn out the corner will enable you to form a seal at the fold of the filter cone to prevent air leakage behind the paper.

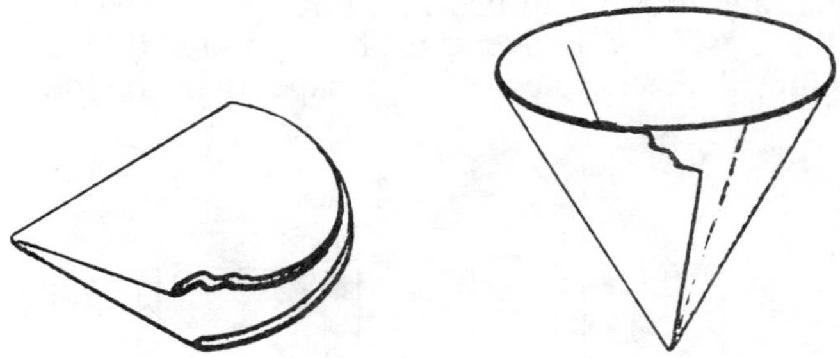

**Figure 10**

Folding filter paper into a cone.

Decant the supernatant liquid into the funnel by pouring it down a stirring rod (see Figure 11). When you have poured off as much of the liquid as you can using this method, transfer the solid using a stirring rod, spatula, or rubber policeman. Use a stream of water (or other appropriate liquid) from a wash bottle to wash the last of the solid into the funnel. Adding the solid last will save you time because most of the liquid will pass through the paper before the paper can become clogged with solid. During the filtration the funnel should never be more than about three-fourths full.

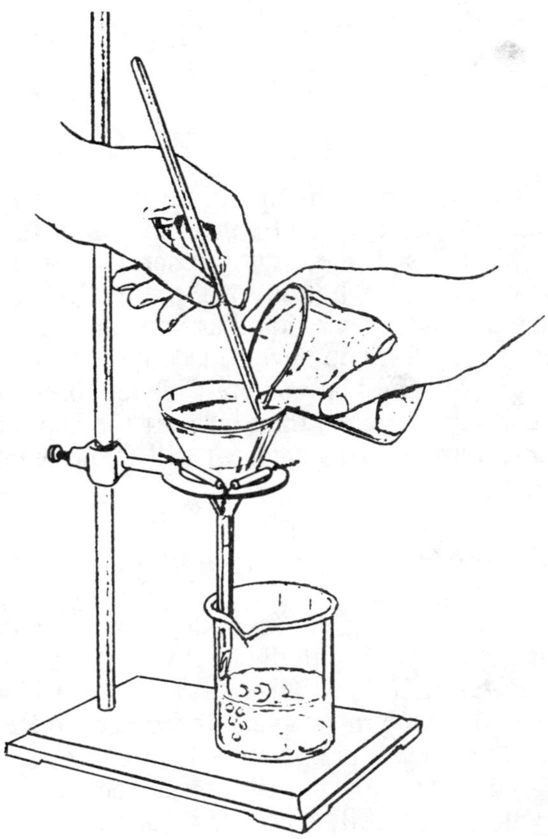

**Figure 11**

Gravity filtration.

Suction filtration requires a thick-walled flask with a sidearm, a Büchner funnel, thick-walled rubber tubing, and a bottle to use as a trap (all shown in Figure 12). The filter flask must be clamped so that it will not tip over if the heavy tubing flops around. After clamping the flask, insert the Büchner funnel. Place a piece of filter paper in the bottom of the funnel; use the size that covers

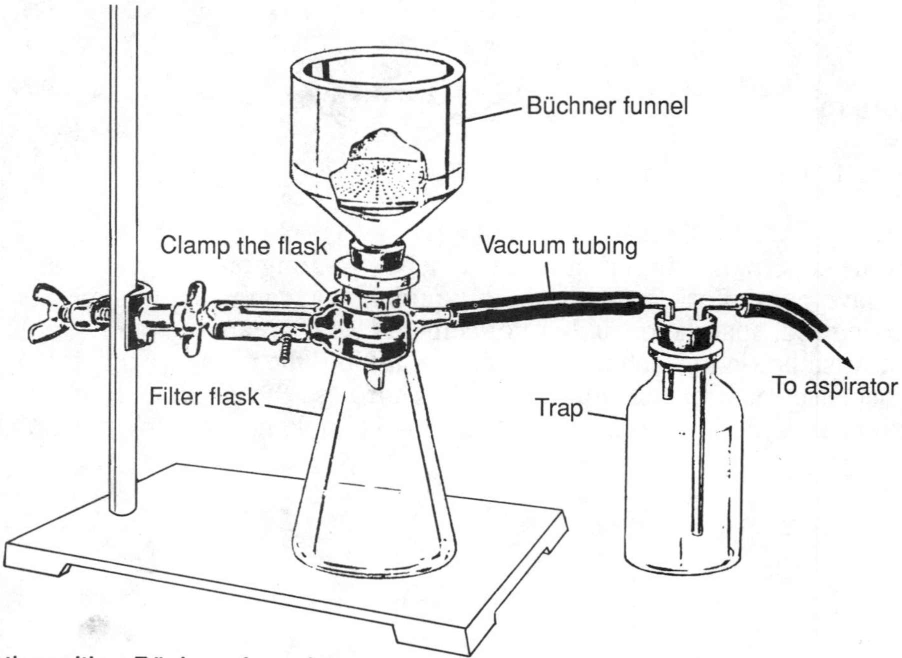


**Figure 12**

Vacuum filtration with a Büchner funnel.

the bottom of the funnel without extending up the sides of the funnel. Connect thick-walled tubing between the flask and trap and between the trap and aspirator; the trap will prevent aspirator water from backing up into the filter flask if the water pressure drops. Moisten the filter paper with a little liquid and press it tightly against the bottom of the funnel. Turn on the aspirator to draw liquid through the paper. Decant as much of the liquid as you can into the funnel and draw it through the funnel before adding the solid. After you have added all the solid, continue the use of the suction to dry the solid as much as possible. When you are finished, break the suction by disconnecting the tubing from the flask, trap, or aspirator; then turn off the aspirator. If you turn off the aspirator first, water might be drawn into the flask where it will dilute or contaminate the filtrate and might even pass into the funnel.

## F. Cleaning Glassware

For most laboratory uses it is sufficient to wash glassware with detergent solution and a brush and to rinse it with tap water and then with distilled water. If more thorough cleaning is necessary your instructor will give you directions. Glassware is clean if water flows smoothly down the sides of the vessel without collecting in droplets. If droplets collect repeat the cleaning.

If a vessel is to be used to transfer or contain a solution of a specific, accurately known concentration it must be rinsed with some of the solution to remove any remaining distilled water. To rinse a beaker or flask, pour a few milliliters of the solution into it; tip and roll the glassware so that the small amount of solution washes over the entire inside wall. Discard the used solution and repeat this rinsing two more times. For this final rinsing of a pipet or buret, put a small amount of solution into it; then hold it horizontally and roll it so that the solution washes over the entire inside wall. Do this three times.

High-accuracy glassware, that is, pipets, burets, and volumetric flasks, should be thoroughly rinsed with tap water and with distilled water after each use. Such glassware is always put away clean, but it is usually permissible to leave a little distilled water in it. A volumetric flask is put away with the stopper in place.

## G. Volume Measurements

You will often need to measure the volumes of liquids in your laboratory work. Several measuring vessels are available and the choice of which to use will depend on the required accuracy of the measurement.

**Beakers and Erlenmeyer flasks.** Some of these have approximate volume markings that can be used when rough estimates of volume are acceptable.

**Graduated cylinders.** These are used if reasonable accuracy is important. A 10-mL graduated cylinder has 1-mL graduation marks; these are numbered. Each 1 mL is divided into five small divisions of 0.2 mL each; these are not numbered. The position of the bottom of the liquid meniscus gives the volume reading. In Figure 13 the meniscus is between the 7.6 and 7.8 mL marks; the volume is more than 7.6 mL but less than 7.8 mL. Mentally divide the distance between the marks into smaller divisions and estimate where the meniscus lies along this scale. Because the 0.2-mL marks are so close together you will probably find it possible to divide the distance only into halves rather than into smaller division such as tenths; that is, you will probably be able to distinguish a distance that is about halfway between the two marks but not a smaller distance. Volumes from about 7.65 mL to 7.75 mL would be indistinguishable and the meniscus would look like it was about halfway between 7.6 mL and 7.8 mL; your reading would therefore be 7.7 mL. Since any true volume from 7.65 mL to 7.75 mL would look the same to you, there is an uncertainty of 0.1 mL in the volume reading of 7.7 mL, a possible error of 1 part in 77, or about 1%. With any other measuring device that has graduation markings you also make the best possible estimate between the markings, but there is always some possible error that cannot be eliminated even with practice, for it arises from the limitations of your eyes and of the equipment.

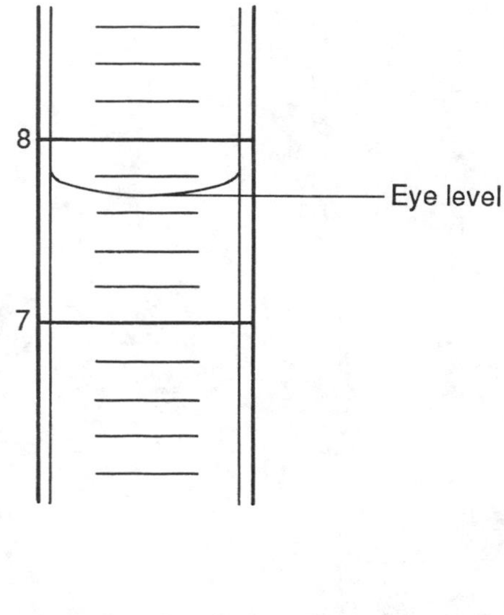


**Figure 13**

Measuring volume with a graduated cylinder.

**Volumetric Glassware.** This term includes pipets, burets, and volumetric flasks. These are constructed and calibrated to be much more accurate than graduated cylinders and are the most accurate glassware you will use to measure volumes.

**Pipets.** Figure 14 illustrates the use of a pipet. Draw liquid into the pipet by suction until the liquid is above the calibration line around the stem; use a bulb for suction. Then control the flow of liquid from the pipet by placing your first finger over the top of the pipet; allow liquid to drain out of the pipet until the bottom of the meniscus just touches the calibration line. Touch the tip of the pipet to the side of a waste beaker to remove any drop of clinging liquid and then touch it to the wall of the vessel into which the liquid is to be transferred. Remove your finger from the top of the pipet to permit the liquid to flow out of the pipet; hold the pipet vertically and tip the receiving vessel. When the flow has stopped, wait about 30 seconds for any remaining liquid to drain down the wall of the pipet. Then roll the pipet tip against the wall of the receiving vessel so that an additional drop flows out. Some liquid will remain in the pipet; that is all right, for the pipet is calibrated *to deliver* the proper volume of liquid when this small amount is retained.

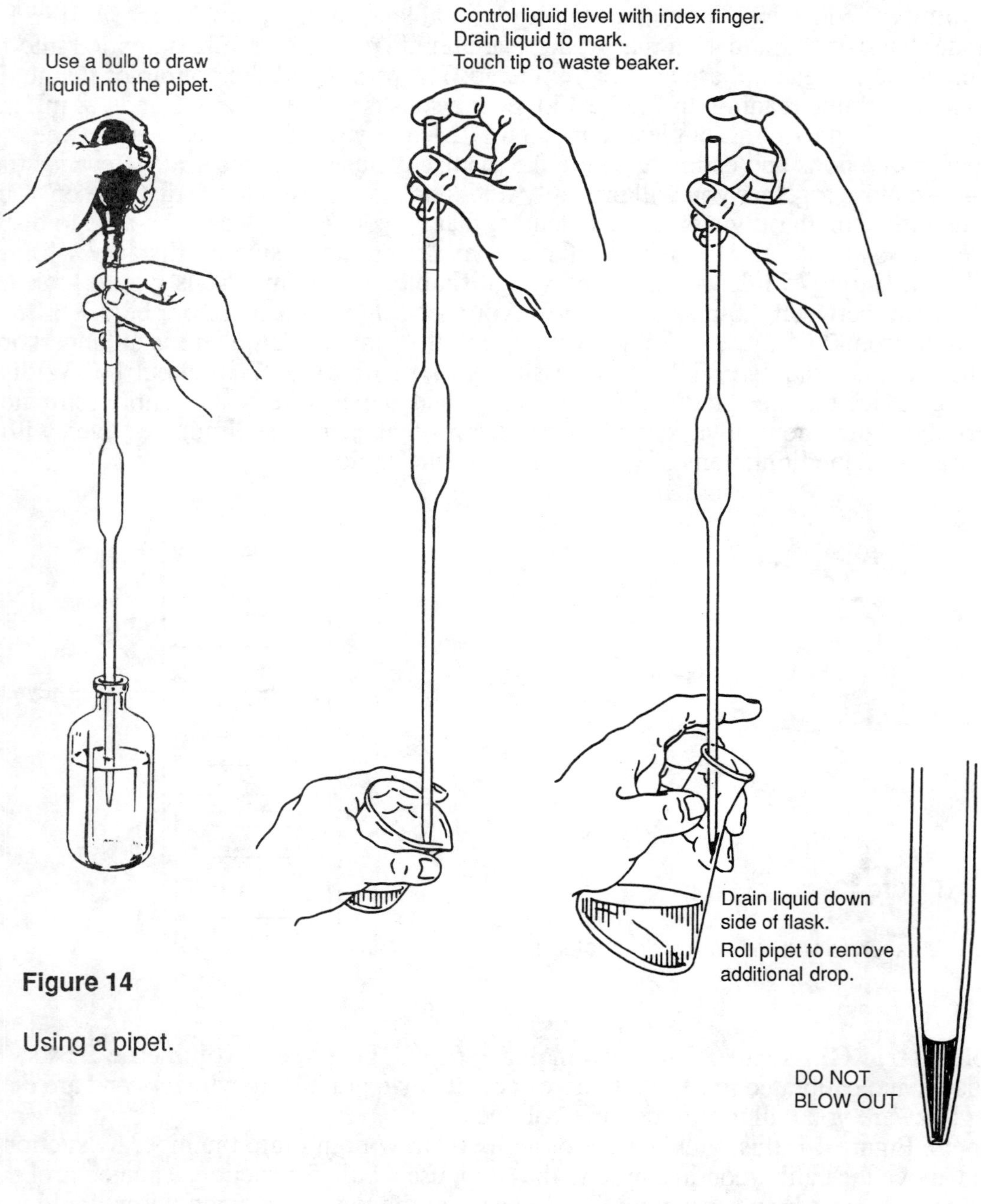


**Figure 14**

Using a pipet.

**Burets.** If a buret's stopcock is made of glass it must be greased before it is used. To do this, place a small ring of grease around the stopcock on each side of the hole; be sure that no grease gets into the hole. A Teflon stopcock never needs to be greased. Clean the buret (see F. Cleaning Glassware). Clamp the buret in place, close the stopcock, and fill the buret with solution. Open the stopcock and drain solution from the buret until the tip of the buret is filled and does not contain an air bubble. If necessary add more solution so that the buret is filled to just below the 0.00 mark.

Each graduation mark on the buret corresponds to 0.1 mL. Estimate the fractional distance between two marks to obtain a reading to the nearest 0.02 mL. The position of the meniscus in Figure 15 is 18.37 mL. In your readings it is important to avoid errors that result from viewing the meniscus from various angles. Note that the meniscus seems to change position depending on whether you look at it from above or below (again, see Figure 15). To make an accurate reading the eye must be level with the meniscus. To ensure this, look at the buret so that the front of the mL graduation mark nearest the meniscus is superimposed on the back of this same mark and your eyes see only a single line; read the meniscus level from this position.

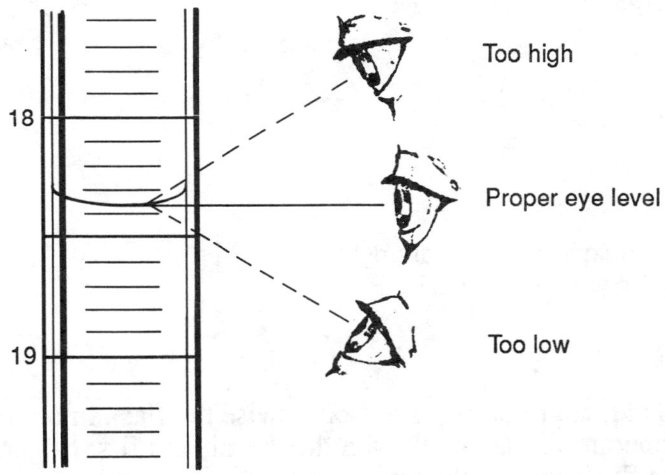


**Figure 15**

Reading a buret.

When you begin to use a buret you might find it difficult or awkward to master the proper technique. However, it will be worthwhile to exert the effort required to use a buret properly. If you are right-handed operate the stopcock with your left hand (see Figure 16a); grasp the stopcock's handle between your thumb on the front of it and your first and second fingers on the back of it.

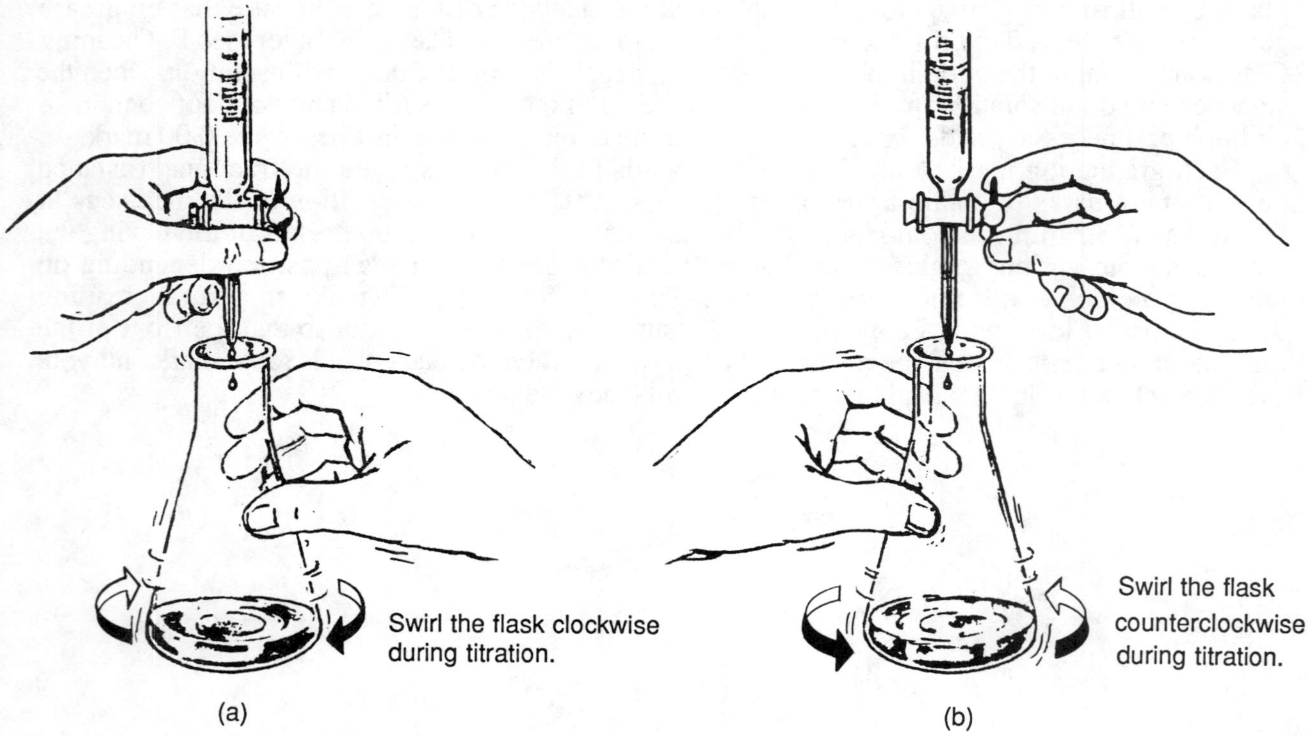


**Figure 16**

Proper buret technique (a) if you are right-handed, and (b) if you are left-handed.

With your right hand, continuously swirl clockwise the flask into which the solution flows. If you are left-handed, operate the stopcock with the thumb and first fingers of your right hand (see Figure 16b); swirl the flask counterclockwise with your left hand.

Before beginning a titration, touch the buret tip against the inside of a waste beaker to remove the drop of liquid that might be clinging to it.

After running solution out of a buret, wait a minute before making a volume reading so that the liquid remaining on the upper wall has time to drain down into the bulk of the liquid.

**Volumetric flasks.** Volumetric flasks (see Figure 17) are calibrated *to contain* a specified volume; they are used to prepare solutions of accurately known molarity (moles of solute per liter of solution). For example, 0.100 mole of NaCl is put into a 1.00-liter volumetric flask and the flask is filled with water to the calibration mark giving 1.00 liter of solution; the concentration is 0.100M. The solid that is to be dissolved must be very carefully weighed, but it is not necessary to measure the amount of solvent that is used; instead, the final volume of solution is accurately known. After you put the solid into the flask, use a little of the solvent to wash out the container in which the solid was weighed and add the washings to the flask; be sure that none of the solid is lost. Add a little more solvent to the flask and swirl the flask to dissolve as much of the solid as possible before the flask is filled with solvent. Then add more solvent until the bottom of the meniscus just touches the calibration line; it is helpful to use a medicine dropper or wash bottle to add the last few milliliters. Stopper the flask and mix the solution by repeatedly inverting and uprighting the flask while holding your hand tightly over the stopper.

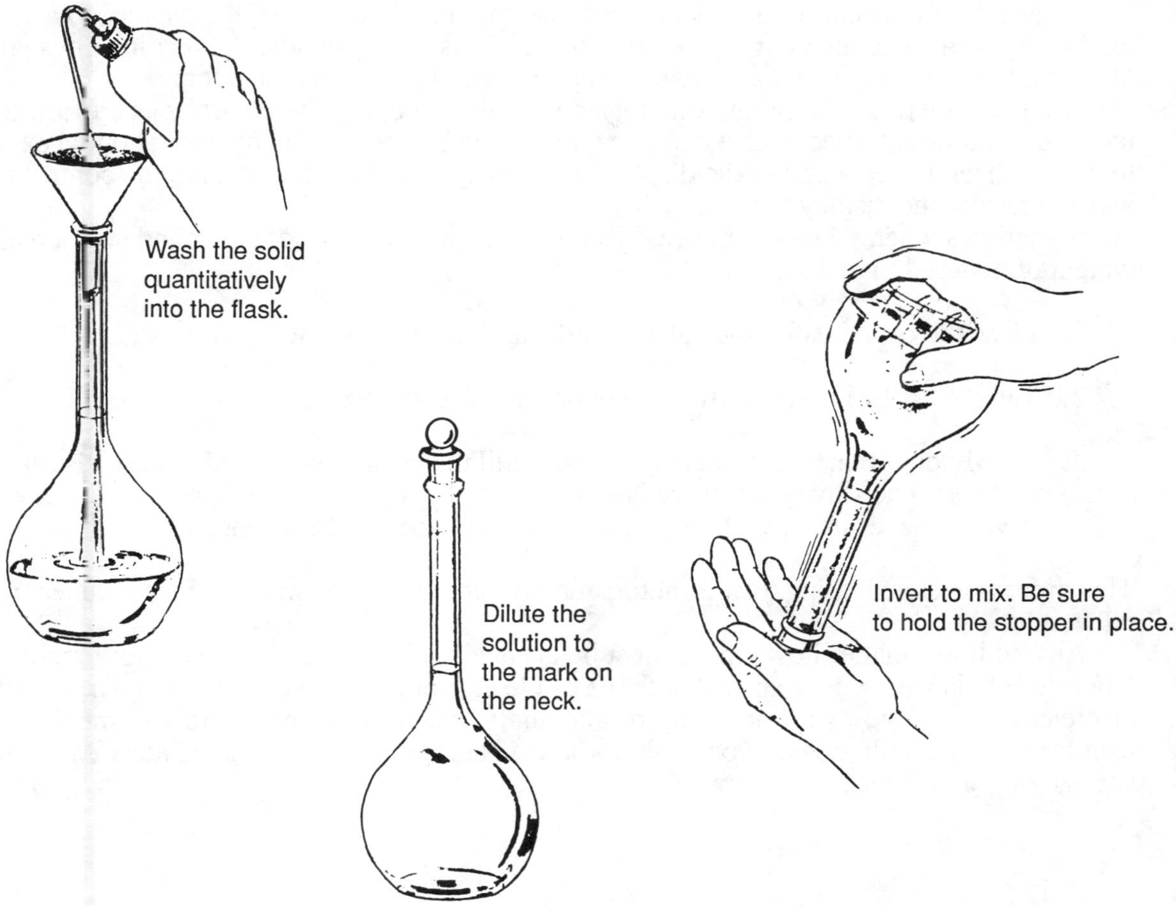


**Figure 17**

Making up a solution using a volumetric flask.

## H. Mass Measurements

In this discussion the words "mass" and "weight" are used synonymously and "to weigh" is used to mean "to determine the mass of".

Two kinds of balances are available to you to use in determining the masses of objects and chemicals. The "laboratory balance" is a mechanical or electronic balance that has a *sensitivity* (ability to measure) of 0.1 g or 0.01 g. In addition, many experiments require an analytical balance, which can be a mechanical or electronic balance; it has a sensitivity of 0.001 g or 0.0001 g.

Any material that is weighed must be in a clean, dry, nonpermeable container such as a beaker or crucible; *do not use paper*. A liquid can be weighed on the analytical balance only if it is in a closed container so that vapor cannot escape, thereby changing its weight and possibly corroding parts of the balance. If a very precise measurement is required, handle the object with tongs or a test tube holder, for your hands can leave moisture, dirt, or oil from your skin on the object. Do not take any chemicals or equipment to the balance except that which is to be weighed.

To turn an electronic balance on, tap the control bar downward. A series of 8's will appear across the display. This allows the user to check that all portions of the display are functioning; if any portion of an 8 does not light up, inform your instructor. When a series of 0's appears, the balance is ready for use.

To use the analytical balance, slide open the door on either side of the balance, place the object on the pan, and close the door. Wait until the weight is displayed and the reading does not change. Do not lean on or bump the table. After removing the object, close the door.

After completing a weighing on an electronic balance, tap the control bar upward to turn the balance off, unless someone else is waiting to use the balance. If the balance is on when you begin to use it, check to see whether the display shows only 0's. If it does not, tap the control bar downward to restore the display to zero.

Sometimes when you use a balance you will weigh an empty container and add some specified weight of material to it. To do this:

1. Place the container on the balance and wait for the weight to be displayed.

2. Tap the control bar downward to return the display to zero.

3. Slowly pour material into the container until the desired weight of material is reached. This *must* be done slowly, for there is a slight time lag in the balance's response and you will go beyond the desired weight if you do not allow time for the balance to "catch up" with you.

This process is called taring, that is, automatically subtracting the weight of a container from the total weight.

Any spill, no matter how small, must be cleaned up immediately. If you spill anything on the laboratory balance, wipe it up with a tissue or clean sponge and then dry the balance with a tissue or clean towel. If you spill anything on the analytical balance, report this immediately to your instructor. If you spill anything on a table where a balance is located, immediately clean it up with a sponge or whisk broom.

# EXPERIMENT 1

# MEASUREMENT OF VOLUME AND MASS

## Purposes

To learn to use balances
To learn to use a pipet
To learn to use a buret
To compare the accuracy of several types of glassware for measuring volume

## Review

Equipment and Techniques F, G, H
Appendix II

## Equipment Needed

balances--laboratory and analytical
150-mL beaker
bottle
buret
crucible and lid
crucible tongs
125-mL Erlenmeyer flask
100-mL graduated cylinder
10-mL pipet
pipet bulb
0.1° thermometer

## Technique and Procedure

### A. Comparisons of Volume Measurements

Most beakers and flasks have approximate volume markings on them. Graduated cylinders are designed to be more accurate. Burets and pipets are more carefully calibrated and are much more accurate. Most glassware that has volume markings also has a statement of how accurate volume measurements are using that piece of glassware. This information might be stated as an *absolute error*, that is, the greatest amount by which a measurement might be in error if the glassware is used correctly, such as 0.02 mL or 5 mL. Or it might by given as a *relative error*, that is, the fraction or percentage that the possible error is of the total measurement, such as 5%. Record this information for each piece of glassware as you use it; if the glassware does not have an error statement, say so in your report. You are to compare experimentally the accuracy of the volume measurements you can make using several kinds of laboratory glassware.

Weigh a clean, dry 150-mL beaker on the laboratory balance. Go back to your desk and use the volume markings on the side of the beaker to measure 100 mL of clear tap water into it as accurately as you can. Remember to have the meniscus at eye level when you read the volume. Weigh the beaker and water on the laboratory balance. Determine the weight of water in the beaker; then use the density of water to determine the volume of this water. This method of determining the volume of water in the beaker is based on the assumption that the balance is indeed accurate to the nearest 0.1 g or 0.01 g (depending on the kind of balance you are using). The error of your measurement is the difference between the volume you actually did measure and the volume you were trying to measure. This might be a fairly large error; if it is, remember that this is mainly the beaker's error and not your error. The purpose of this exercise is for you to determine how large the glassware's errors are so that you will know what glassware is appropriate when you must have a certain degree of accuracy.

Repeat this procedure with a 125-mL Erlenmeyer flask to measure 100 mL of clear tap water.

Repeat this procedure with a 100-mL graduated cylinder to measure 100 mL of clear tap water.

(Any time you use tap water, collect a little of it and look at its color. If it has a reddish or brownish or yellowish color, there are probably iron deposits in it from the water pipes. If this occurs, let the water run until it is colorless. Use only colorless and clear tap water.)

## B. Analytical Balance

The purpose of this part of the exercise is to make you aware of

(1) the *sensitivity* of the analytical balance, that is, the smallest mass it can detect;

(2) the *precision* of the analytical balance, that is, the agreement among repeated measurements of the mass of a single object.

Take your balance instructions with you when you go to the balance.

Practice using tongs to pick up and put down a crucible, to pick up and put down a crucible lid, to pick up and put down a crucible with the lid on it, and to carry a crucible with the lid on it. When you can perform these activities without dropping anything, clean and dry the crucible and lid. After drying them as thoroughly as possible with a towel, let them stand for several minutes before you weigh them so that any remaining traces of moisture can evaporate. After the crucible and lid are clean and dry, handle them only with tongs so that you do not leave fingerprints on them, which can change their masses.

Weigh the crucible on the analytical balance. Remove it from the balance pan, close the doors of the weighing chamber, and wait for the reading to be removed from the readout. Look at the readout. Is it all zeros? If the readout remains at some quantity other than zero, tap the control bar down gently to rezero the balance. Then weigh the crucible lid. Remove the lid and close the doors. Rezero the balance if necessary, and weigh the crucible with the lid on it.

Repeat this sequence of weighings: crucible, lid, crucible and lid. After each weighing, remove the object(s) from the balance pan, close the doors of the weighing chamber, and wait for the reading to be removed. When you begin each new weighing check the readout and rezero the balance if necessary.

When you complete the weighings, close the door of the weighing chamber and turn the balance off.

## C. Pipet

Get about 30 mL of distilled water in a small, clean beaker. After you have the water, check out a 0.1° thermometer. Record the water's temperature to the nearest 0.1°C. Immediately return the thermometer. Be careful! These thermometers are quite expensive.

Obtain a 10-mL pipet, pipet bulb, and small bottle. Place the bulb securely on the end of the pipet. If the bulb has a firm tip that is tapered and soft on the inside, press it gently against the end of the pipet. Practice pipetting ten mL of water until you can control the flow of liquid into and out of the pipet. Try not to draw liquid into the bulb, but if you do, inform your instructor or stockroom clerk when you return the bulb, so that it can be dried before it is used again. Look on the upper stem of the pipet for an error statement; if there is none, look at other pipets until you do find such a statement.

Weigh the bottle (with its lid on it) on the analytical balance; be sure that it is dry on the outside. Pipet ten mL of distilled water into the bottle and weigh the bottle (with its lid on it) on the analytical balance. Use the density of water to calculate the actual volume of water in the bottle. What is the error of your measurement? If it is as great as 0.1 mL, you are not using the pipet as accurately as possible, but this might be expected if you are inexperienced in using a pipet. Later, when you use a pipet in a chemical analysis, an error of 0.1 mL can noticeably decrease the accuracy of your results and might affect your grade. If your error is 0.2 mL or more you might not be using the pipet correctly; ask your instructor to watch you to see whether you are making a mistake.

Pipet another ten mL of distilled water into the bottle and weigh the bottle (with its lid on). What are the weight and actual volume of this second portion of water?

When you have completed this work, return the pipet and bulb. Empty the water from the bottle into the beaker; keep the bottle and the beaker of water to use with a buret.

## D. Buret

Set up a buret in a clamp; tighten the clamp slowly until the buret is held securely. Close the stopcock of the buret and fill the buret to the top with distilled water from the beaker. Open the stopcock and drain all the water back into the beaker. Look at the inside wall of the buret: does any water cling in droplets? If so, the buret is not clean. If your buret needs to be cleaned, take it to a sink, close its stopcock, and scrub it with a few milliliters of detergent solution and a long-handled brush. Then rinse it repeatedly with tap water to remove all the detergent. Then rinse it three times with five-mL portions of distilled water poured from the beaker or squeezed from a wash bottle; tip and roll the buret so that each portion of water washes over the entire inside surface.

Set up the buret and fill it nearly to the top with distilled water. Open the stopcock and let water fill the tip of the buret. Look at the top of the tip (right next to the stopcock) for an air bubble; if you see one, open the stopcock all the way and allow water to flow rapidly from the buret until this bubble is gone. Adjust the water level to slightly below the zero mark; there is no need to have it exactly at the zero mark.

Weigh the bottle (with its lid on) on the analytical balance; it doesn't matter if it is wet inside, for you will be measuring the weight of water that is added to it.

Read the buret to the nearest 0.01 mL by estimating between marks; remember that you are reading downward from the top. Drain about ten mL of water into the bottle. Allow the buret to stand for about a minute before you read it again so that water on the inside wall will have time to flow down into the remaining water. Read the buret to the nearest 0.01 mL. (Note that you don't have to try to measure exactly ten mL, simply a volume that is close to ten mL). Weigh the bottle and water (with the lid on the bottle). Use the density of water to determine the actual volume of water. If the difference between the volume you measured by reading the buret and the actual volume (calculated from mass and density) is as great as 0.1 mL you are not using the buret as accurately as possible, but you will improve with practice. If the difference is more than 0.1 mL, you might be reading the buret incorrectly; ask your instructor for assistance.

Read the buret again; this reading should be the same as the previous one. Drain about ten mL more water into the bottle, read the buret after about a minute, and weigh the bottle and water (with the lid on the bottle). Determine the weight and actual volume of this second portion of water. Is your technique improving?

When you have completed this work, return the drained buret and the empty bottle with the lid off.

## E. Density of Water

To complete your report, you must know the density of water to four significant figures. Many handbooks and dictionaries of chemistry have tables that give the density of water at 0.1° intervals. Use such a table to report the density of water at the temperature you measured for it. Note that if the table has more than four figures in the density value, you are to round it off to four figures.

# EXPERIMENT 1

# REPORT SHEET

Name ______________________________________________ Date__________________

Section ____________________________________________

## A. Comparisons of Volume Measurements

Mass of 150-mL beaker ______________________

Mass of 150-mL beaker + 100 mL water ______________________

Mass of water ______________________

Actual volume of water ______________________

Error of your measurement ______________________

Mass of 125-mL flask ______________________

Mass of 125-mL flask + 100 mL water ______________________

Mass of water ______________________

Actual volume of water ______________________

Error of your measurement ______________________

Mass of 100-mL graduated cylinder ______________________

Mass of 100-mL cylinder + 100 mL water ______________________

Mass of water ______________________

Actual volume of water ______________________

Error of your measurement ______________________

**B. Analytical Balance**

| | Trial 1 | Trial 2 | Difference |
|---|---|---|---|
| Mass of Crucible | | | |
| Mass of Lid | | | |
| Calculated mass of crucible + lid | | | |
| Measured mass of crucible + lid | | | |
| Difference between calc. mass and meas. mass of crucible + lid | | | |

What is the sensitivity of the analytical balance, that is, what is the smallest amount of mass that it can detect?

According to your measurements, what is the precision of the analytical balance, that is, what is the largest difference between repeated measurements of a particular object (or objects)?

## C. Pipet

| | |
|---|---|
| Water temperature | ____________ |
| Water's density at this temperature | ____________ |
| Mass of bottle | ____________ |
| Mass of bottle + ten mL water | ____________ |
| Mass of water | ____________ |
| Actual volume of water | ____________ |
| Error of your measurement | ____________ |
| Mass of bottle + twenty mL water | ____________ |
| Mass of second ten mL water | ____________ |
| Actual volume of second ten mL water | ____________ |
| Error of your measurement | ____________ |

## D. Buret

| | Trial 1 | Trial 2 |
|---|---|---|
| Initial buret reading | ________ | ________ |
| Final buret reading | ________ | ________ |
| Measured volume of water | ________ | ________ |
| Mass of bottle | ________ | ________ |
| Mass of bottle + water | ________ | ________ |
| Mass of water | ________ | ________ |
| Actual volume of water | ________ | ________ |
| Error of your measurement | ________ | ________ |

## Questions

1. Give the error statement (which might be absolute, relative, or verbal) that you found on each of these pieces of glassware, or write "none" if you did not find one. Then state the error of your measurement with the glassware.

   150-mL beaker

   125-mL flask

   100-mL cylinder

   10-mL pipet

   25-mL buret

2. Which of the three pieces of glassware that you used to measure 100 mL of water did you use most accurately? Least accurately?

3. Which of the two pieces of glassware that you used to measure 10 mL of water did you use more accurately?

# EXPERIMENT 2

# PHYSICAL PROPERTIES OF SUBSTANCES

## Introduction

Chemistry is the study of matter. One aspect of this study is the characterization of materials, that is, the determination of properties that distinguish materials from one another. This includes the determination of both physical and chemical properties of materials. Some physical properties, such as physical state and color, are obvious without an observer's having to do any work to determine them. The identification of an unknown substance is usually based on measuring its physical properties. Some of the properties that might be considered are:

1. *color*
2. *melting point*: temperature at which a substance changes from solid to liquid
3. *boiling point*: temperature at which a liquid's vapor pressure equals atmospheric pressure
4. *density*: mass per unit of volume
5. *solubility*: amount of substance that dissolves in a specified quantity of a particular liquid

Several of these properties, not just one, are measured to confirm identification of a substance. For example, suppose that a substance is a yellow solid with a melting point of 95°C and a boiling point of 347°C. This information enables us to decide that the substance is benzil. Of course many substances are yellow solids; more than one of these might melt around 95°C and more than one of these might boil around 347°C. However, it is quite unlikely that there are two substances that are yellow, and melt at 95°C, and boil at 374°C. The combination of these properties leaves no doubt that the substance is benzil.

It is most meaningful to determine physical properties that have numerical values, because there can be no doubt about the result (assuming that the limits on the accuracy of the measurements are clearly stated). In contrast, the word description of a property like color can be quite ambiguous; what one person describes as "pink", another might call "light red", or "rose", or even "coral". Nonetheless, such a qualitative, subjective description still has value; as least we know that the "pink" material has color and it is not black or green!

In this experiment you will measure some physical properties of pure substances and of mixtures of substances. You will then use these measured values to identify an unknown solid and an unknown liquid. This experiment will give you experience in (1) making numerical measurements of physical properties of substances and (2) identifying pure substances by their physical properties.

## Review

Equipment and Techniques A, C, D
Appendix II

## Equipment Needed

- beakers
- bottle
- capillary tubes
- clamps
- "Glassine" paper
- glass rod
- microburner
- 5-mL pipet
- ring
- ring stand
- rubber band
- scissors
- slit rubber stopper
- spatula
- 20-mL test tube
- thermometer
- wire gauze

## Chemicals Needed

- acetoacetanilide
- methanol
- naphthalene
- unknown liquid (from list on page 31)
- unknown solid (from list on page 31)

## Technique and Procedure

### A. Liquid Density

Density is defined as mass per unit of volume. Measurements of density can be quite simple to make, especially for liquids. You simply take a carefully measured volume of the liquid and determine its mass by weighing it. A pipet or a buret is usually used to make an accurate measurement of a liquid's volume. In this experiment you will use a pipet.

Obtain a liquid unknown. Immediately record its identification code.

Obtain a small, dry bottle. Weigh this bottle and its lid on the analytical balance. Pipet 5.00 mL of your unknown liquid into this bottle. (It might be helpful to place the sample bottle of unknown liquid in a small beaker so that you won't knock it over as you pipet from it). Weigh this bottle, lid, and liquid. Calculate the weight of liquid. Leave the liquid in this weighing bottle.

Pipet a second 5.00 mL of liquid into the weighing bottle. Reweigh the bottle, lid, and liquid. Calculate the weight of this second 5.00 mL of liquid, that is (weight of bottle, lid, and 10.00 mL of liquid) minus (weight of bottle, lid, and 5.00 mL of liquid). If the weights of 5.00 mL of liquid for the two trials differ by more than 1%, carry out a third trial. That is, an acceptable difference is

$$\frac{\text{higher value - lower value}}{\text{average value}} \times 100 = 1\% \text{ or less} \qquad (2\text{-}1)$$

If the third trial gives a weight that is not within 1% of either previous trial, consult your instructor. You are probably not using the pipet or balance correctly, or you are carrying out one of the calculations incorrectly.

When you complete these weighings, pour the liquid back into the original sample bottle and save it to use to measure its boiling point. When you finish all your work with the liquid, dispose of it by pouring it into the waste container that is provided. Return the sample bottle and weighing bottle and their lids with their lids off.

| Compound | m.p., °C | b.p., °C* | density, g/mL** |
|---|---|---|---|
| acenaphthene | 95 | 278 | 1.024 |
| acetoacetanilide | 85 | decomposes | 1.103 |
| acetone | -95 | 56.3 | 0.784 |
| acetone oxime | 60-61 | | |
| biphenyl | 69-71 | 254-255 | 1.6 |
| *p*-bromochlorobenzene | 67 | 196 | 1.575 |
| 1-bromo-2, 2-difluoroethane | -74.5 | 57 | 1.817 |
| 2-butanone | -87 | 79.6 | 0.800 |
| *n*-butyl chloride | -123 | 78 | 0.881 |
| cyclohexane | 6.5 | 80.7 | 0.774 |
| *p*-dichlorobenzene | 53 | 174 | 1.458 |
| diphenylamine | 53-54 | 302 | 1.159 |
| diphenylcarbonate | 80-81 | 306 | 1.12 |
| ethyl acetate | -84 | 77 | 0.894 |
| *n*-heptane | -91 | 98 | 0.680 |
| methanol | -98 | 64.7 | 0.787 |
| methyl iodide | -66 | 42 | 2.279 |
| N-methyl-*o*-toluamide | 69-71 | | |
| naphthalene | 80 | 218 | 1.145 |
| phenyl benzoate | 68-69 | 314 | 1.235 |
| 1-propanol | -126 | 97 | 0.800 |
| 2-propanol | -88 | 82.3 | 0.781 |
| salicylaldoxime | 57 | | |
| *o*-terphenyl | 56-58 | | |
| triphenylphosphate | 50 | 245 | |

*at 760 torr
**at 20°C or 25°C

## B . Boiling Point

**Cautions!** (1) **Work under a hood.** The vapor of your unknown liquid might be irritating or toxic. (2) **The unknowns are flammable;** do not pour them near a flame. To avoid the possibility of igniting vapor as it leaves the test tube, use the lowest possible flame to heat the water bath around the liquid. **Do not leave your burner on after you stop heating the water bath.**

1. Known compound. The purpose of working with a pure liquid of known boiling point is to enable you to assure yourself that your technique is good. When you are confident you can make this measurement accurately, proceed to Part 2. and make the measurement on your unknown liquid.

Obtain one capillary tube and snap it in half with your fingers. Seal the smooth (unbroken) end of each piece of tubing in the flame of a microburner. To do this, put the smooth end of the tube at the edge of the burner flame and slowly rotate the tube. Leave the tube in the flame until you see a definite "glob" of melted glass sealing the end of the tube.

Place a 250-mL beaker on wire gauze on a ring on a ring stand with a microburner beneath the beaker. Clamp a 20-mL test tube to the ringstand so that the bottom of the test tube is close to, but not touching, the bottom of the beaker. Drop one of the sealed capillary tubes into the tube, open end down and closed end up. Put methanol into the test tube to a great enough depth that the closed end of the capillary tube is a little above the liquid level; this will probably be about five mL. Put clean tap water into the beaker to a great enough depth that the water level in the beaker is slightly above the liquid level in the test tube.

Put a thermometer into a slit rubber stopper and clamp the stopper onto the ring stand. Have the thermometer immersed in the liquid up to the immersion mark on the thermometer; if the thermometer does not have an immersion mark, have the thermometer close enough to the bottom of the test tube that its bulb will be covered by liquid even after some of the liquid has boiled away. Have the bottom of the stopper at least one cm above the top of the test tube so that vapor from the boiling liquid can easily escape from the test tube without having much contact with the stopper. Your equipment should look like Figure 2-1.

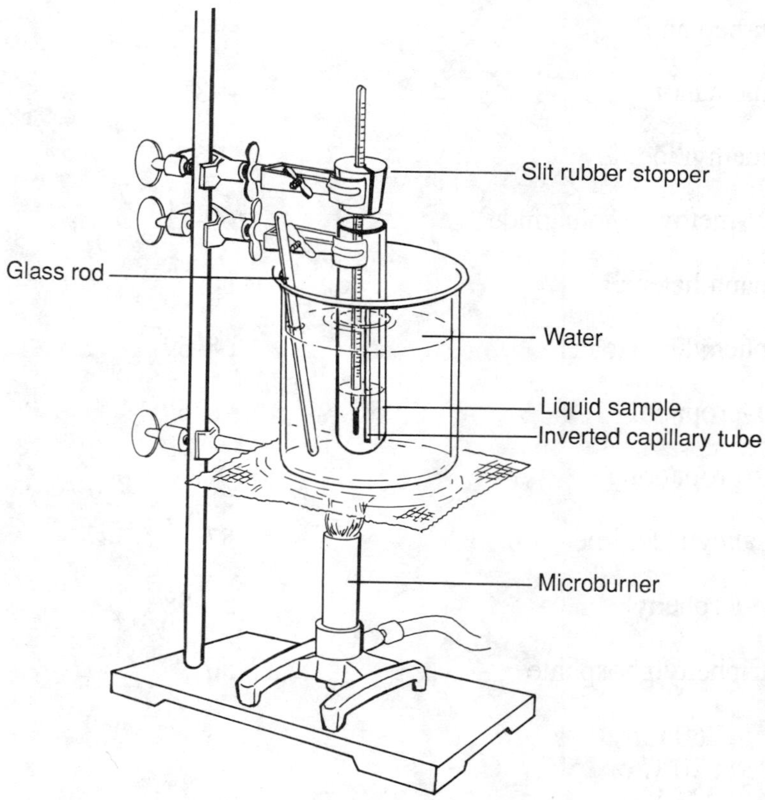


**Figure 2-1**

Determining a boiling point.

Begin to heat the water bath, stir it gently and continuously, and watch what happens in the capillary tube. It is initially filled with air at atmospheric pressure. As the bath is heated, this air expands and begins to escape from the open end of the capillary tube as tiny bubbles. Vapor from the sample pushes into the tube, expelling even more air. When the liquid starts to boil, the broken capillary tube tip provides a rough surface on which bubbles of vapor form, and so boiling will probably be most obvious at the tip of the capillary tube.

Remember that the boiling point is *defined* as the temperature at which a liquid's vapor pressure equals the atmospheric pressure on the surface of the liquid. For bubbles of vapor to rise to the liquid surface against the downward pressure of the atmosphere and the pressure exerted by the liquid that is above the bubbles, the gas inside a bubble must exert a pressure that is slightly greater than atmospheric pressure. Consequently, the temperature at which the liquid begins to boil might be slightly higher than its defined boiling point.

Because the liquid's vapor pressure might be slightly higher than atmospheric pressure when the liquid starts to boil, the vapor present in the capillary tube will force liquid out of the tube; the liquid level will be higher in the test tube than it is in the capillary tube. At this point turn off the burner and stop heating the bath. The temperature will begin to drop and approach the boiling point of the liquid from above. When the liquid's boiling point is reached, the pressure from vapor in the capillary tube will equal the pressure of the atmosphere; the liquid levels will be the same in the capillary tube and in the test tube. Record this temperature as the boiling point of the liquid. It might be that, as the liquid begins to cool, you cannot distinguish the instant at which the two liquid levels are the same; if this happens, record the temperature at which you can discern that liquid is moving into the capillary tube.

Repeat the measurements until two consecutive values differ by no more than 1°; be sure to estimate to the nearest 0.5°. Use the same capillary tube for all the measurements. If necessary, put more liquid into the test tube; if you do so, **take the test tube out of the hood away from all burners while you add the liquid.**

From the definition of boiling point it is clear that the temperature at which a liquid boils depends on the pressure of the atmosphere. In tables, the boiling points of substances are usually given for a pressure of 760 torr; this is called *normal* (or standard) *boiling point.* If a boiling point is measured at some pressure other than 760 torr, the pressure at which the boiling point was determined is reported, for example, b.p. 102°C/600 torr. You are measuring boiling points at a pressure that is probably not 760 torr; read the current atmospheric pressure from a barometer and record it.

2. Unknown compound. Use the procedure described in Part 1. to determine the boiling point of your unknown liquid. Use a dry test tube and a fresh capillary tube. Repeat the measurements until two consecutive values differ by no more than 1°.

After completing this part of the experiment, pour the remaining liquid back into the sample bottle and save it if you still need to use it to measure its density. When you finish all your work with the liquid, dispose of it by pouring it into the waste container that is provided; return the sample bottle and its lid with its lid off.

When you are trying to pick your unknown from the list of compounds, keep in mind two factors that might cause your experimental value to be different from those in the list even if you have worked as carefully as possible: (1) An error in your thermometer might cause consistently low or consistently high values. (2) Standard boiling points are listed and your measurement was probably made at an atmospheric pressure different from 760 torr. After you record the barometric pressure, think about whether the variation in atmospheric pressure would cause your measured boiling point to be lower than or higher than the standard boiling point. Also use the value you obtained for the known liquid to decide whether you should expect the value you obtained for the unknown liquid to be slightly lower or slightly higher than those in the list.

## C. Melting Point

1. Pure, known compound. First, you are to determine the melting point of a known compound to assure yourself that your technique is good.

Place a small beaker on wire gauze on a ring on a ringstand with a microburner beneath the beaker.

Obtain three capillary tubes and snap each one in half with your fingers. Seal the broken end of each piece of tubing in the flame of a microburner as described in Part B.1. A fresh tube must be used for each measurement. Obtain, break, and seal more tubes if you need them.

Place a few crystals of naphthalene on a piece of "Glassine" paper; use a spatula to crush them to a fine powder. Push the powder into a mound. Press the open end (which is smooth) of the capillary tube into the mound of powder *once*. Turn the tube upright and repeatedly tap the sealed end against the desktop until *all* the powder falls to the *bottom* of the tube. Again, press the open end of the tube into the heap of powder once and tap the bottom of the tube against the desktop until all the powder falls to the bottom of the tube. Repeat this process until there is about five mm (or about one-half the length of the thermometer bulb) of the solid in the bottom of the tube. Cut a narrow strip off the end of a piece of thin-walled rubber tubing. Use this rubber band to hold the capillary tube onto the thermometer. Have the solid right beside the thermometer bulb so that solid and bulb will be subjected to the same intensity of heat simultaneously.

Put the thermometer into a slit rubber stopper and clamp the thermometer onto the ring stand. Put clear tap water into the beaker to a great enough depth that the thermometer is immersed in the water up to the immersion mark on the thermometer; if the thermometer does not have an immersion mark, have enough water in the beaker that the thermometer is covered to a point slightly above the bulb. Be sure that the open end of the capillary tube is above the water level. Also, have the rubber band above the water level. Your equipment should look like Figure 2-2.

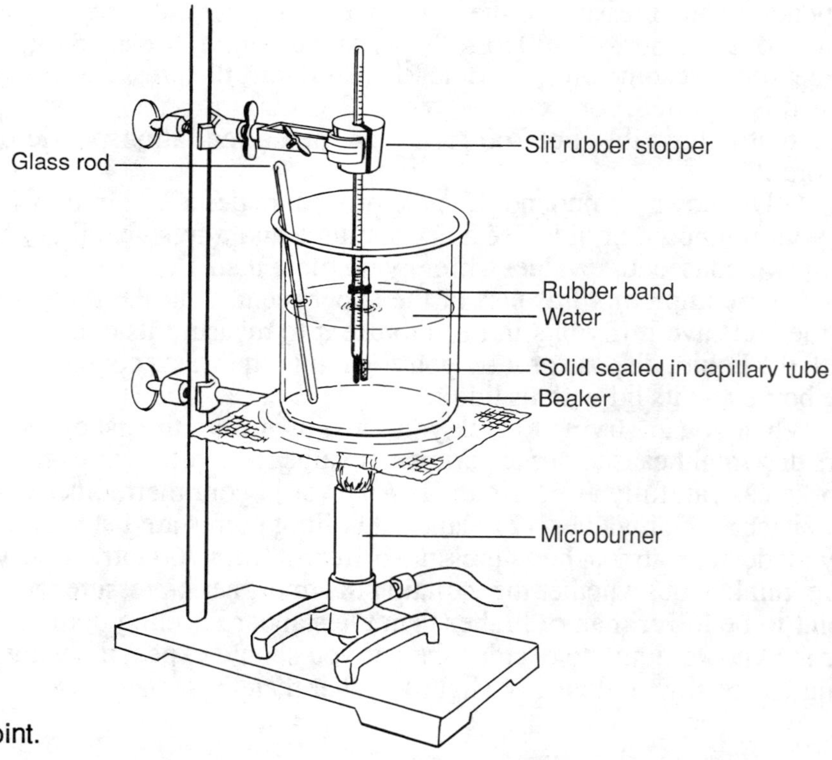


**Figure 2-2**

Determining melting point.

Begin to heat the bath and stir it gently and continuously. Since you know the melting point of naphthalene, you can rapidly heat the bath to a temperature about 20° below the expected melting temperature. Then reduce the heat so that the temperature rises no more than 2° per minute as you approach the melting point. Record the temperature at which you first see solid change to liquid; the solid will seem to collapse as it begins to melt and will be translucent or transparent, rather than opaque. Record the temperature at which all solid finally changes to liquid. Your sample of naphthalene should melt within 3° of 80°C. If it does not, or if the melting range is more than 2°, you were probably heating the water bath too rapidly. Repeat the measurement with a fresh sample of naphthalene in a fresh capillary tube. Be sure the initial bath temperature is several degrees below 80°C; heat at a rate of about 1° per minute. Repeat measurements (a new sample in a fresh capillary tube for each one) until your value is within 3° of 80°C and has a range of no more than 2°.

2. Known mixture. On the "Glassine" paper with the naphthalene place an equal volume of acetoacetanilide; crush it into a powder and thoroughly mix the two powders together. Put some of this mixture into a capillary tube and determine its melting range. Make two trials, the first one with rapid heating to determine the approximate range, the second one with slow heating. Note that although these separate pure substances melt at similar temperatures, the mixture melts at a lower temperature and over a much wider temperature range. This lowering and spreading of melting point can also be used for compound identification. A material that is thought to be substance A can be mixed with a sample of substance A. If the melting point of the mixture is the same as the melting point of pure A, the unknown material is indeed substance A. If the unknown material is not substance A, the mixture will melt over a fairly wide temperature range and at a temperature different from that of pure A. This procedure is called taking a *mixture melting point.*

3. Unknown compound. Obtain an unknown solid; record its identification code. Determine its melting point (or a range of not more than 2°). Repeat trials (with a fresh sample and capillary tube for each trial) until consecutive values agree within 1°. Tentatively identify the compound from the list on page 31. Keep in mind that your measured value might be in error by at least two or three degrees even if you thought you were being very careful. This is partly because of your lack of experience making this kind of measurement, but it is also because your thermometer might be in error by one or two degrees.

Obtain a small amount of the compound having a melting point closest to that of your unknown. Crush this on the "Glassine" paper with your unknown and thoroughly mix the two together. Measure the melting point (or range) of this mixture.

If the unknown and known are different compounds, the melting point of the mixture will be lower than that of the unknown and will be spread over a wide temperature range. If this occurs you do not need to repeat the measurement. Find a different known to mix with some of your unknown; prepare this mixture on a new sheet of "Glassine" paper with a new sample of your unknown; be sure to crush thoroughly and mix thoroughly.

If the unknown and known are the same compound, the melting point of the mixture will be sharp and within a degree or two of your original measurement for the unknown. If this occurs, you have identified your unknown and you do not have to make any more measurements.

When you complete all your work with knowns and unknown, put all left-over solids, the "Glassine" paper on which you crushed them, and used capillary tubes into the plastic bag (or other container) that is provided in the hood. This is to prevent an accumulation of vapors from volatile compounds in the laboratory atmosphere or in the waste can.

## Results

### A. Liquid Density

Calculate density for each of your trials. Report an average of all trials or of your two best trials. Be sure to report your results to the correct number of significant figures; include units.

### B. Boiling Point

Report your final results for boiling points of the known and unknown.

Identify your unknown liquid on the basis of your density and boiling point data.

### C. Melting Point

Report your final results for melting points of known, unknown, and mixtures. Be sure to state the melting range for each compound and for each mixture.

Identify your unknown solid on the basis of your melting point data.

For each of your unknowns put the following information on the back of the appropriate Report Sheet. Use the name of the unknown compound to look it up in a chemical dictionary, chemical encyclopedia, chemical handbook, or organic chemistry textbook. Write the molecular formula of the compound; if you find a formula that is intended to convey information about structure, rewrite it to group all atoms of one kind together and put them in the order C, H, and other atoms alphabetically. Find and state one other intrinsic, physical, quantitative property of the compound that is not given in this experiment and that cannot be determined from the formula of the compound. Give the definition of this property. Cite a reference for the property.

## Questions

1. In your determination of density, which measurement limits the number of significant figures in your calculated density, mass or volume?

2. Why is a pipet used instead of a graduated cylinder to make the volume measurement for the density determination?

# EXPERIMENT 2

# REPORT SHEET

Name ______________________________ Date ______________

Section ______________________________

## A. Liquid Density

Unknown identification code ______________

Weight of bottle and lid ______________

Weight of bottle, lid, and 5.00 mL liquid ______________

Weight of first 5.00 mL of liquid ______________

Weight of bottle, lid, and 10.00 mL liquid ______________

Weight of second 5.00 mL liquid ______________

Weight of bottle, lid, and 15.00 mL liquid ______________

Weight of third 5.00 mL liquid ______________

Density for first 5.00 mL liquid ______________

Density for second 5.00 mL liquid ______________

Density for third 5.00 mL liquid (if needed) ______________

Average value for density ______________

## B. Boiling Point (or Range)

Methanol ______________

Unknown ______________

Identity of unknown liquid ______________________________

(On the back of this Report Sheet provide the added information about your unknown and answer the Questions).

# EXPERIMENT 2

# REPORT SHEET

Name ______________________________ Date ______________

Section ______________________________

**C. Melting Point (or Range)**

Naphthalene ______________

Mixture of Naphthalene and acetoacetanilide ______________

Unknown compound ______________

Mixtures

Compound added to unknown ______________ m.p. __________

Compound added to unknown ______________ m.p. __________

Compound added to unknown ______________ m.p. __________

Identity of unknown solid ______________________

Unknown identification code ______________________

(On the back of this Report Sheet provide the added information about your unknown).

# EXPERIMENT 3

# COMPOUND STOICHIOMETRY

## Introduction

It is of great importance to chemists to know the formulas of compounds. That is, chemists want to know how many atoms of one element combine with how many atoms of another element. For example, we can cause magnesium to react with oxygen by heating magnesium in oxygen or air. One gram of magnesium will produce 1.66 grams of magnesium oxide. The weight of oxygen in the oxide is the difference between these two weights, 0.66 g. Knowing the molar weights of magnesium and oxygen, we can calculate how many moles of each element reacted to form the oxide and the relative numbers of atoms involved in this reaction.

$$\text{moles Mg} = \frac{1.00\ \text{g}}{24.3\ \text{g/mole}} = 0.0412\ \text{mole}$$

$$\text{moles O} = \frac{0.66\ \text{g}}{16.0\ \text{g/mole}} = 0.41\ \text{mole}$$

For the compound, the mole ratio of the two elements is 0.0412/0.041, or 1.0/1.0. Since one mole of atoms of any element is the same number of atoms as one mole of atoms of any other element, we know that the atom ratio is also 1.0/1.0; the simplest, or *empirical, formula* is MgO.

## Technique

In this experiment you will determine the formula of a compound that forms when a metal reacts with sulfur. You will heat an accurately weighed sample of a metal with an excess of sulfur to form a metal sulfide. Whatever sulfur does not react with the metal burns in air producing gaseous sulfur dioxide.

The weight of sulfur that reacts with the metal can be determined as the weight of the oxygen was in the MgO example. You can determine the empirical formula of the metal sulfide using the molar weights of the metal and sulfur.

## Review

Equipment and Techniques A, H
Appendix II

## Equipment Needed

- balances, analytical and laboratory
- bottles or test tubes
- burner
- ceramic mat or wire gauze
- clay triangle
- crucibles and lids (2)
- crucible tongs
- ring
- ring stand

## Chemicals Needed

copper wire or metal shavings
powdered sulfur

## Procedure

Wash and thoroughly dry a crucible and lid. Place them on a clay triangle on a ring and ring stand as shown in Figure 3-1; be sure the lid is slightly ajar. Apply the full heat of a burner to the crucible and lid for at least 10 minutes to burn off or vaporize any possible contaminants or volatile materials; if your burner flame leaves a deposit of soot on the crucible or lid, be sure this is burned off. Turn off the burner and allow the crucible to cool for a minute or two on the clay triangle. Use crucible tongs to place the crucible and lid on a ceramic mat or wire gauze. Throughout the rest of the experiment handle the crucible and lid only with tongs. When the crucible and lid are at room temperature (after 15-20 minutes), weigh them on the analytical balance. If, during the weighing, the weight readout changes continuously, the crucible and lid are not cool; immediately remove them from the balance, return to your desk while they cool, and wait at least five minutes before you weigh them again.

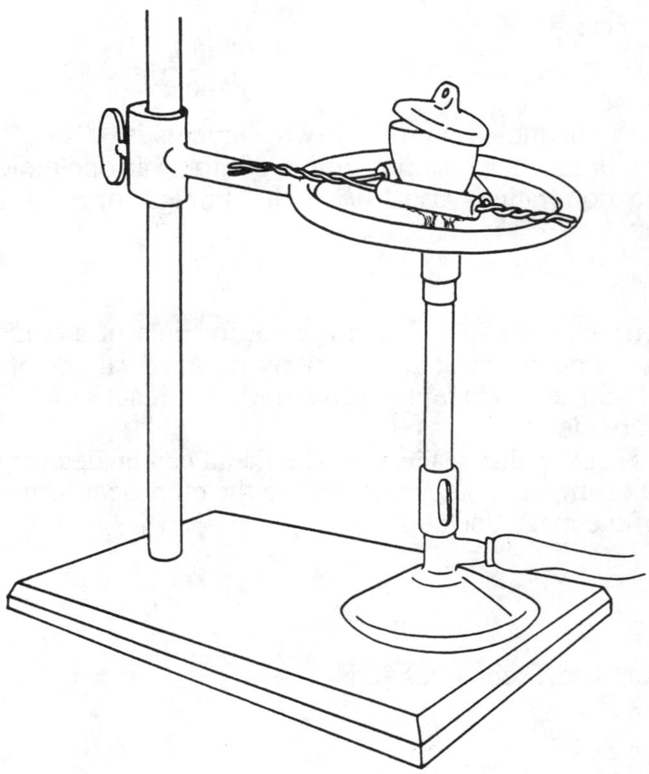

**Figure 3-1**

Heating a crucible.

While the crucible is cooling weigh out about one gram of copper wire or metal shavings on the laboratory balance. If you use copper wire, coil it around a pencil and compress the coil so that it will fit into the crucible. After you have weighed the crucible and lid, place the metal in the crucible and weigh the crucible, lid, and metal on the analytical balance.

Weigh about one gram of sulfur into a beaker on the laboratory balance; since an excess of sulfur is used it does not need to be weighed any more accurately than to the nearest 0.1 g. Transfer this weighed sulfur to the crucible that contains the metal.

Move your ring stand, ring, clay triangle, and burner to a hood. Place the covered crucible on the clay triangle and heat it with a low flame. The sulfur will melt (m.p.<120°C) and the solid metal and liquid sulfur will react. Some sulfur will vaporize and the hot vapor will push out of the crucible; it will then burn in air with a blue flame. (**Caution! This heating must be done in a hood**, for sulfur vapor and sulfur dioxide are irritating and can be harmful to your health.) When you see a blue flame around the lid of the crucible, continue to heat the crucible with a low flame. When you no longer see a blue flame of burning sulfur, turn up the burner and strongly heat the crucible until the bottom of it is dull red. After the clay triangle and crucible bottom are dull red, heat the crucible for an additional five minutes; *keep the crucible covered during all of this heating*, so that the metal and metal sulfide cannot react with oxygen in the air. Be sure to burn off any soot that might have been deposited on the crucible or lid. Turn off the burner and allow the covered crucible to cool for a minute or two on the clay triangle. Then place the covered crucible on a mat or gauze to continue cooling. After about five minutes you may open the crucible; if you see any sulfur inside the crucible or on the crucible lid, return the covered crucible to the hood and heat it strongly for another five minutes. Cool the crucible to room temperature. Weigh it with the metal sulfide it contains on the analytical balance.

At this point there might be a trace of unreacted sulfur in the crucible even though you can't see it. It is also possible that not all the metal was in contact with, and reacted with, sulfur. Weigh 0.1 g of sulfur into a beaker and put it into the crucible. Cover the crucible, take it to a hood, and repeat the heating in the same way that you did it originally. This repeated heating should ensure complete reaction of metal and complete removal of excess sulfur. Cool and weigh the covered crucible and metal sulfide. If this second weight is within 0.002 g of the first weight, you have satisfactory precision. If the two weights do not agree within 0.002 g, add another 0.1 g of sulfur to the crucible; repeat the heating, cooling, and weighing. If the second and third weights are within 0.002 g, you have satisfactory precision. If they do not, repeat the addition of 0.1 g of sulfur, heating, cooling until two consecutive weights agree within 0.002 g; only then can you be sure of complete reaction of metal and complete removal of excess sulfur. In your calculations, use the last weight.

Turn in your product in a bottle or test tube that is labelled with your name, the date, the name of the product, and the Trial number.

Carry out a second determination of the formula of this metal sulfide. Do it in exactly the same way that you did the first trial, except put about 2.0 g of sulfur into the crucible for the first heating. This will allow you to determine whether or not the initial amount of sulfur used in the reaction affects the formula of the compound that forms. While the first crucible is cooling, heat the second crucible. Then heat the reaction mixture of metal and sulfur in the first crucible, while the second crucible is cooling. By dovetailing your operations for the two crucibles, you will not waste time. One crucible can be heating while you are cooling and weighing the other crucible. Before you begin the experiment, look for and record any identifying marks or numbers on the two crucibles so that you can be sure of distinguishing them one another throughout the experiment. Turn in both products in separate containers.

## Notebook

In your notebook, prepare a data table like the one on the Report Sheet and record your data there as you collect them. Also write down any pertinent observations you make in the course of the experiment. For example, you should record the color and appearance of both reactants and of the product. Be sure to record any difficulties you have or any mistakes you make that might affect your results.

## Results

Use the weight of metal sulfide and weight of metal to calculate weight of sulfur in the metal sulfide.

Use molar weights of metal and sulfur to calculate to the correct number of significant figures how many moles of each is present in the amount of metal sulfide that formed (that is, moles of metal and sulfur that reacted).

Calculate to the correct number of significant figures the ratio of moles of metal to moles of sulfur present in the metal sulfide that formed. Divide this out; do not leave it as a fraction.

Change this mole ratio to the simplest whole number ratio. Write the empirical formula of the metal sulfide.

## Questions

1. What evidence (both qualitative and quantitative) was there that a reaction occurred between the metal and sulfur?

2. Write an equation for the reaction that occurred between the metal and sulfur.

3. You are told that consecutive weights of crucible, lid, and metal sulfide that agree within 0.002 g are sufficiently precise. Consider your last value for weight of metal sulfide in Trial 1. Add 0.002 g to this weight, calculate the weight of sulfur that would be in this weight of metal sulfide, and calculate the compound's formula using this weight of sulfur. Subtract 0.002 g from your last value for weight of metal sulfide in Trial 1, calculate the weight of sulfur that would be in this weight of metal sulfide, and calculate the compound's formula using this weight of sulfur. Compare the two mole ratios when written to four significant figures. Is the empirical formula affected?

4. What difference, if any, did you observe in your results for Trial 1 and Trial 2 in which different initial amounts of sulfur were used? What is the reason for this behavior?

# EXPERIMENT 3

## REPORT SHEET

Name ______________________________ Date ______________

Section ______________________________

Trial 1

**Data**

Weight of crucible and lid ______________

Weight of crucible, lid, and metal ______________

Weight of metal reacted ______________

Identity of metal ______________

Weight of crucible, lid, and metal sulfide
Give all values; use last value in calculations.

__________ __________ __________ ______________

Weight of metal sulfide formed ______________

Weight of sulfur in product ______________

**Calculation of formula**

Moles of metal reacted ______________
Show calculation.

Moles of sulfur reacted ______________
Show calculation.

Mole ratio, metal/sulfur ______________

Simplest whole number ratio, metal/sulfur ______________

Empirical formula of metal sulfide ______________

Trial 2

**Data**

Weight of crucible and lid ____________

Weight of crucible, lid, and metal ____________

Weight of metal reacted ____________

Identity of metal ____________

Weight of crucible, lid, and metal sulfide
Give all values; use last value in calculations.

__________ __________ __________ ____________

Weight of metal sulfide formed ____________

Weight of sulfur in product ____________

**Calculation of formula**

Moles of metal reacted ____________
Show calculation.

Moles of sulfur reacted ____________
Show calculation.

Mole ratio, metal/sulfur ____________

Simplest whole number ratio, metal/sulfur ____________

Empirical formula of metal sulfide ____________

Turn in both of your products in separate labelled containers.

(Attach a separate sheet with the answers to the Questions).

# EXPERIMENT 4

# REACTION STOICHIOMETRY

## Introduction

The net result of a reaction (a chemical change) is summarized by a chemical equation. In order to write a chemical equation, a chemist must determine experimentally what the reactants are, in what ratio they react (the reaction stoichiometry), what the products are, and the amount of each product. This experiment is designed to determine the relative amounts of reactants (reaction stoichiometry) for a particular reaction.

When two or more substances are mixed, it is possible to determine whether a reaction has occurred by noting whether the mixture has properties that are different from properties of any of the original components of the mixture. Further, by noting how an observed property varies with the ratios in which reactants are mixed, the stoichiometry of the reaction can be determined.

Consider an example in which solutions of A and B are mixed and a precipitate P forms. In a series of experiments, solutions of A and B are mixed in a manner such that the sum of the moles of A and B remains constant and the amount of precipitate that forms is determined. If A is in excess of the amount needed for the reaction, an increase in the amount of B will result in the formation of more precipitate, and vice versa. The amount of precipitate formed is a maximum when A and B are mixed in the correct stoichiometric amounts. This *stoichiometric point* can be determined by plotting the amount of precipitate formed vs. the amount of A used in the experiments, Figure 4-1. In this example the maximum amount of precipitate forms with the ratio 1A:1B.

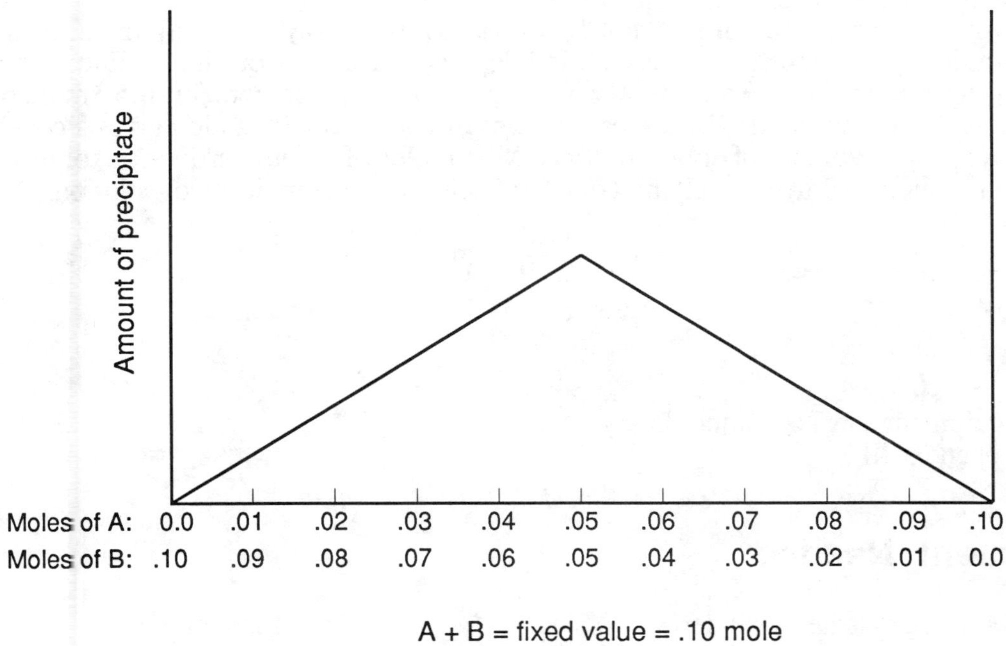


**Figure 4-1**

Amount of precipitate as a function of relative amounts of reactants.

## Technique

The formation of a precipitate is just one of the many properties that could be used to determine the stoichiometry of a reaction. In this experiment you will base your decision about the proper ratio between reactants on the amount of heat that is evolved during a reaction of NaOCl (sodium hypochlorite). For example, with $Na_2S_2O_3$ (sodium thiosulfate)

$$x\text{NaOCl} + y\text{Na}_2\text{S}_2\text{O}_3 \rightarrow \text{products} + Q \tag{4-1}$$

$Q$ stands for the heat that is evolved. This heat causes the temperature of the reaction mixture to rise by an amount

$$\Delta T = T_{\text{final}} - T_{\text{initial}} \tag{4-2}$$

The greater the amount of heat that is released by the reaction, the larger $\Delta T$ will be. Consequently, you can use the temperature change to monitor the amount of heat released during a reaction.

How does the quantity of heat released relate to the reaction stoichiometry? Suppose we know the values of $x$ and $y$ in Equation (4-1). If we mix exactly $y$ moles of $Na_2S_2O_3$ with $x$ moles of NaOCl, a certain amount of heat, $Q$, is produced. However, if we mix fewer then $y$ moles of $Na_2S_2O_3$ with more than $x$ moles of NaOCl, not all the NaOCl will react because not enough $Na_2S_2O_3$ is present to react with all of it. Consequently, a smaller amount of heat is evolved because fewer moles of $Na_2S_2O_3$ react; a smaller $\Delta T$ results. This argument can also be applied when $Na_2S_2O_3$ is in excess. It is essential to keep the total number of moles (moles $Na_2S_2O_3$ + moles NaOCl) constant; if we increase the number of moles of $Na_2S_2O_3$, then, at the same time, we must decrease the number of moles of NaOCl by the same amount, and vice versa. The maximum amount of heat is released and $\Delta T$ is a maximum when the reactants are mixed in exactly the correct stoichiometric ratio. This experiment makes use of the preceding argument by having you mix reactants in various ratios and observe the rise in temperature. You can conclude that the ratio producing the largest $\Delta T$ is the correct stoichiometric ratio.

You will study the reaction of NaOCl with $Na_2S_2O_3$, or $Na_2SO_3$ (sodium sulfite), or KNCS (potassium thiocyanate), or KI (potassium iodide), as your instructor directs. Each reactant is in a 0.50M aqueous solution. The term 0.50M means that the solution contains 0.50 mole of the compound in one liter of solution. The use of solutions of known concentration gives a convenient way to obtain a specific volume of solution. Since M is moles of solute per liter of solution, moles of solute can be calculated by multiplying volume of solution times molarity of solution:

$$\text{L}\left(\frac{\text{moles}}{\text{L}}\right) = \text{moles} \tag{4-3}$$

## Review

Equipment and Techniques G
Appendix III

## Equipment Needed

- 100-mL graduated cylinder
- graph paper (provided by the student)
- ruler
- Styrofoam cup
- 0.1° thermometer

## Chemicals Needed

0.50M KI; potassium iodide
0.50M KNCS; potassium thiocyanate
0.50M NaOCl; sodium hypochlorite
0.50M $Ma_2SO_3$; sodium sulfite
0.50M $Na_2S_2O_3$; sodium thiosulfate

## Procedure

Put 60.0 mL of 0.50M NaOCl into a Styrofoam cup and measure its temperature using a 0.1° thermometer. Measure the temperature of 40.0 mL of 0.50M KI (or other reactant, as directed by your instructor) in a graduated cylinder. These temperatures should be the same; if they are not the same, record both of them and use their average as $T_{\text{initial}}$. Pour this second reactant into the cup containing the NaOCl, quickly mix the reactants thoroughly with the thermometer, and record the maximum temperature that is reached, $T_{\text{final}}$. The mixture will warm up as reaction occurs and then it will gradually cool as heat is lost to the surroundings. However, Styrofoam is a fairly good insulator, and so this cooling will occur slowly and you should have no trouble observing the maximum temperature with sufficient accuracy.

After you complete this first trial, the solution in the Styrofoam cup can be washed down the drain. Rinse the cup with tap water and distilled water; dry it thoroughly. Rinse and dry the graduated cylinder.

Repeat the preceding procedure using different amounts of reagents having a total volume of 100 mL. Use the volume combinations 40:60, 80:20, and 20:80.

After you have measured and recorded the temperature changes for these four combinations of reactants, make a graph of volume of NaOCl solution used ($x$ axis) vs. temperature change, $\Delta T$ ($y$ axis). Your graph should include points for the volume "combinations" 0:100 and 100:0, as well as the four experimental points.

Your plot should consist of two straight-line segments, a line at the left having a positive slope and a line at the right having a negative slope, similar to Figure 4-1. The intersection of these two straight-line segments corresponds to maximum $\Delta T$ at the stoichiometrically correct amount of NaOCl solution. Draw each of these straight-line segments only if at least three points form the straight line. If there are not at least three points for each straight-line segment, you must make one more measurement using a different volume combination of reactants.

Consider ratios of 50:50, 70:30 (or 30:70), and 90:10 (or 10:90). Pick one combination that clearly will give a third point in the straight-line segment that does not already have three points. If you have any doubt about an appropriate choice, consult your instructor so that you will not waste time and materials on a trial that does not give you helpful data. Run a sixth trial if your instructor advises it because of considerable experimental error that decreases the usefulness of your plot.

## Results

Make a graph on which you plot moles of NaOCl ($x$ axis) vs. $\Delta T$ ($y$ axis). (Note that this is not the same as the graph you made while you were doing your laboratory work.) Below the scale for moles of NaOCl, give a second scale for moles of the other reactant; identify the reactant. Use this graph to decide on the stoichiometry of the reaction. Write the equation for the reaction $x$NaOCl + $y$Y $\rightarrow$ products + $Q$ using the correct coefficients $x$ and $y$ determined from your data plot. (Y refers to the material with which NaOCl reacted. You are not expected to know identities and coefficients of products.)

## Questions

1. What is the purpose of keeping the total volume of the reacting solutions the same when determining the stoichiometric ratio?

2. Imagine that a reaction occurs with the absorption of heat, and so the temperature of the reaction mixture drops as the reaction progresses. Sketch a graph that shows $\Delta T$ vs. number of moles of a reactant for a reaction of this type. Is $\Delta T$ a negative or positive quantity?

# EXPERIMENT 4

## REPORT SHEET

Name ______________________________ Date ______________

Section ______________________________

**Data**

| 0.50M NaOCl (mL) | 0.50M reactant Y (mL) | $T_{initial}$ | $T_{final}$ | $\Delta T$ |
|---|---|---|---|---|
| | | | | |
| | | | | |
| | | | | |
| | | | | |
| | | | | |
| | | | | |

Attach your graph of $\Delta T$ vs. number of moles of NaOCl. Explain how you determined the coefficients. Write the equation for the reaction.

(On the back of the Report Sheet answer the Questions).

# EXPERIMENT 5

# SYNTHESIS OF AN ALUM, $KAl(SO_4)_2 \cdot 12H_2O$

## Introduction

Potassium aluminum dodecahydrate, $KAl(SO_4)_2 \cdot 12H_2O$, belongs to a class of inorganic compounds called *alums* (which is accented on the first syllable). The composition of an alum is represented by the general formula $M^+M^{3+}(SO_4)_2 \cdot 12H_2O$. In this formula, $M^+$ is a monoatomic ion such as $Na^+$, $K^+$, $Tl^+$, or $Ag^+$; or is $NH_4^+$. Although the name "alum" sounds as if the compound must contain aluminum, the $M^{3+}$ might be from any of several main-group metals (p block elements) and transition metals (d block elements) that form triply charged ions: Al, Ti, V, Cr, Mn, Fe, Co, Ga, In, Fe, and Ir. Some examples are $NaAl(SO_4)_2 \cdot 12H_2O$, which is a component of some baking powders; $NH_4Al(SO_4)_2 \cdot 12H_2O$, which is used in dyeing textiles; and $KCr(SO_4)_2 \cdot 12H_2O$, which is used in tanning leather.

The term *double salt* also describes $KAl(SO_4)_2 \cdot 12H_2O$. A double salt is an ionic compound that contains two different cations or two different anions. In aqueous solution, double salts give positive tests for all three of the ions they contain. $KAl(SO_4)_2 \cdot 12H_2O$ would give positive tests for $K^+$, $Al^{3+}$, and $SO_4^{2-}$.

*Hydrate* is another term that describes $KAl(SO_4)_2 \cdot 12H_2O$. A hydrate is a substance that contains a specific number of water molecules per formula unit. Hydration is common with ionic compounds, for example $Na_2SO_4 \cdot 10H_2O$ and $CoCl_2 \cdot 6H_2O$. The water of hydration is included when calculating the compound's formula weight. The water molecules are part of the crystal of a compound along with cations and anions, but they can break out of the crystal when it is heated. This dehydration might occur at a specific temperature characteristic of the particular compound. If the compound is heated in a relatively closed container, the escaping water might dissolve the remaining salt forming a liquid solution. Consequently, this combination of dehydration and dissolving looks like melting; the temperature at which it occurs can be measured in the same way that a melting point is measured.

## Reactions Involved in the Synthesis

Metals dissolve in water or aqueous solutions only by reacting with water or some component of the solution. The reaction converts the metal to a positive ion. Aluminum will dissolve in acids by reacting with $H_3O^+$ and will dissolve in bases by reacting with $OH^-$. The reaction you will use to dissolve aluminum involves $OH^-$ from KOH:

$$2Al(s) + 2OH^-(aq) + 6H_2O \rightarrow 2Al(OH)_4^-(aq) + 3H_2(g) \qquad (5\text{-}1)$$

Hydrogen is flammable, and so this reaction must be carried out in a hood to remove $H_2$ from the laboratory atmosphere.

Next, you will add sulfuric acid to the solution containing $Al(OH_4)^-$. As $H_3O^+$ reacts with $Al(OH)_4^-$, insoluble $Al(OH)_3$ forms and precipitates. As more acid is added, the precipitated Al $(OH)_3$ dissolves:

$$Al(OH)_4^-(aq) + H_3O^+(aq) \rightarrow Al(OH)_3(s) + 2H_2O \quad (5\text{-}2)$$

$$Al(OH)_3(s) + 3H_3O^+(aq) \rightarrow Al^{3+}(aq) + 6H_2O \quad (5\text{-}3)$$

The product forms and precipitates when the solution containing $Al^{3+}$, $K^+$ (from KOH), and-$SO_4^{2-}$ (from $H_2SO_4$) is cooled:

$$K^+(aq) + Al^{3+}(aq) + 2SO_4^{2-}(aq) + 12H_2O \rightarrow KAl(SO_4)_2 \cdot 12H_2O \quad (5\text{-}4)$$

## Review

Equipment and Techniques A, C, D, E, G, H
Experiment 2C

## Equipment Needed

- balance
- beakers
- bottle or test tube
- burner
- clamp
- Erlenmeyer flasks
- filter paper
- funnel
- glass rod
- "Glassine" paper
- graduated cylinder
- ice
- melting point capillary tubes
- ring
- ring stand
- slit rubber stopper
- spatula
- thermometer
- wire gauze

## Chemicals Needed

- acetone
- Al foil
- 6M $H_2SO_4$; sulfuric acid
- 3M KOH; potassium hydroxide
- thymol blue solution

## Procedure

1. Dissolve Al by reaction with KOH.

Tear a piece of aluminum foil approximately three inches long from the roll that is provided. Weigh it on the laboratory balance. Remove some Al from it or add more Al to it until you have 1.0-1.1 grams; record the weight. Tear the Al into pieces about the size of large postage stamps and put these pieces into an Erlenmeyer flask that is labelled with your name.

To the Al in the flask add 20.0 mL of 3M KOH. **Now take the flask to the hood.** Within a few minutes the Al and KOH will begin to react producing hydrogen; this reaction will become progressively more vigorous. Do not handle the flask; the reaction is so exothermic that the entire flask, including the neck, will become extremely hot. After the obvious evolution of hydrogen ceases, let the flask stand for a few minutes more to ensure complete reaction of the Al. Then cool the flask slightly by wiping the neck with a wet sponge. When the flask is cool enough that you can safely pick it up, take it back to your desk and cool it further under cold, running water until you can handle the entire flask comfortably.

If the solution in the flask is *absolutely clear*, proceed to the next step in the synthesis. If the solution is not absolutely clear, filter it by gravity filtration. Collect the filtrate in another Erlenmeyer flask. Discard the filter paper in the regular trash container.

2. Convert $Al(OH)_4^-$ to $Al^{3+}$ by reaction with $H_2SO_4$.

To the clear solution in an Erlenmeyer flask, add 16.0 mL of 6M $H_2SO_4$ a few milliliters at a time. After each addition, swirl the flask to mix the contents. Initially, $Al(OH)_3$ will precipitate; then some of it will dissolve as more $H_2SO_4$ is added. If solid sticks to the inside wall of the flask, wash it down with a few drops of $H_2SO_4$. After you have added all the $H_2SO_4$, gently heat the flask for several minutes; have the liquid *barely bubbling*. If the amount of $H_2SO_4$ you used was sufficient for the amounts of Al and KOH you used, all the solid in the flask should dissolve.

If solid remains, continue the gentle heating for several more minutes; do not increase the intensity of the heating. If solid still remains, gradually add more $H_2SO_4$ dropwise with gentle heating until all the solid dissolves and you have a clear solution; this should not require more than one mL (about 20 drops) of $H_2SO_4$.

3. Precipitate $KAl(SO_4)_2 \cdot 12H_2O$ by cooling.

Cool the flask to about room temperature under cold, running water. Then put it into a beaker of ice water; be sure the water in the beaker is at least as deep as the solution in the flask. Occasionally stir the solution with a glass rod. After crystals of the product begin to form, cool the mixture for at least another ten minutes; continue to stir it occasionally. While you wait for the product to precipitate, set up the equipment for filtration; use gravity or suction filtration as your instructor directs.

4. Collect, wash, and dry $KAl(SO_4) \cdot 12H_2O$.

Allow the precipitated product to settle to the bottom of the flask. Decant as much as possible of the liquid into the funnel and leave as much as possible of the solid in the flask. To the material remaining in the flask, add about 10 mL of distilled water that has been chilled in a flask or small beaker in an ice bath. Swirl the flask to mix the contents, allow the solid to settle again, and decant as much as possible of the liquid into the funnel.

The *washing* of the precipitate is done to remove the solution containing excess starting materials so that the product will be pure. To monitor the progress of the washing, put two or three drops of distilled water into a 4-mL test tube, add one drop of thymol blue, and collect a few drops of the filtrate in this test tube. The thymol blue is an acid-base indicator that is pink in a strongly acidic solution, such as the sulfuric acid you are using; it is yellow in a solution that is weakly acidic or neutral, such as a solution of pure $KAl(SO_4) \cdot 12H_2O$ or distilled water. The first filtrate that you collect should give a pink color with the indicator showing that an excess of acid was used; when all this excess acid is washed out of the product, you can be confident that any other excess starting material has also been washed away. After testing this first filtrate, keep the test tube of indicator and filtrate to use to make color comparisons, so that you can monitor your progress in washing out the acid.

Repeat the washing with another 10 mL of ice-cold water; test a few drops of the *last of the filtrate* with a fresh water-and-thymol-blue solution in a clean test tube. Repeat this washing and testing (using a fresh tube of indicator solution for each test) until the filtrate gives a yellow, or perhaps a pale orange, color with thymol blue. This will probably take at least four or five washings. Then use a little more ice-cold water to transfer the remaining solid to the funnel. Allow the solution to drain from the solid as thoroughly as possible.

Obtain 20 mL of acetone in a dry container. Pour about five mL of acetone over the entire surface of the solid in the funnel; then allow all the liquid to drain from the funnel. Repeat this procedure three more times. The acetone washes away the water; acetone is much more volatile than water and it will evaporate fairly rapidly to give a dry product. When you complete the filtering and washing, the filtrate can be discarded by washing it down the drain. The thymol blue solutions can also be discarded by washing them down the drain.

Obtain a large piece of filter paper and fold it in half. Lift the wet filter paper and product from the funnel and use a spatula to scrape the solid onto the larger piece of dry filter paper. Use a spatula to spread the solid out as much as possible; cut through it repeatedly with a spatula to expose new solid surface and hasten drying. When the large piece of filter paper is quite wet, transfer the solid to another large piece of filter paper folded in half; continue cutting through and spreading the solid.

Transfer a small amount of the solid to a piece of "Glassine" paper. Spread it out completely so that it can dry thoroughly in a few minutes. While it dries, set up the equipment for measuring melting point.

5. Determine "Solid to Liquid" transition temperature of $KAl(SO_4)_2 \cdot 12H_2O$.
   Wait until you are sure that the small sample of the product is completely dry.
   Follow the procedure in Experiment 2C for determining melting points. Record a temperature range, but make repeated measurements to determine a range that is as narrow as possible; use a fresh capillary tube and sample for each measurement. The narrowness of this range and the accuracy of your result depend on the purity, including dryness, of your product.

6. Determine yield.
   While you are carrying out the preceding step, you should occasionally cut through and redistribute your remaining product to facilitate drying.
   Weigh a bottle or test tube that is labelled with the compound's formula, the date, and your name; put your product into this and reweigh it with the product. Turn in your product as part of your report.

## Report

Calculate the number of moles of each reactant you put into the reaction mixture: Al, KOH, and $H_2SO_4$. On the basis of the composition $KAl(SO_4)_2 \cdot 12H_2O$, determine which reactant limits the amount of product that can form.

Calculate the theoretical yield of product, your actual yield, and your percentage yield.

Report the product's "Solid to Liquid" transition temperature.

Turn in your product.

## Questions

1. The way this experiment is carried out it is impossible to recover 100% of the product that forms, no matter how carefully you work. Explain why; consider a particular physical property of the product.

2. Consider the weight of Al foil that you need. Is this actually the weight of Al that reacts? Explain. How will this affect your calculated percentage yield?

3. In the fourth step of the synthesis you washed the product with ice-cold water. If this were not done, what ions, that were used in excess of the stoichiometric amounts, would contaminate the product?

4. In many syntheses, yield must be sacrificed for purity. For this synthesis, what particular step in the procedure increased the purity of the product, but decreased the yield of the product?

5. Describe any mistakes you made, accidents that occurred, or sloppy technique that might have lowered your percentage yield.

# EXPERIMENT 5

# REPORT SHEET

Name ______________________________ Date ______________

Section ______________________________

Mass of Al foil used ______________________________

Moles of Al used ______________________________

Volume of 3M KOH used______________________________

Moles of KOH used ______________________________

Mass of KOH used______________________________

Volume of 6M $H_2SO_4$ used ______________________________

Moles of $H_2SO_4$ used______________________________

Mass of $H_2SO_4$ used ______________________________

Limiting reactant ______________________________

Theoretical yield of $KAl(SO_4)_2 \cdot 12H_2O$, moles ______________________

Theoretical yield of $KAl(SO_4)_2 \cdot 12H_2O$, grams ______________________

Actual yield of $KAl(SO_4)_2 \cdot 12H_2O$, grams ______________________

% yield of $KAl(SO_4)_2 \cdot 12H_2O$ ______________________

"Solid to Liquid" transition temperature range ______________________

(On the back of the Report Sheet answer the Questions).

# EXPERIMENT 6

# A SIMPLE COVALENT COMPOUND: SYNTHESIS AND PROPERTIES

## Introduction

Many compounds can be synthesized by the interaction of elements with another. This is especially true of *binary* compounds, that is, compounds that contain only two elements. The formula of the compound that results and the type of bonding within the compound depend on the natures of the atoms of the two elements. A fairly great difference in the electronegativities of the atoms results in a transfer of electrons from one atom to the other; *ions* form. Atoms of similar electronegativities share electrons in covalent bonds; *molecules* form. In this experiment you will allow tin and iodine to react with one another. The electronegativities of 1.8 for tin and 2.5 for iodine correctly suggest that the bonding should be covalent. An iodine atom, with seven valence electrons, attains a noble-gas configuration by the formation of one covalent bond. A tin atom, which has four valence electrons, can form four covalent bonds. Consequently, the formula of the resulting compound is $SnI_4$; the name is tin(IV) iodide (commonly, tin tetraiodide or stannic iodide).

Molecular and ionic compounds can often be distinguished from one another by their physical properties. The attractive forces between molecules that hold them to one another in the solid and liquid states are generally much weaker than the strong attractive forces between the oppositely charged ions of an ionic compound. Consequently, compounds composed of molecules have much lower melting points and boiling points; they are also much more volatile. Another major difference is the solubilities of ionic and molecular compounds. Ionic compounds are more likely to be soluble in water than in other liquids that are commonly used as solvents; however, some ionic compounds are not soluble even in water. Molecular compounds are often soluble in many liquids that are composed of polar or nonpolar molecules, but are less likely to be soluble in water (unless they react with it). In this experiment you will compare these properties for some molecular and ionic compounds.

## Review

Equipment and Techniques D, H

## Equipment needed:

- balance
- beakers
- burner
- 125-mL Erlenmeyer flask
- graduated cylinder
- rubber stopper
- sample bottle or test tube
- spatula
- 4-mL test tubes
- watch glass

## Chemicals needed:

$CH_2Cl_2$; dichloromethane
$C_6H_{14}$; hexane
$C_{10}H_8$; naphthalene
$CaSO_4$; calcium sulfate
$I_2$; iodine
NaCl; sodium chloride
Sn; tin (foil or granular)

## Procedure

### A. Synthesis of $SnI_4$

Into a clean and dry 125-mL Erlenmeyer flask, weigh 0.5 ± 0.1 g of Sn using the laboratory balance; record the weight you actually take. To this Sn, add 10 mL of $CH_2Cl_2$; stopper the flask loosely. (Because **$CH_2Cl_2$ is quite volatile and is hazardous to your health** if a large amount is inhaled, you should uncover it only when necessary and for as brief a time as possible. When you pour $CH_2Cl_2$, place the two containers as close to one another as possible to minimize vaporization.) Weigh 1.0 to 1.2 g of $I_2$ into a beaker; record the weight you actually take. Cover this beaker with a watch glass so that $I_2$ vapor will not pass into the air in the laboratory. Add the $I_2$ to the mixture of Sn and $CH_2Cl_2$; immediately restopper the flask.

The color of $SnI_4$ is much different from the color of $I_2$, and so you can monitor the progress of the reaction by observing the color of the solution. Gently swirl the flask until the color of the solution no longer changes. If you know the color has changed, but you are not sure whether the color change is complete, write down your description of the color in as much detail as possible. Then observe the color again after two or three minutes and decide whether it has changed or is still the same. If it has changed, reaction was still occurring; if it has remained the same, reaction is complete. Complete reaction of the iodine will take about 15 to 25 minutes if everything has been done correctly.

**In a hood** decant the $CH_2Cl_2$ solution of $SnI_4$ into a 50-mL beaker labelled with your name; leave the unreacted tin in the flask. If any solid $SnI_4$ remains in the flask, put two to three mL of $CH_2Cl_2$ into the flask to dissolve the $SnI_4$; decant this solution into the beaker. Evaporate the $CH_2Cl_2$ to obtain the solid product. Several ways to do this are described here. Your instructor will tell you which method to use.

Method 1: Lower the hood window so that just enough space remains to place the beaker under the window. Position the beaker so that about half of it extends in front of the window and half of it extends behind the window. As air is drawn into the hood it will pass over the solution and speed the evaporation of the $CH_2Cl_2$. This is the simplest and most convenient method if a hood that draws well is available.

Method 2: If compressed air is available, direct a stream of it across the surface of the solution (as shown in Figure 6-1). This must be done **in a hood**.

Method 3: Use a water aspirator to pull a stream of air across the surface of the solution (see Figure 6-2). This should be done **in a hood**.

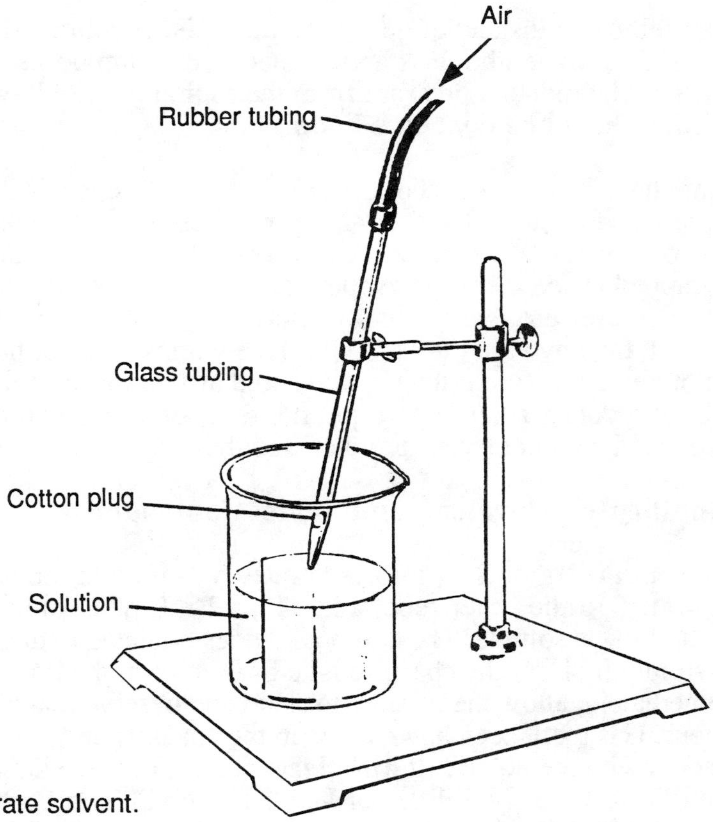


**Figure 6-1**

Using compressed air to evaporate solvent.

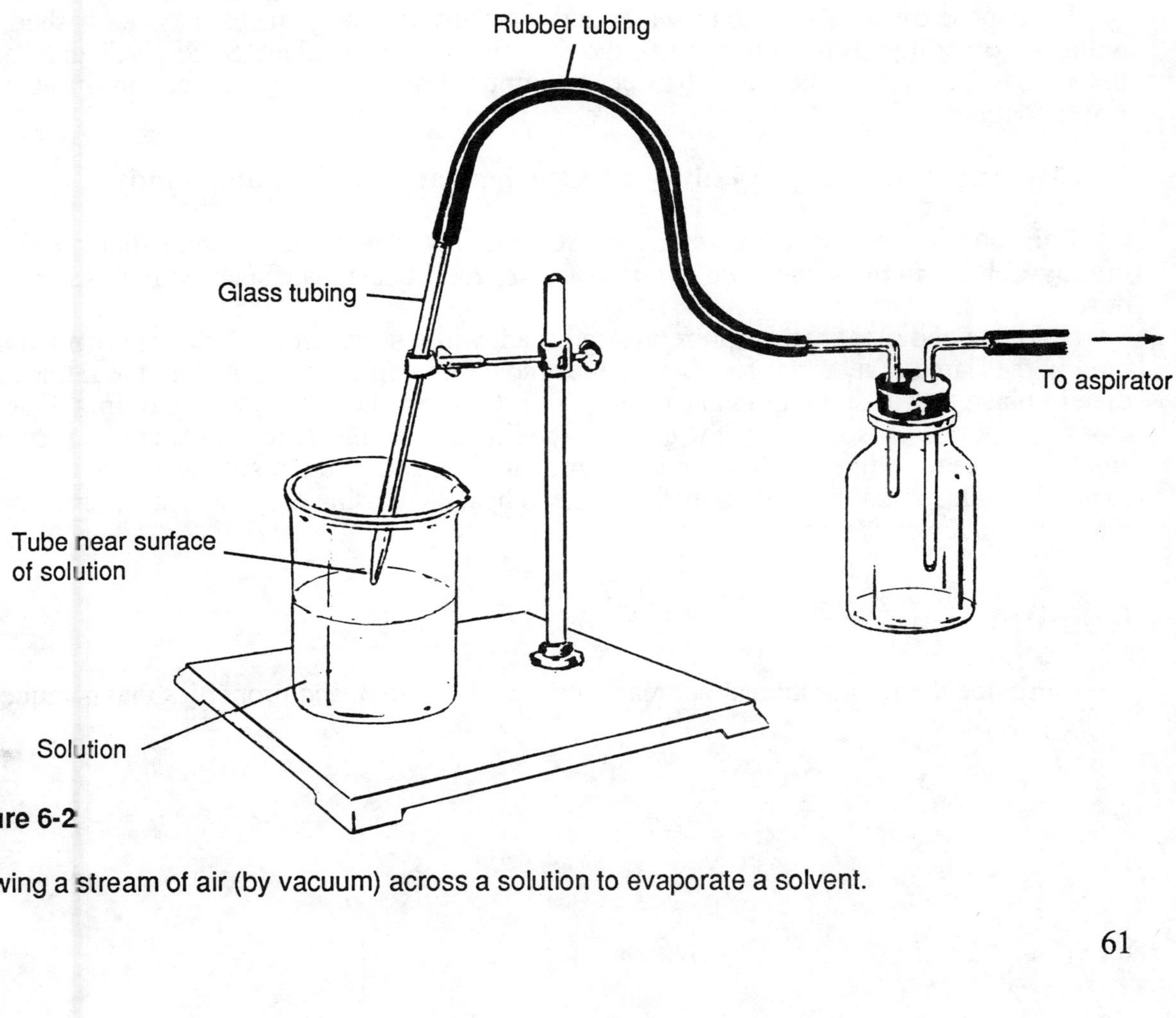


**Figure 6-2**

Drawing a stream of air (by vacuum) across a solution to evaporate a solvent.

Method 4: Instead of decanting the $CH_2Cl_2$ solution into a beaker, decant it into a filter flask. Stopper the flask and connect it to a trap bottle that is connected to a water aspirator. The reduced pressure from the aspirator will allow the $CH_2Cl_2$ to boil away. Again, this should be done **in a hood**.

Drying the $SnI_4$ will take about 15 to 45 minutes, depending on the method. While you wait for it to dry, proceed to parts B and C with other substances. When the $SnI_4$ is dry, scrape it out of the beaker onto a piece of filter paper. Use a spatula to break up large pieces of the solid so that any $CH_2Cl_2$ trapped inside them can evaporate. Do *not* continue your investigations of the properties of $SnI_4$ until you are sure that it is completely dry. (**Smell it cautiously**; hold it a few inches from your face and fan any vapor toward your face with your other hand. If you can detect any odor, the solid is not yet dry.) Weigh the dry $SnI_4$ so that you can calculate your percentage yield. Turn in your $SnI_4$ with your written report; put it into a bottle or test tube that is labelled with your name, the formula and weight of your product, and the date.

### B. Solubility of Covalent and Ionic Compounds

Place about 0.1 g of $SnI_4$ into each of two 4-mL test tubes. To one tube, add about 2 mL of distilled water; to the other tube, add about 2 mL of hexane ($C_6H_{14}$). For each liquid, record whether the $SnI_4$ is soluble. The compound does not have to dissolve completely to be classified as soluble; simply look for the characteristic $SnI_4$ color in the liquid. If the $SnI_4$ does not seem to dissolve immediately, allow the mixture to stand and observe it again after five minutes and again after half an hour. Has there been any *change* in the appearance (color, texture) of the $SnI_4$ that suggests it has reacted with the liquid? (It *will react* with one of these liquids.)

Similarly, test the solubilities of $I_2$, $C_{10}H_8$, NaCl, and $CaSO_4$ in water, $C_6H_{14}$, and $CH_2Cl_2$. Classify each of these solids as soluble or insoluble in each of these three liquids.

To dispose of these solutions, water and water-soluble substances may be washed down the drain. All other liquids must be put into the specified waste container. Solid NaCl and $CaSO_4$ (*only* these) may be put into the waste basket. Any unused naphthalene must be put into the specified waste container.

### C. Melting Point and Volatility of Covalent and Ionic Compounds

**This portion of the experiment must be done in a hood.** Be sure that no one is evaporating $CH_2Cl_2$ in the same hood. Your instructor might designate specific hoods for each procedure.

Hold one end of a spatula (preferably a broad, white porcelain one; definitely not an all-metal one) in the flame of a burner for about 20 seconds. Remove the spatula from the flame and immediately place a very small amount of $SnI_4$ on it. Observe the behavior of the $SnI_4$. Does it melt? Does it vaporize? If so, what is the color of the liquid or of the vapor? Repeat this procedure with small samples of iodine, naphthalene, sodium chloride, and calcium sulfate. (Note that you are not to put the solids into the burner flame; you are to heat them using only the heat retained by the spatula after it is removed from the flame.)

## Results

Provide the information about reactants, reaction, yield, and properties that is requested on the Report Sheet.

## Questions

1. Why was $CH_2Cl_2$, rather than water, chosen as the solvent for the reaction between tin and iodine? (There is more than one reason.)

2. Why was excess tin used in the reaction, rather than excess iodine? (Think about what you do after the reaction to separate the components of the reaction mixture and isolate the pure product.)

3. When the five substances were heated as described in the experiment, which substances behaved similarly? Account for similarities in terms of the type of particles (ions or molecules) of which each substance is composed. Account for differences in behavior.

4. Draw Lewis electron dot formulas for $SnI_4$, $I_2$, naphthalene, hexane, and $CH_2Cl_2$. Is each of these molecules polar or nonpolar? (You will have to look up naphthalene in a chemical handbook or encyclopedia, or in an organic chemistry textbook; do not try to guess at its structure.)

5. What similarities in solubilities did you observe for the five substances? Account for similarities in terms of the type of particles (ions or molecules) of which each substance is composed. Account for differences in behavior. In your discussion include the name for the kind of attraction between particles in each solid, the name for the kind of attraction between particles in each liquid, and the name for the kind of attraction between solute and solvent particles in each solution.

# EXPERIMENT 6

## REPORT SHEET

Name ______________________________ Date ______________

Section ______________________________

Write a balanced equation for the reaction between tin and iodine to produce $SnI_4$.

Mass of Sn used ______________________

Moles of Sn used ______________________

Mass of $I_2$ used ______________________

Moles of $I_2$ used ______________________

Limiting reactant ______________________

Theoretical yield of $SnI_4$, moles ______________________

Theoretical yield of $SnI_4$, grams ______________________

Show calculation.

Actual yield of $SnI_4$, grams ______________________

Actual yield of $SnI_4$, moles ______________________

Percentage yield of $SnI_4$ ______________________

Show calculation.

Turn in your product in a labelled bottle or test tube.

Describe the physical appearance of $SnI_4$.

Solubilities:

| Substance | In water | In $C_6H_{14}$ | In $CH_2Cl_2$ |
|---|---|---|---|
| $SnI_4$ | | | |
| $I_2$ | | | |
| $C_{10}H_8$ | | | |
| NaCl | | | |
| $CaSO_4$ | | | |

Describe the *change* in appearance of $SnI_4$ in the liquid with which it *reacted.*

Behavior when heated:

$SnI_4$

$I_2$

$C_{10}H_8$

NaCl

$CaSO_4$

(Attach a separate sheet with the answers to the Questions).

# EXPERIMENT 7

# SYNTHESIS OF $[Co(NH_3)_6]Cl_3$

## Introduction

One activity that occupies chemists is the synthesis or preparation of chemical compounds. A goal of synthetic chemistry is to make substances that have never been previously prepared. Some chemists work to find ways of preparing in the laboratory compounds that occur in nature and are useful. An example is the development of methods for synthesizing penicillin. Other chemists strive to make totally unknown compounds in the expectation that they may have useful properties or that their behavior can tell us something fundamental about the workings of nature. In order to perfect their grasp of the techniques involved in making compounds, chemists must spend a good deal of time duplicating syntheses of known compounds in the course of their training.

The purpose of this experiment is to gain experience in using the laboratory techniques involved in synthetic chemistry. The compound you will prepare is called a *coordination compound.* The name of this compound is *hexaammine cobalt(III) chloride.* In $[Co(NH_3)_6]Cl_3$, the cation $[Co(NH_3)_6]^{3+}$ has six molecules of ammonia bonded to $Co^{3+}$ in a regular geometric fashion (see Figure 7.1). All the groups bonded together are written inside square brackets. The three $Cl^-$ ions serve to balance the cation charge just as they do in a simple salt such as $CaCl_2$. The compound will be prepared by the reaction of cobalt chloride, ammonia and hydrogen peroxide as in Equation (7-1):

$$2CoCl_2{\cdot}6H_2O) + 2NH_4Cl + 10NH_3 + H_2O_2 \rightarrow 2[Co(NH_3)_6]Cl_3 + 14H_2O \qquad (7\text{-}1)$$

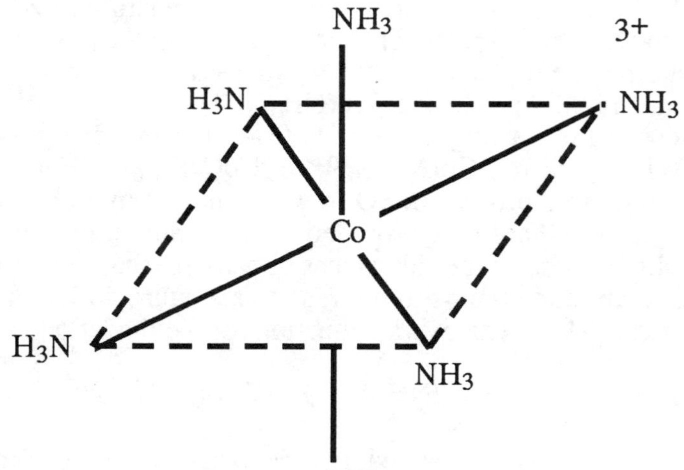


Figure 7-1 Structure of $[Co(NH_3)_6]^{3+}$

Almost all synthetic reactions are carried out in solution by dissolving the reactants and allowing them to form the desired product. After the product forms in solution, the experimenter must isolate it (get it out of solution) and purify it (rid it of all traces of the unused reactants and any side products). (Side products are any reaction products other than the one of interest.)

Since the solubility of most compounds decreases with temperature, a common technique for isolating a product is to cool the solution until the amount of compound dissolved exceeds its solubility at that temperature. The excess then crystallizes out of solution. Unfortunately, as the temperature drops, compounds other than the desired product (such as unreacted starting material or side products) also become less soluble and may crystallize out of solution, too.

These contaminants can be removed by *recrystallization*. Recrystallization involves dissolving the impure product in a minimum volume of hot solvent, filtering the solution to remove insoluble impurities, and then cooling the solution to crystallize the desired compound while unwanted materials remain in solution. Sometimes the recrystallization must be repeated many times in order to obtain a pure product. In this experiment, however, you can stop after only one recrystallization since $[Co(NH_3)_6]Cl_3$ can be purified fairly easily by this method.

## Reactions Involved in the Synthesis

As in most syntheses, we start out with compounds that are conveniently available commercially, in this case with $CoCl_2{\cdot}6H_2O$. This is the most expensive reagent in the synthesis, so all other reagents are added in slight molar excess to make sure that all the cobalt chloride hexahydrate is consumed.

Ammonia ($NH_3$) must be added, of course, since it is the group attached to cobalt in the product. Ammonia is commonly available in the laboratory as an aqueous solution. Sometimes these solutions of ammonia are referred to as "ammonium hydroxide" because the following reaction takes place to a very limited extent when $NH_3$ is dissolved in water:

$$NH_3 + H_2O \rightarrow NH_4^+ + OH^- \qquad (7\text{-}2)$$

Experimental measurements show that more than 99% of the material is still present as $NH_3$; only about 1% has reacted to form ammonium and hydroxide ions. Hence, it is certainly more accurate to refer to the solutions as aqueous ammonia, written as $NH_3(aq)$.

Actually, the cobalt chloride hexahydrate has a structure similar to that of the product with the six waters directly attached to $Co^{2+}$. Its formula is more correctly written as $[Co(H_2O)_6]Cl_2$. The first step in the reaction is the replacement of the attached waters by $NH_3$. Charcoal catalyzes this process so that it occurs rapidly producing $[Co(NH_3)_6]^{2+}$.

The cobalt present in the starting material is $Co^{2+}$, but the final product contains $Co^{3+}$. The function of the $H_2O_2$ is to oxidize $[Co(NH_3)_6]^{2+}$ to $[Co(NH_3)_6]^{3+}$. $H_2O_2$ is reduced to $H_2O$ in the reaction. Since there are more protons per O in water than in hydrogen peroxide, some source of additional protons is needed. These are provided by the ammonium ion of $NH_4Cl$. Removing the proton leaves $NH_3$ which is one of the substances already present in the solution. So, we avoid the addition of many different compounds to the reaction mixture which could potentially remain to contaminate the desired product. The overall equation for the oxidation is:

$$2[Co(NH_3)_6]^{2+} + H_2O_2 + 2NH_4^+ \rightarrow 2[Co(NH_3)_6]^{3+} + 2H_2O + 2NH_3 \qquad (7\text{-}3)$$

Since hydrogen peroxide has been added in large excess, the unreacted material is now decomposed by heating:

$$2H_2O_2 \xrightarrow{\Delta} 2H_2O + O_2\,(g) \qquad (7\text{-}4)$$

It is now possible to understand why the reaction mixture is cooled when the oxidation with $H_2O_2$ is done. The oxidation reaction releases heat. This heat could speed up the decomposition reaction (7-4). Reaction (7-4) itself releases heat which would further speed up the decomposition. Much $O_2$ gas would be released causing the mixture to foam out of the flask.

The crude product is precipitated by cooling the reaction mixture in an ice bath until no more precipitate seems to be forming. Because charcoal is not soluble in water, black charcoal will be in the precipitate with your yellow compound, and the entire mixture will look black.

Recrystallization is carried out to remove charcoal. Once the product has been dissolved in hot HCl solution and the charcoal filtered off, the solution is cooled to reprecipitate the hexaamminecobalt(III) chloride. Concentrated HCl is added to precipitate the compound as completely as possible. The use of excess chloride in this circumstance is an example of the *common ion effect*. The addition of a substance (HCl) with an ion common to a compound in solution ($Cl^-$) drives the precipitation reaction farther toward completion:

$$[Co(NH_3)_6]^{3+}(aq) + 3Cl^-(aq) \rightarrow [Co(NH_3)_6]Cl_3(s) \quad (7\text{-}5)$$

## Review

Equipment and Techniques A, B, D, E, H

## Equipment Needed

- beakers (50 mL, 100 mL, 250 mL)
- Büchner funnel
- burner
- Erlenmeyer flask (125 mL)
- filter flask (125 mL)
- filter paper
- graduated cylinders (10 mL, 100 mL)
- ice bath
- medicine dropper
- ring stand with ring
- spatula
- stirring rods
- test tube with cork
- thermometer
- vacuum tubing
- wire gauze

## Chemicals Needed

- C; activated charcoal
- $CH_3C(O)CH_3$; acetone ice
- $CoCl_2{\cdot}6H_2O$; cobalt chloride hexahydrate
- 12M HCl; conc. hydrochloric acid
- 0.5M HCl; hydrochloric acid
- 30% $H_2O_2$; hydrogen peroxide
- 15M $NH_3(aq)$; conc. ammonia
- $NH_4Cl$; ammonium chloride

## Procedure

**Caution! All work must be performed under the hood.**

### A. Making the Compound

(If you should remove more reagent from the stock bottles than you need, rinse the excess down the drain with plenty of water. Charcoal is not soluble in water and can be thrown in the wastebasket.)

Weigh out 9.0 ± 0.1 g $CoCl_2{\cdot}6H_2O$. Record the weight you actually take. This weight must be known to the indicated degree of accuracy since cobalt chloride is the limiting reagent. The exact weights of other materials used in this synthesis are less important since they are all present in molar excess.

Weigh out about 6 g $NH_4Cl$ into a clean 100-mL beaker. Add 15 mL of distilled water. Place the beaker on wire gauze on a ring over your burner. Use a low flame to heat the solution gently until you bring it to a boil. Stir it with your stirring rod during the heating.

When the solution is boiling, add the $CoCl_2 \cdot 6H_2O$ to it gradually. Continue stirring during the addition; be sure to heat and stir gently so that drops of the hot solution do not spatter out of the beaker. After all the $CoCl_2 \cdot 6H_2O$ has dissolved, turn off the burner. Use beaker tongs or a towel to pick up the hot beaker and pour the hot solution into the Erlenmeyer flask containing the charcoal. Rinse the beaker with a little distilled water to transfer all the cobalt chloride to the flask. Cool the flask to room temperature under running water. To this cooled mixture add 23 mL conc. $NH_3$. Cool the flask in an ice bath until the temperature of the mixture is 10°C or lower. (Two students should share an ice bath.)

While the mixture is cooling, use a graduated cylinder to measure about 13 mL of 15% $H_2O_2$ and pour it into your 50 mL beaker. **Caution! Hydrogen peroxide can cause severe burns. If you spill even a drop of it on your skin, wash it off immediately with plenty of running water. If you spill any on the bench top, wipe it up with a sponge and then wash out the sponge repeatedly under running water. Hydrogen peroxide can also damage your clothing.**

When the temperature of the reaction mixture falls to 10°C or lower, carry out the oxidation by adding hydrogen peroxide a few drops at a time with the medicine dropper. During the addition, leave the flask in the ice bath; swirl the flask gently to mix the $H_2O_2$ with the other reactants.

After all the $H_2O_2$ has been added, return the flask to the wire gauze and heat the mixture gently to 50-60°C. Swirl the flask occasionally to keep the contents mixed. Continue heating and swirling for about 20 minutes, or until the original pink color characteristic of $CoCl_2 \cdot 6H_2O$ turns to a golden yellow. A good way to test the color of the solution (which will look black because of the suspended charcoal) is to dip a stirring rod into the flask and transfer a drop of solution from the tip of the rod to a piece of white filter paper. The solution will spread out on the paper, enabling you to see its color.

After the proper color change has occurred, the flask and its contents are cooled in an ice bath. The product will precipitate and be mixed with insoluble charcoal. Continue cooling until the volume of precipitate no longer seems to be increasing. Filter the solution using a Büchner funnel containing two pieces of filter paper. It does not matter if a little charcoal passes into the filtrate, but be sure you don't lose any of the crystals of the product. Wash the precipitate in the funnel with two 5-mL portions of ice-cold distilled water. To perform the washing operation, disconnect the suction and add the wash solvent; then reconnect the vacuum tubing and suck the solvent through the filter paper. If you must interrupt your work, this is a good stopping point. Place the filter paper and precipitate in your evaporating dish and store in your drawer until the next laboratory period. Discard the filtrate down the drain with lots of water.

## B. Recrystallization of the Crude Product

Pour 75 mL of 0.5M HCl into a 250-mL beaker and bring the solution to a boil. Transfer the crude product (filter paper and all) to this beaker. Stir the mixture and continue heating until nothing else seems to be dissolving. Filter the hot solution through a Büchner funnel to remove the insoluble charcoal. In this filtration, you must be sure that the filtrate is free of charcoal. If any charcoal passes through into the filtrate, pour the mixture back into the beaker, reheat it, and refilter using clean filter paper. The filter paper containing the charcoal can be discarded in the wastebasket.

Pour the golden yellow, charcoal-free solution into a clean 125-mL Erlenmeyer flask; if any crystals have formed in the filter flask, scrape them all out into the Erlenmeyer. Add 20 mL conc. HCl to the solution. Cool the solution in an ice bath until no more crystals seem to be forming. Filter the mixture through a Büchner funnel containing two pieces of filter paper. Wash the crystals with two 5-mL portions of ice cold distilled water to remove any HCl. Then wash them with 10 mL of acetone to wash away the water. Acetone dissolves the water and carries it away. Since acetone is much more volatile than water, the crystals dry faster than they would if they were wet with water. Dry the precipitate as thoroughly as possible by suction. Then scrape the crystals onto a large piece of dry filter paper and spread them out so that any solvent still remaining can evaporate. The filtrate containing the washings may be washed down the drain with plenty of water.

Weigh the dry compound by difference into a clean, dry test tube and stopper the tube. Label the tube with your name and with the identity and weight of the compound. Turn it in to your instructor.

## Results

The equation for the preparation of $[Co(NH_3)_6]Cl_3$ is given as Equation (7-1). Perform the following calculations involving this preparation:

1. Calculate the number of moles of $CoCl_2{\cdot}6H_2O$ that you used in the synthesis.

2. Using Equation (7-1), calculate the number of moles of $[Co(NH_3)_6]Cl_3$ that your synthesis could have produced.

3. Use the result of 2. to calculate the expected number of grams of the product. This is the *theoretical yield* of the product.

4. Use the theoretical yield and your actual yield to calculate your *percentage yield*.

## Questions

1. If the product had not been recrystallized, would your percentage yield have been larger or smaller? Explain.

2. Calculate the number of moles of $NH_4Cl$ you used in this synthesis. Would all of this react? Explain.

# EXPERIMENT 7

## REPORT SHEET

Name ______________________________ Date ______________

Section ______________________________

**Data:**

Weight of $CoCl_2 \cdot 6H_2O$ used ______________________

Weight of $[Co(NH_3)_6]Cl_3$ isolated ______________________

**Results:**

Moles $CoCl_2 \cdot 6H_2O$ used ______________________

Show calculation:

Moles $[Co(NH_3)_6]Cl_3$ possible ______________________

Weight $[Co(NH_3)_6]Cl_3$ possible ______________________

Show calculation:

Percentage yield of $[Co(NH_3)_6]Cl_3$ ______________________

Show calculation:

Turn in your product in a labelled test tube or bottle.

(On the back of the Report Sheet answer the Questions).

# EXPERIMENT 8

# LINE SPECTRA OF ELEMENTS

## Introduction

The bright glow of neon in advertising signs is a striking example of a common phenomenon: many elements emit colored light when their atoms are excited (provided with excess energy) by an external energy source such as an electric current or heat. When we look at a neon sign we see just one color, bright red-orange. However, the light emitted by excited neon atoms is actually polychromatic, that is, it consists of many colors.

When polychromatic light passes through a prism, the prism separates it into many rays, each of a single wavelength and color. This ability of a prism to separate light of different wavelengths depends on the refraction of light; there is a change in the direction of travel of light as it enters the prism and again as it leaves the prism, and these changes in direction of travel are different for different wavelengths of light. In Figure 8-1 this is illustrated for polychromatic light that consists of four colors. This series of colors is a *spectrum.* When the light emitted by excited atoms passes through a prism a series of discrete lines of color is produced. Hence, the spectrum is called a *line spectrum*; it is also called an *atomic emission spectrum*, since it is produced by the emission of light from excited atoms. Each element has a unique atomic emission spectrum with different numbers and colors of lines. For example, neon's atomic emission spectrum consists of dozens of closely spaced lines, many of which are red or yellow. In contrast, only four widely separated lines are visible in hydrogen's atomic emission spectrum.

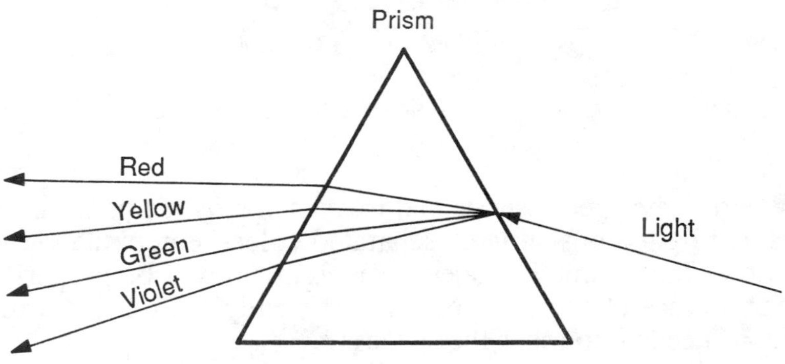


**Figure 8-1**

Dispersion of light by a prism.

When white light passes through a prism, a "rainbow" results. The spectrum is a *continuous* one consisting of bands of color gradually blending into one another with no breaks between colors. It shows that white light consists of *all colors* of visible light. In contrast, the line spectrum of an element shows that excited atoms emit only some *specific colors* of light.

In this experiment you will observe the line spectra of four elements using a Bunsen *spectroscope*, Figure 8-2. For one of these elements, hydrogen, you will make calculations that relate the spectrum to the atom's internal structure.

## Technique

**Use and Calibration of a Spectroscope.** Look at Figure 8-2. The spectroscope has three "arms". One arm has a narrow slit at the end of it through which light enters the spectroscope. The widths of spectral lines depend partly on the width of this slit; the narrower the slit, the narrower the lines. However, a very narrow slit results in faint lines because very little light enters the spectroscope. The best slit width must be a compromise between lines as narrow as possible and light of sufficient intensity for you to see the lines easily.

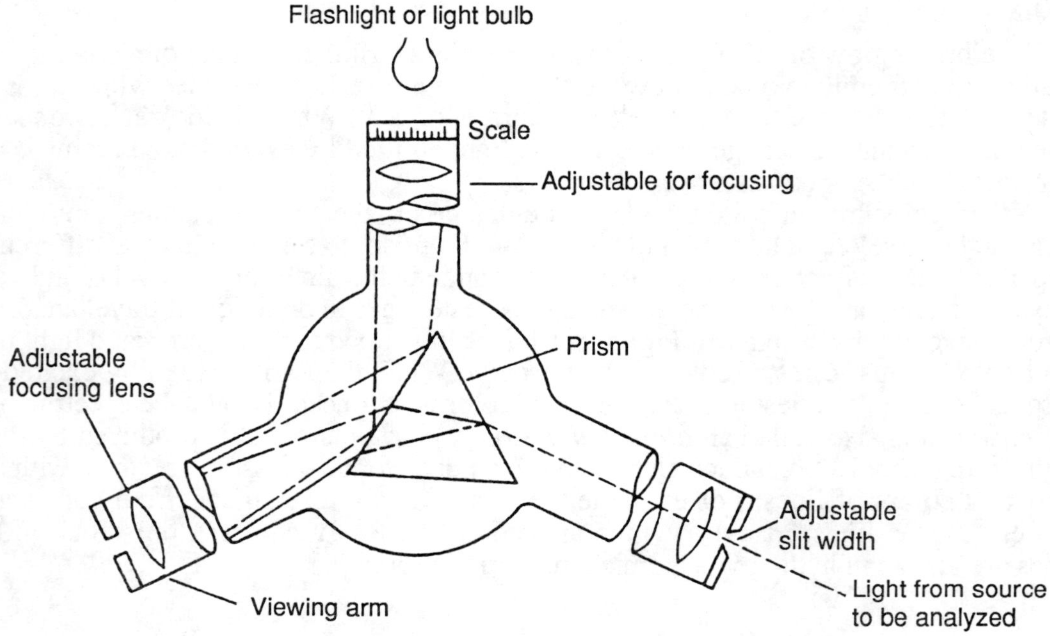


**Figure 8-2**

Schematic drawing of a Bunsen spectroscope.

Light enters the spectroscope and is *refracted* by the prism in the center of the spectroscope. The refracted light, consisting of separated colors, enters the viewing arm; this requires proper positioning of the prism. The spectroscope prism will be properly positioned before you begin your work; do not change its position. The endpiece on this arm can be pulled farther out or pushed in, as needed, to focus the spectrum.

The third arm of the spectroscope has a numerical scale, the image of which is projected into the viewing arm when the scale is illuminated. This scale is unitless; the numbers are not wavelengths or any other particular distance. You must relate the numbers on the scale to known wavelengths; this is called *calibrating* the spectroscope.

To calibrate the spectroscope, you will look at the line spectrum of helium. Wavelengths for the lines in this spectrum are given on the Report Sheet. You will record the scale reading for each visible line. Then you will make a calibration curve, a graph on which you plot wavelength vs. scale reading. You can then use this calibration curve to determine the wavelength of a line in the spectrum of any other element if you have viewed the spectrum and determined the scale reading for the line. Because the prism position can be different for each spectroscope, the correlation between wavelength and scale reading can be different for each spectroscope. Your calibration curve applies to just one spectroscope, and all your data must be taken from that spectroscope.

**Relating Hydrogen's Spectrum and Electronic Structure.** A light wave is characterized by its wavelength and frequency. Its velocity is the product of these

$$c = \lambda \nu \tag{8-1}$$

where c = velocity of light, $\lambda$ = wavelength, and $\nu$ = frequency. The energy of a light wave is directly proportional to its frequency

$$E = h\nu \tag{8-2}$$

The proportionality constant, h, Planck's constant, is $6.626 \times 10^{-34}$ J • s. The relationship between a light wave's energy and its wavelength is obtained by solving Equation (8-1) for $\nu$ and substituting into Equation (8-2):

$$E = \frac{hc}{\lambda} \tag{8-3}$$

An atomic emission spectrum contains light of only specific wavelengths. This means that the excited atoms emit light of only *specific energies*. In contrast, white light has *all energies* of visible light.

The energy of electrons in atoms is quantized; that is, only certain quantities of energy are possible. The energy emitted by an excited atom is equal to the difference between two of these possible levels of energy:

$$\Delta E = E_{II} - E_{I} = \frac{hc}{\lambda} \tag{8-4}$$

where $E_{II} > E_{I}$.

For a hydrogen atom, the wavelength of light that is emitted by an excited atom is related to the principal quantum numbers of the energy levels between which a transition occurs by the equation

$$\frac{1}{\lambda} = 1.097 \times 10^{7}\ \text{m}^{-1} \left(\frac{1}{n_{I}^{2}} - \frac{1}{n_{II}^{2}}\right) \tag{8-5}$$

where $n_{II} > n_{I}$.

As an example of the use of Equations (8-4) and (8-5), consider an electron in hydrogen's fifth energy level ($n_{II} = 5$) losing energy down to hydrogen's first energy level ($n_{I} = 1$):

$$\frac{1}{\lambda} = 1.097 \times 10^{7}\ \text{m}^{-1} \left(\frac{1}{1^{2}} - \frac{1}{5^{2}}\right)$$

$$= 1.053 \times 10^{7}\ \text{m}^{-1}$$

$$\lambda = 9.497 \times 10^{-8}\ \text{m}$$

$$\Delta E = E_{II} - E_{I} = \frac{hc}{\lambda}$$

$$= \frac{(6.626 \times 10^{-34} \text{J} \cdot \text{s})(3.00 \times 10^{8}\ \text{m} \cdot \text{s}^{-1})}{9.497 \times 10^{-8}\ \text{m}}$$

$$= 2.09 \times 10^{-18}\ \text{J}$$

The energy difference between hydrogen's fifth and first energy levels is $2.09 \times 10^{-18}$ Joule; when this electronic transition occurs, the emitted photon has wavelength of $9.497 \times 10^{-8}$ m.

## Review

Appendix II
Appendix III

## Equipment Needed

Bunsen spectroscope
clamp
flashlight
gas discharge tubes
graph paper (provided by the student)
power supply (high-voltage transformer)
ring stand

## Procedure

### A. Setting up the Equipment

You will work in pairs for the entire experiment.

Your instructor will show you how to connect the gas discharge tubes to the power supply. Begin by placing a tube of hydrogen in the power supply. Position the tube at the end of the slit arm of the spectroscope; have the tube very close to, but not quite touching, the arm so that as much as possible of the light from the tube enters the spectroscope and as little as possible of the room's light enters the spectroscope. After the hydrogen tube is properly in place, plug in and turn on the power supply.

Look through the viewing arm at hydrogen's spectrum. Turn the screw that is beside the slit to make the slit wider or narrower until you see four lines that are as narrow as possible. You might see only three lines. The fourth line is violet and is not easy to see against a black background; also, people's eyes vary in their sensitivity to violet. If you do not see the violet line, try placing the tube closer to the slit or ask your instructor for assistance in adjusting the viewing arm. Focus the spectrum by adjusting the viewing lens.

Clamp a flashlight onto a ring stand and position it so that its light shines into the scale arm. Position the scale horizontally; focus it by adjusting it farther into or out of the arm.

If you have any doubt about any of these adjustments, have your instructor check your equipment before you proceed.

### B. Calibrating the Spectroscope

Turn off the power supply and replace the hydrogen tube with a helium tube. Record the color of helium's glow. The Report Sheet gives wavelengths of light in the helium spectrum and the relative intensity of lines on a scale that assigns a value of 1000 to the most intense line. You will not be able to see all these lines with the equipment you are using; you probably cannot see lines assigned an intensity value less than about 30.

Before you come to the laboratory, make a data table for helium like the one on the Report Sheet. Use it to record scale readings for the visible lines; you should record scale readings to three figures. One student should look at the spectrum and call out readings for the other student to record in the notebook of the student who is observing the spectrum. The student who is making readings must be careful not to change position; the spectrum and the scale image enter the prism and viewing arm from different directions and their relative positions change if the viewing angle changes. After one student completes the readings, the two students should trade jobs, so that both students have their own data; the two sets of data might be different, but each will be internally consistent. After recording the helium data, leave the spectroscope so that another team of students can use it.

Use your helium data to make a calibration graph of wavelength ($y$ axis) vs. scale reading ($x$ axis). The plot should be a *smooth curve*. If there are many points that do not lie along a smooth curve, or any points that are far from a smooth curve you have misassigned some lines. For example, you might have assigned the scale reading that actually correlates with the red line at 706.5 nm to the red line at 728.1 nm. If the plot is not fairly smooth, look at helium again (when your spectroscope is available to you again) and try to improve the assignments.

### C. Hydrogen Spectrum

Put a hydrogen tube into the power supply and record the color of its glow. Look at hydrogen's spectrum and assign a scale reading (three figures) to each of the four lines.

### D. Other Elements

Your instructor will assign two elements, other than hydrogen and helium, to you. Some possibilities are argon, krypton, mercury, and neon. Record the color of the glow for each of these elements. Look at the spectrum of each of these two elements; decide which is simpler (has fewer lines), which is more complex.

For the element that has the more complex spectrum, write a qualitative description of its spectrum. State what colors are present and what colors are absent. Note the colors of the most intense lines and of the most abundant colors. Describe the occurrence of "bundles" (groups of closely spaced lines of a particular color). Your description should be sufficiently clear and detailed that anyone who read it would know what to expect if they looked at the spectrum.

For the element that has the simpler spectrum, write a qualitative description of its spectrum. Then determine scale readings for two to six of its most prominent lines. If many lines seem similarly prominent, there are several things you can do to distinguish the ones that are truly the most intense: Make the slit narrower; move the tube farther from the slit; slightly close your eyes. Each of these can cause the less intense lines to fade out.

## Results

### A. Setting Up the Equipment

No report is required for this part.

### B. Calibrating the Spectroscope

On the Report Sheet, give helium's composite color (its "glow") when electricity is passed through it. Give scale readings for the lines you could see. Make a new graph of wavelength vs. scale reading and attach it to the Report Sheet.

### C. Hydrogen Spectrum

1. Record hydrogen's composite color.

2. Use your helium calibration curve and your hydrogen scale readings to determine the wavelength for each line in hydrogen's spectrum.

3. Use Equation (8-5) to identify the energy levels, $n_I$ and $n_{II}$, between which the transition occurs or each of the four spectral lines. This equation cannot be "solved" in a straight-forward algebraic fashion; instead you must try *possible* solutions. Use an experimentally determined $\lambda$ to calculate $1/\lambda$. Pick values for $n_I$ and $n_{II}$; use them to calculate the value of $1.097 \times 10^7 \text{ m}^{-1} (1/n_I^2 - 1/n_{II}^2)$. If your value for this term and your value for $1/\lambda$ are the same (within slight experimental

error), you have picked the values of n that correspond to the energy levels between which a transition occurred. If not, try another set of n values.

In deciding what n values to try, keep in mind the following:

(a) All the n values will be fairly small. As n increases, energy levels are closer to one another; therefore, large, consecutive values of n give small values for $\Delta E = E_{II} - E_I$, large values for $\lambda$, and small values for $1/\lambda$. The lines produced by these transitions are of lesser energy than visible light. If $n_I$ and $n_{II}$ are very much different, $\Delta E$ is large, $\lambda$ is small, and $1/\lambda$ is large. The lines produced by these transitions are of greater energy than visible light.

(b) The difference between $n_I$ and $n_{II}$ will be less for lines toward the red (low energy) end of the spectrum and greater for lines toward the violet (high energy) end of the spectrum.

(c) Begin with the simplest combinations: $n_I = 1$ and $n_{II} = 2$; $n_I = 1$ and $n_{II} = 3$; etc. If none of these values matches a value of $1/\lambda$ that you calculated from your experimentally determined wavelength, try values of $n_I = 2$ and $n_{II} = 3$; $n_I = 2$ and $n_{II} = 4$; etc.

4. Use your experimentally determined wavelengths to calculate the energy emitted by each transition, Equation (8-3).

5. Construct an energy level diagram for hydrogen. Use graph paper so that you will have an accurate scale for the energy levels. As the base line for the energy scale, use the lowest energy level involved in the transitions that produced lines in the spectrum. Along the energy scale (vertical), show the amount by which the energy of each higher level is greater than the energy of the lowest level. Label each level with the proper quantum number. Draw arrows between energy levels indicating the transitions that produce lines in the spectrum. Label each arrow with the wavelength of the light that is produced by the transition.

**D. Other Elements**

Give the glow color and a qualitative description of each element's spectrum. Use the helium calibration curve to assign wavelengths to the most prominent lines in the spectrum of one of the elements.

## Questions

1. Use Equation (8-5) to show that a transition between two energy levels with large n values results in the emission of light of very great wavelength. (Hint: Choose similar values of 20 or higher for $n_I$ and $n_{II}$ to calculate the wavelength of the emitted light.) Note that you are to calculate $\lambda$, not just $1/\lambda$.

2. Use Equation (8-5) to show that a transition between energy levels of very small $n_I$ and very large $n_{II}$ emits light of very small wavelength.

# EXPERIMENT 8

## REPORT SHEET

Name ______________________________ Date ______________

Section ______________________________

### B. Helium Spectrum for Calibrating the Spectroscope

Composite color ______________________________

| λ, nm | Line color | Relative intensity | Scale reading |
|---|---|---|---|
| 728.1 | red | 30 | ______________ |
| 706.5 | red | 70 | ______________ |
| 667.8 | red | 100 | ______________ |
| 587.6 | yellow | 1000 | ______________ |
| 504.8 | light green | 15 | ______________ |
| 501.6 | light green | 100 | ______________ |
| 492.2 | dark green | 50 | ______________ |
| 471.3 | blue-green | 40 | ______________ |
| 447.1 | blue-violet | 100 | ______________ |
| 443.8 | blue-violet | 10 | ______________ |
| 438.8 | violet | 30 | ______________ |
| 414.3 | violet | 15 | ______________ |
| 412.1 | violet | 25 | ______________ |
| 402.6 | violet | 70 | ______________ |
| 400.9 | violet | 10 | ______________ |
| 396.4 | violet | 50 | ______________ |

Attach the calibration curve.

## C. Hydrogen Spectrum

1. Composite color ______________________

2.

| Line color | Scale reading | $\lambda$, nm |
| --- | --- | --- |
| | | |
| | | |
| | | |
| | | |

3. Principal Quantum Numbers for Hydrogen Transitions

| $\lambda$, m | $\frac{1}{\lambda}$, $m^{-1}$ | $n_I$ | $n_{II}$ | $1.097 \times 10^7 m^{-1}\left(\frac{1}{n_I^2} - \frac{1}{n_{II}^2}\right)$ |
| --- | --- | --- | --- | --- |
| | | | | |
| | | | | |
| | | | | |
| | | | | |
| | | | | |

Show a sample calculation. Include all steps.

4. Energy Values for Hydrogen Energy Level Diagram

| λ, m | E = hc/λ, Joules |
|---|---|
| __________ | __________ |
| __________ | __________ |
| __________ | __________ |
| __________ | __________ |

Show a sample calculation. Be sure to include all units.

5. Attach the energy level diagram.

In the space below, answer the questions.

## D. Other Elements

Spectrum of ________________________________

Give the composite color and a qualitative description.

Spectrum of ________________________________

Give the composite color and a qualitative description.

| Line color | Scale reading | λ, nm |
|---|---|---|
| ________ | __________ | ______ |
| ________ | __________ | ______ |
| ________ | __________ | ______ |
| ________ | __________ | ______ |
| ________ | __________ | ______ |
| ________ | __________ | ______ |

# EXPERIMENT 9

# SALT IDENTIFICATION USING VISIBLE and INFRARED SPECTRA

## Introduction

Experiment 8, "Line Spectra of Elements", showed you (among other things) that gaseous atoms excited in an electric discharge emit light of characteristic colors, for example red for Ne, etc. By using the Bunsen spectroscope, this colored light can be resolved into lines, each of a single wavelength. This behavior of excited atoms reveals interesting facts about the arrangement of electrons in atoms.

Characteristic light colors can also be put to a simpler, more practical use: they can be used to determine the presence of particular elements. Given the elements which may possibly be present in an unknown mixture, you could observe the light emission from each one separately and compare it with the light emission from the unknown. For example, if Ne and Ar are both present in an unknown, then its emission spectrum will contain lines from both elements; if only Ne is present, then only lines from Ne will be observed, etc.

During the nineteenth century, the Bunsen spectroscope was used very extensively to make determinations of this kind. Even before the reasons for the existence of line spectra were known, chemists relied on observations of the wavelengths of emitted light as evidence for the presence or absence of particular elements. If the light emitted is in the visible region of the electromagnetic spectrum, an observation of its color is often sufficient to identify the emitting element. The color is, of course, determined by the wavelengths of emitted light.

Determining which elements are present in a sample is called *qualitative analysis*. The usual method of testing for the presence or absence of elements is to observe a property that is unique to the element being tested for. Some ions when excited by the heat of a flame emit characteristic electromagnetic colors of light in the visible region (4000 - 7000 Å) of the electromagnetic spectrum as energetically excited electrons drop back into lower energy levels. Among the ions which can be identified by the characteristic colors of their emission spectra are $Na^+$, $K^+$, $Ca^{2+}$, $Sr^{2+}$ and $Ba^{2+}$.

For the most part, the electronic spectra of simple *anions* do not exhibit emission or absorption in the visible region. (Exceptions are yellow $CrO_4^{2-}$, orange-red $Cr_2O_7^{2-}$and deep purple $MnO_4^-$.) However, many polyatomic anions do absorb light in the infrared (IR) region of the electromagnetic spectrum (1-100 μ: 1 micron = $10^{-6}$ m). These infrared absorptions result from excitation of stretching and bending vibrations for the ions.

The exact number of such absorptions depends on the number and kinds of atoms in the polyatomic species and on how they are arranged with respect to one another. As one example, the $NO_2^-$ ion has a bent planar structure and may undergo two kinds of stretching vibrations. (These are known as the two *stretching modes* of the ion.) A *bending mode* is also possible. The three vibrational modes for $NO_2^-$ are depicted in Figure 9-1.

In the symmetric mode (Figure 9-1a) both N-O bonds stretch and then compress in unison; energy absorbed in the 1325-1380 $cm^{-1}$ region (~7.5 μ) will cause the ion to vibrate in this fashion. In the antisymmetric mode (Figure 9-1b) one N-O bond stretches as the other compresses; this vibration is excited by energy in the 1230-1275 $cm^{-1}$ (~8.0 μ) region. In addition, a bending motion in which the O-N-O angle changes is depicted in Figure 9-1c; this vibration is excited by energy in the 810-850 $cm^{-1}$ region (~12 μ).

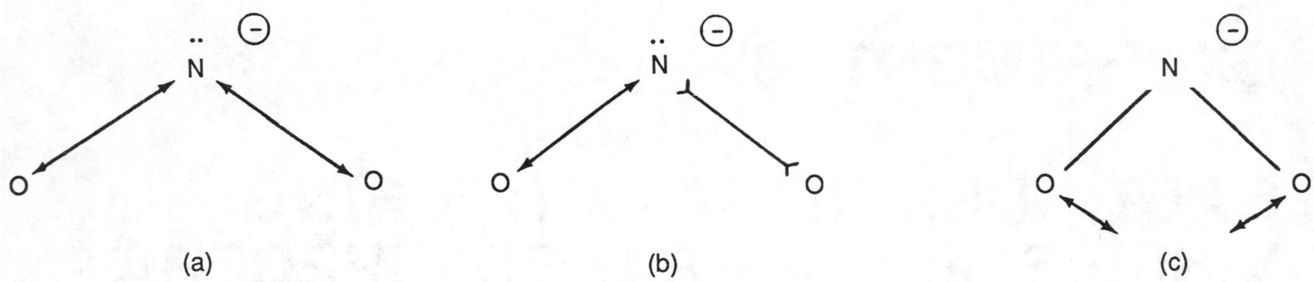

**Figure 9-1**

Stretching (a,b) and bending (c) modes of the nitrite ion.

The IR spectrum of $NaNO_2$ (Figure 9.2) shows how strongly the salt absorbs infrared radiation at each wavelength. You can see that there is a strong absorbance which peaks at 830 $cm^{-1}$ due to excitation of the O-N-O bending vibration. The two stretching vibrations are quite close energetically, and so only one rather broad band resulting from the overlap of two closely spaced bands appears in the spectrum of $NaNO_2$.

Another feature of Figure 9-2 bears mention. It turns out to be experimentally more convenient to measure transmittance of light in the infrared region as a function of wavelength. Hence, at wavelengths where absorbance of light is a maximum, the transmittance is a minimum and the peaks point downward rather than upward.

You are not expected to know exactly how many absorptions a particular anion should have. The purpose of the preceding discussion is simply to acquaint you with the reasons why more than one band can appear for a simple anion. The number of possible vibrations (and so the number of absorption bands) increases with the complexity of the anion.

Even when an experimenter does not know the particular atomic motion causing each absorption band, the infrared spectrum of a polyatomic anion serves as a property by which the anion can be identified. (The spectrum is often referred to as the "fingerprint" of the anion.) Several spectra of inorganic salts are depicted at the end of this experiment. You may notice that the appearance of some spectra depends not only on the anion, but also on the identity of the cation—at least to some

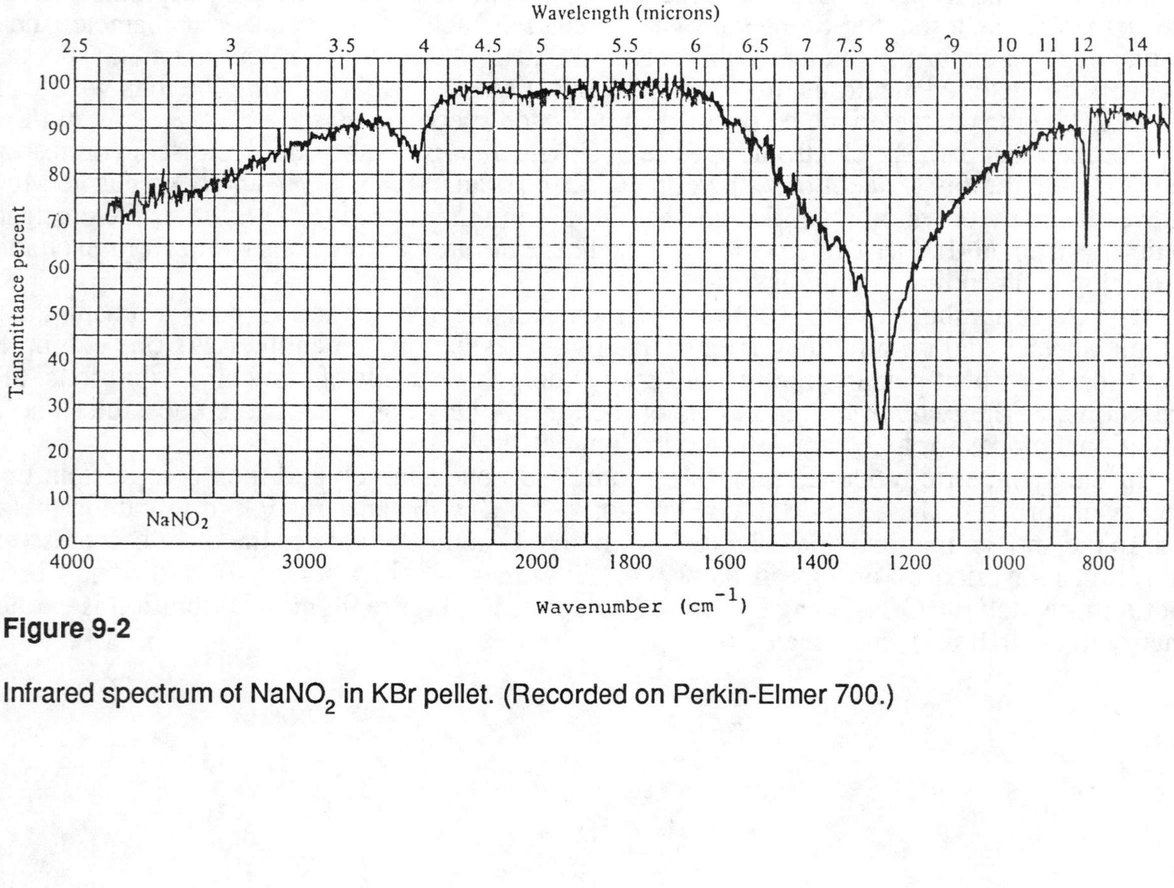


**Figure 9-2**

Infrared spectrum of $NaNO_2$ in KBr pellet. (Recorded on Perkin-Elmer 700.)

extent. This arises because the solid-state packing of anions and cations differs in different salts. Nevertheless, it should be obvious that infrared spectra are characteristic properties by which a particular salt can be identified.

Your instructor will provide you with an unknown inorganic salt from the group whose infrared spectra are given on pages 90-99. These salts are listed in Table 9-1. You will then identify the cation present using flame tests and the anion present using infrared spectroscopy.

Table 9-1

Bromates: $KBrO_3$, $NaBrO_3$

Carbonates: $(NH_4)_2CO_3$, $BaCO_3$, $Li_2CO_3$, $K_2CO_3$, $Na_2CO_3$, $SrCO_3$

Nitrates: $NH_4NO_3$, $Ba(NO_3)_2$, $LiNO_3$, KNO, $NaNO_3$, $Sn(NO_3)_2$

Nitrites: $KNO_2$, $NaNO_2$

Phosphates: $(NH_4)H_2PO_4$, $K_2HPO_4 \cdot 3H_2O$

Sulfates: $(NH_4)_2SO_4$, $BaSO_4$, $K_2SO_4$, $Na_2SO_4$, $SrSO_4$

Sulfites: $BaSO_3$, $K_2SO_3 \cdot 2H_2O$, $Na_2SO_3$, $SrSO_3$

Thiocyanates: $NH_4NCS$, $Ba(NCS)_2 \cdot 2H_2O$, KNCS, NaNCS

## Review

Equipment and Techniques D

## Equipment Needed

burner
cobalt glass
glass rod (4")
micro test tubes
mortar and pestle
pellet press
nichrome wire
infrared spectrophotometer
 with beam attenuator

## Chemicals Needed

$BaCl_2$; barium chloride
$CH_3C(O)CH_3$; acetone
conc. HCl; hydrochloric acid
KBr; potassium bromide (spectro grade)
KCl; potassium chloride
LiCl; lithium chloride
NaCl; sodium chloride
$NH_4Cl$; ammonium chloride
$SrCl_2$; strontium chloride
unknown salt (from Table
 9-1) dried overnight at 70°C.

## Procedure

### A. Flame Tests

Obtain a piece of nichrome wire and make a small loop in one end by wrapping the wire around a pen or pencil. Seal the wire into the end of a 4-inch piece of glass rod (which will serve as a handle). This wire will be used to introduce samples into the burner flame. Obviously you will want to be sure that the wire itself is free of any material which could color the flame. You can clean the wire by dipping the loop into some concentrated HCl in a small test tube and then holding it in the hottest part of the burner flame. It will be necessary to clean the wire each time you change salts. (Why?) The purpose of the HCl is to convert any compounds on the wire to their chlorides, which are quite easily volatilized. The HCl will eventually become contaminated and should be changed before working with your unknown salt.

Obtain a small sample of a known salt of each cation (LiCl, NaCl, KCl, $NH_4Cl$, $BaCl_2$, $SrCl_2$). For each sample, dip the wet loop of the clean wire into the solid and thrust it into the flame. Observe the characteristic flame color of each sample and also any pertinent features of the color. (Does the color appear immediately when the sample is inserted into the flame? Does it last for a long time or a short time? Is it a light or dark shade?) Record these observations in your notebook. You should observe each flame several times. Then identify your unknown by its flame behavior.

The yellow color produced by $Na^+$ emission will appear in all flames. This is because $Na^+$ impurities are impossible to remove. They are introduced from glass storage bottles. In deciding whether your unknown is really a sodium salt, you will have to compare the intensity of the observed yellow color with that from a known sodium salt. In some cases, the yellow color from $Na^+$ masks emission due to other cations present. This is particularly true for $K^+$. So it is advisable to view the flame test for any unknown suspected of containing $K^+$ through a double thickness of blue cobalt glass. This glass will filter out the yellow light from $Na^+$.

Dispose of any excess salts and of the HCl by washing down the drain with plenty of water.

### B. Infrared Spectra

A convenient way to record the infrared spectrum of a solid is to use a solid solution in KBr. (Water absorbs strongly in the infrared region and so cannot be used as the solvent. Absorptions from water would mask absorptions belonging to any other solution component.) Potassium bromide does not absorb in the infrared region above 650 $cm^{-1}$. This solid solution is pressed into a translucent pellet.

To make a pellet, you should take only your solid sample and a spatula to the spectrophotometer area. A small mortar and pestle will be provided there. With your spatula add 10 to 12 mg of your solid to the mortar and grind it thoroughly with the pestle to a very fine powder. You should stop occasionally to scrape the powder into a pile in the mortar and then begin grinding again. Remove the KBr from the desiccator, where it is kept to exclude water. Now add about 200 mg of KBr to the mortar. Mix thoroughly with your sample by grinding slightly with the mortar.

When the sample is a fine powder, obtain the pellet press which consists of a metal block with a threaded hole and two polished end bolts (see Figure 9-3). Screw one of these bolts about halfway into the block. With your spatula, sprinkle a thin, even coating of the powder to cover the surface of the bolt. This will probably not require all the powder you prepared. Now, insert the other bolt and tighten it as much as you can with your fingers. Place one of the bolts in the hexagonal opening mounted on the wooden block. Using the hexagonal wrench provided, tighten the other bolt as tightly as you can to provide maximum pressure on the powder inside. Maintain the pressure for several seconds, and then let the block set for about two minutes. At the end of this time, loosen the top bolt with the wrench and the bottom bolt with your fingers. Remove both bolts. You should have a translucent pellet inside the threaded cavity of the block. If the pellet broke or is not translucent, repeat the procedure.

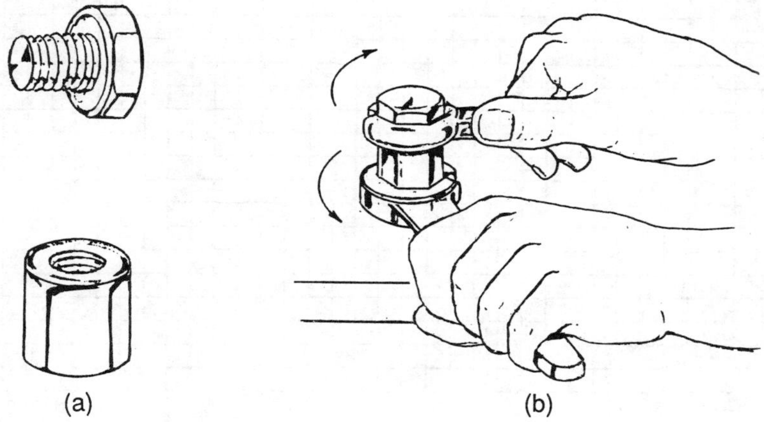

**Figure 9-3**

(a) pellet press block and bolt; (b) Making a KBr pellet.

The block is placed in the sample beam of the infrared spectrophotometer. Record the spectrum following the instructions supplied by your instructor. Figure 9-4 shows infrared spectra of salts.

Wash the mortar and pestle with water, flushing the powder you prepared down the drain with plenty of water. Also wash the block and bolts. Rinse all of these with distilled water followed by acetone and place them in the drying oven. It is a good idea to keep any of your leftover unknown salt until your laboratory report is graded in case any dispute arises over its correct identity. When you are ready to dispose of it, rinse down the drain with plenty of water.

## Results

On the Report Sheet record your observations on the flame tests you conducted.

Identify your unknown salt from the flame test and the infrared results. (If your sample has absorbed water from the air or if it contains water of crystallization, you will see bands around 2.9 μ and 6.25 μ on account of vibrations of the $H_2O$ molecule in addition to those due to the presence of the anion of your unknown salt.)

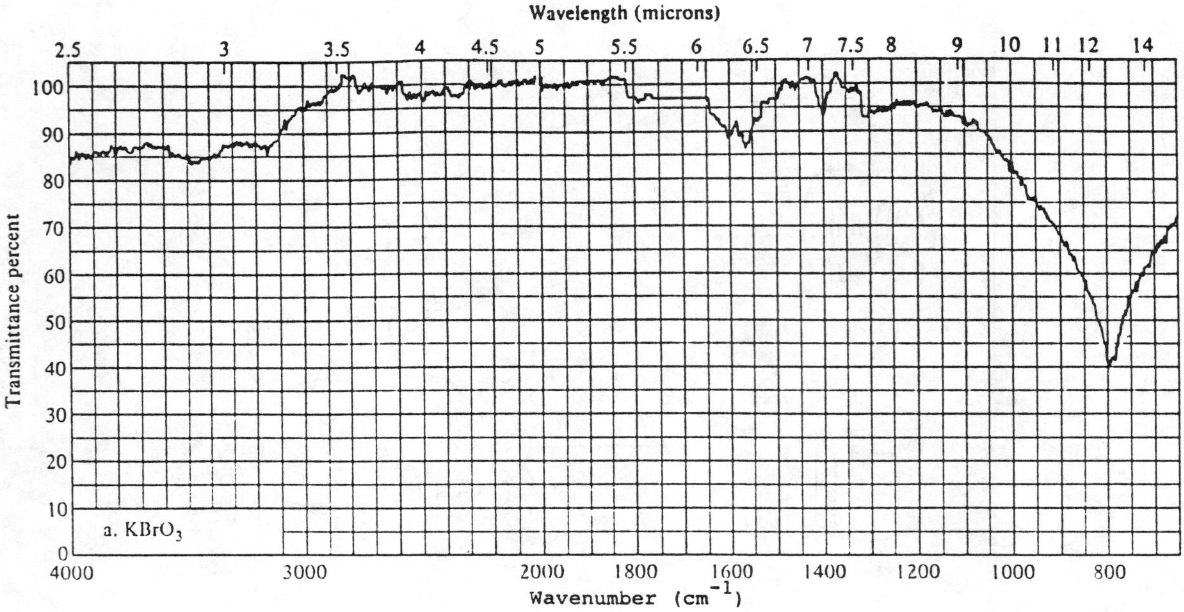


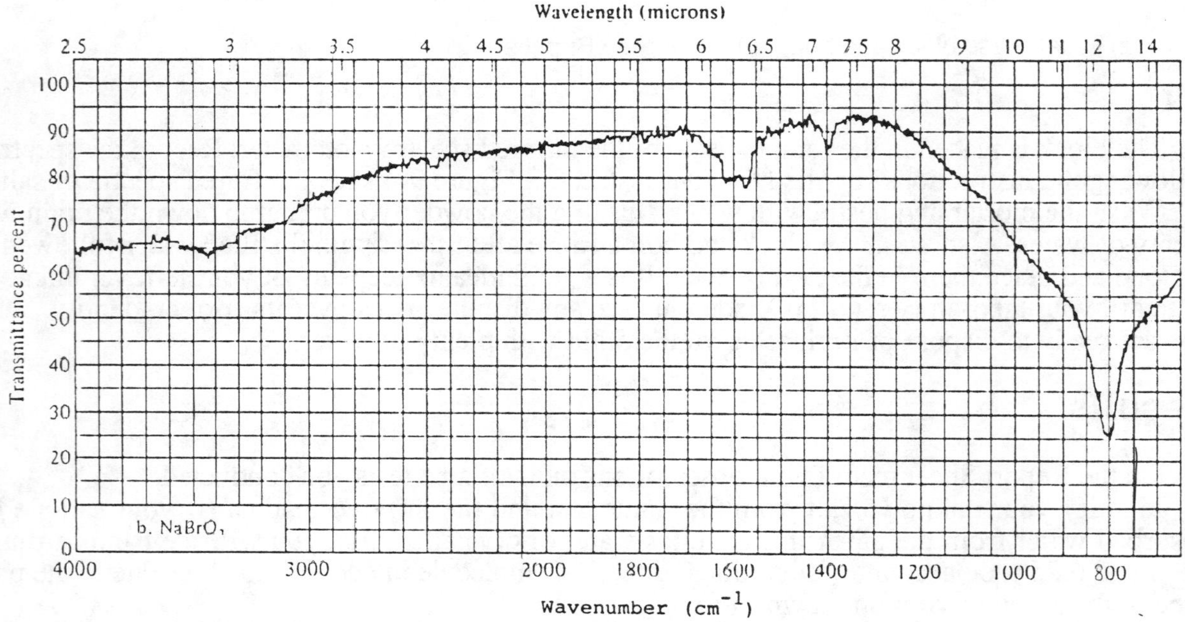


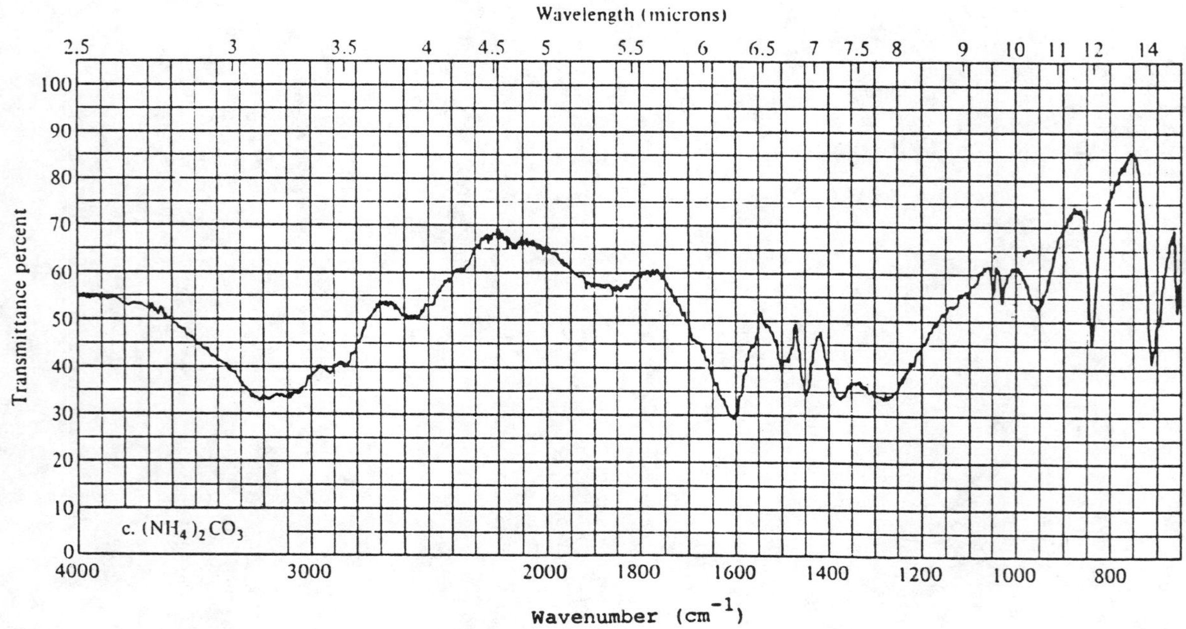


**Figure 9-4** Infrared spectra of unknown salts in KBr pellets. (Recorded on Perkin-Elmer 700.)

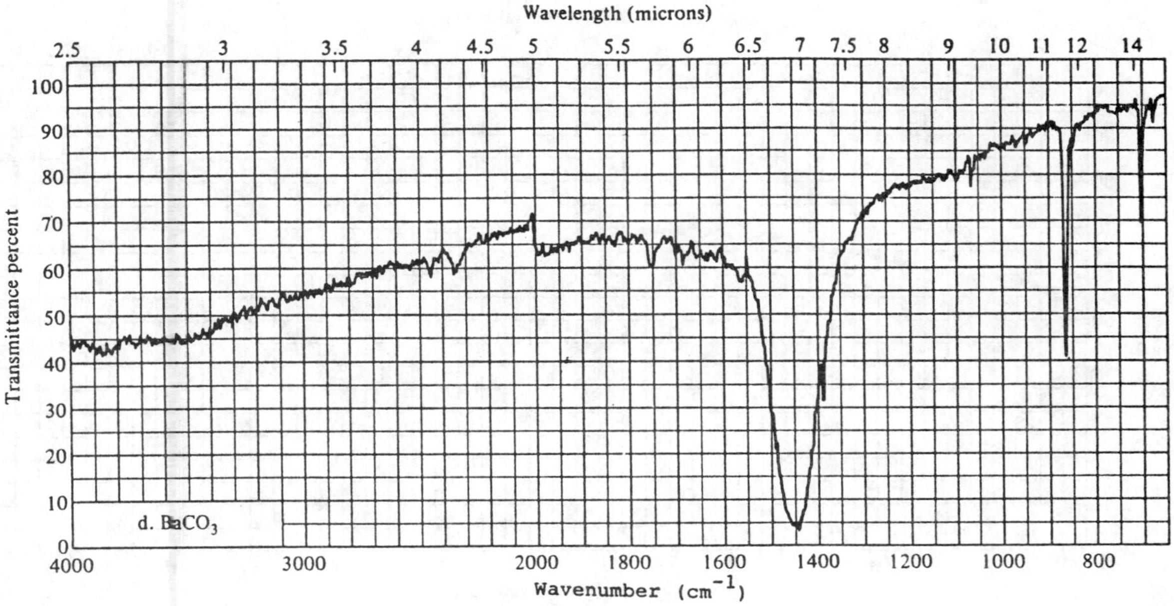
Wavelength (microns)
Transmittance percent
Wavenumber (cm$^{-1}$)
d. $BaCO_3$

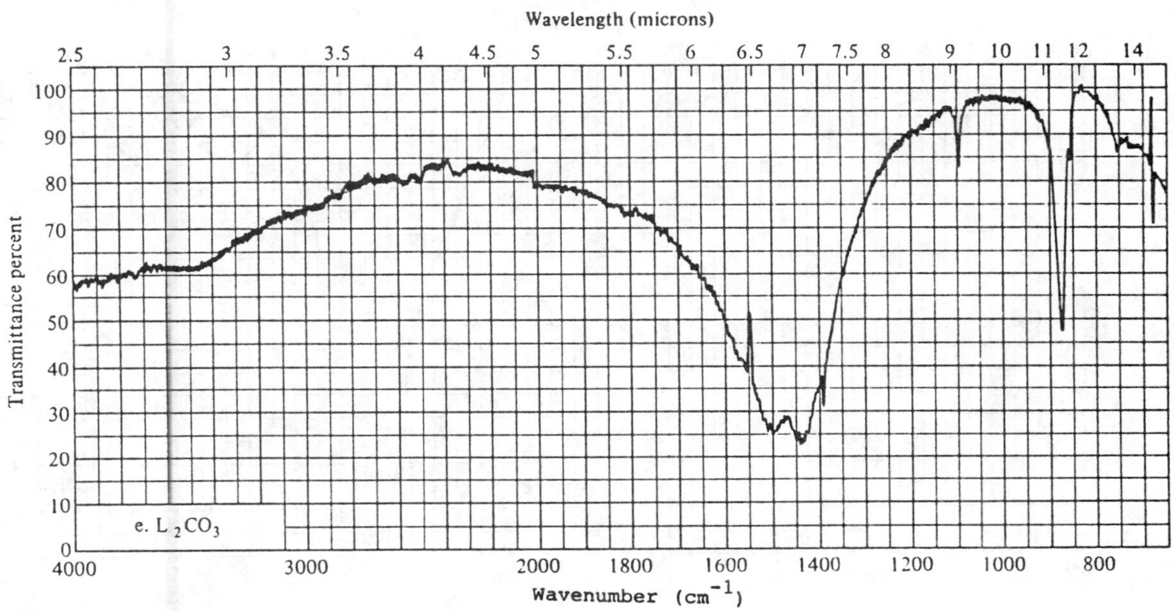
Wavelength (microns)
Transmittance percent
Wavenumber (cm$^{-1}$)
e. $L_2CO_3$

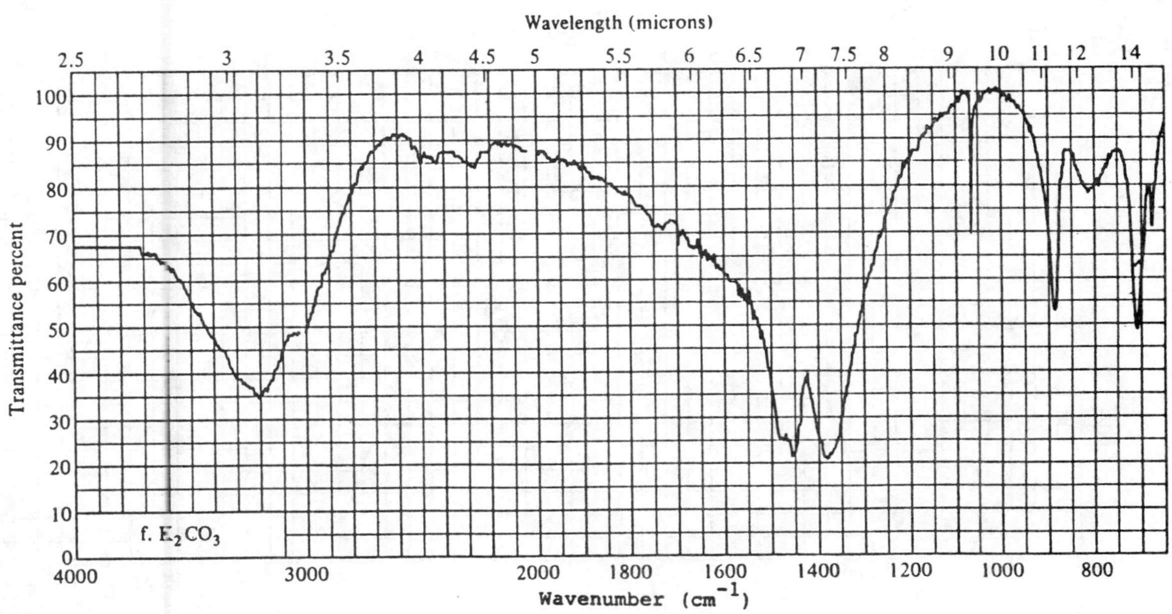
Wavelength (microns)
Transmittance percent
Wavenumber (cm$^{-1}$)
f. $E_2CO_3$

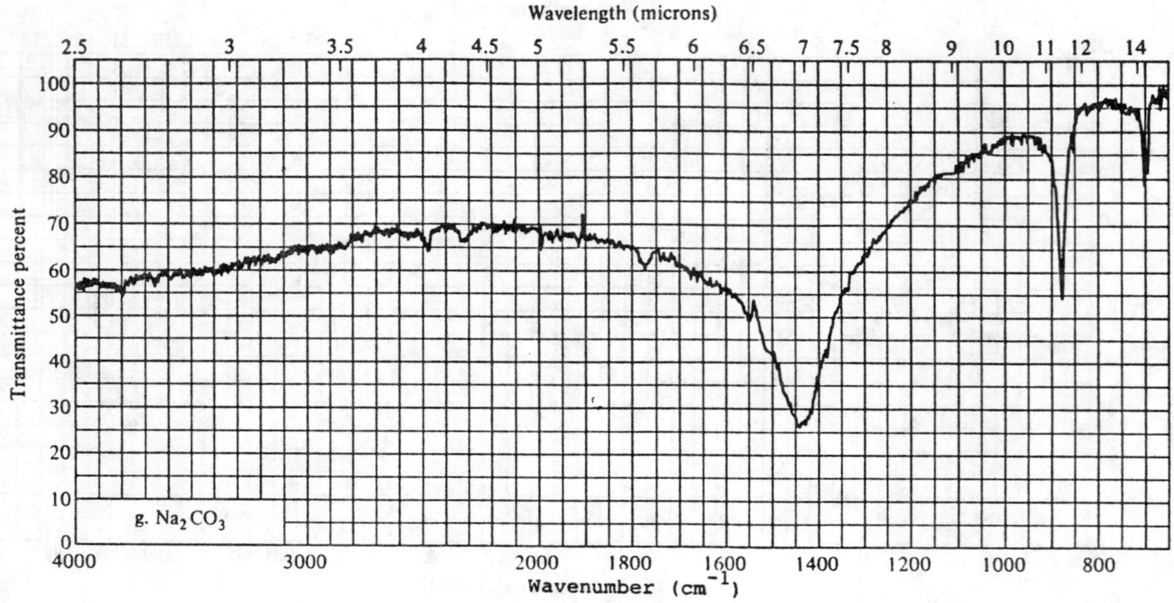
Wavelength (microns)
Transmittance percent
g. $Na_2CO_3$
Wavenumber (cm$^{-1}$)

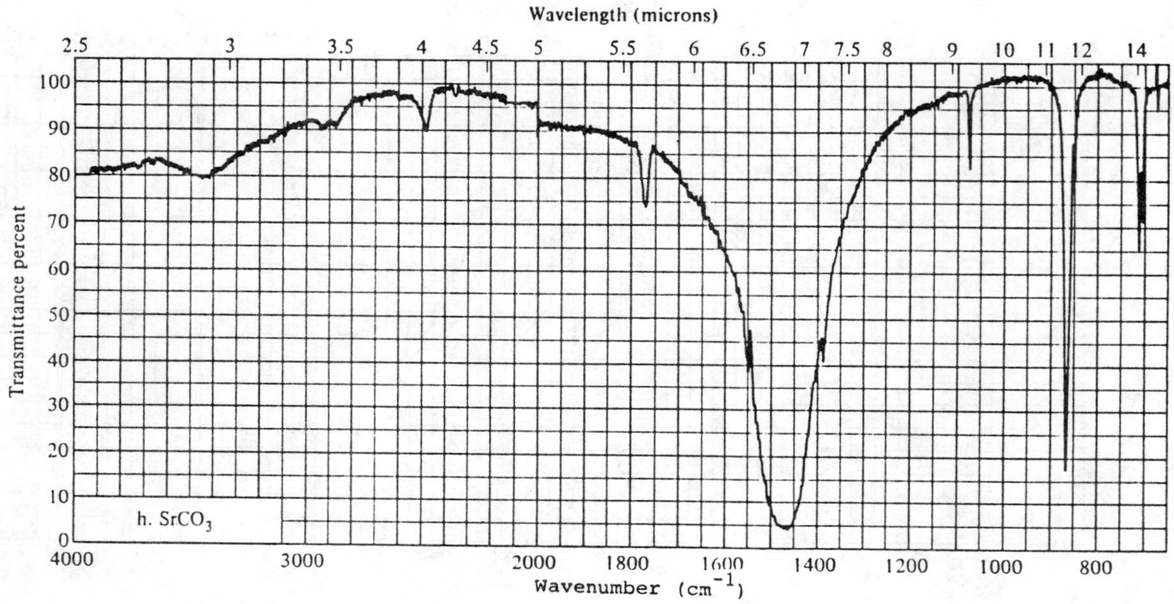
Wavelength (microns)
Transmittance percent
h. $SrCO_3$
Wavenumber (cm$^{-1}$)

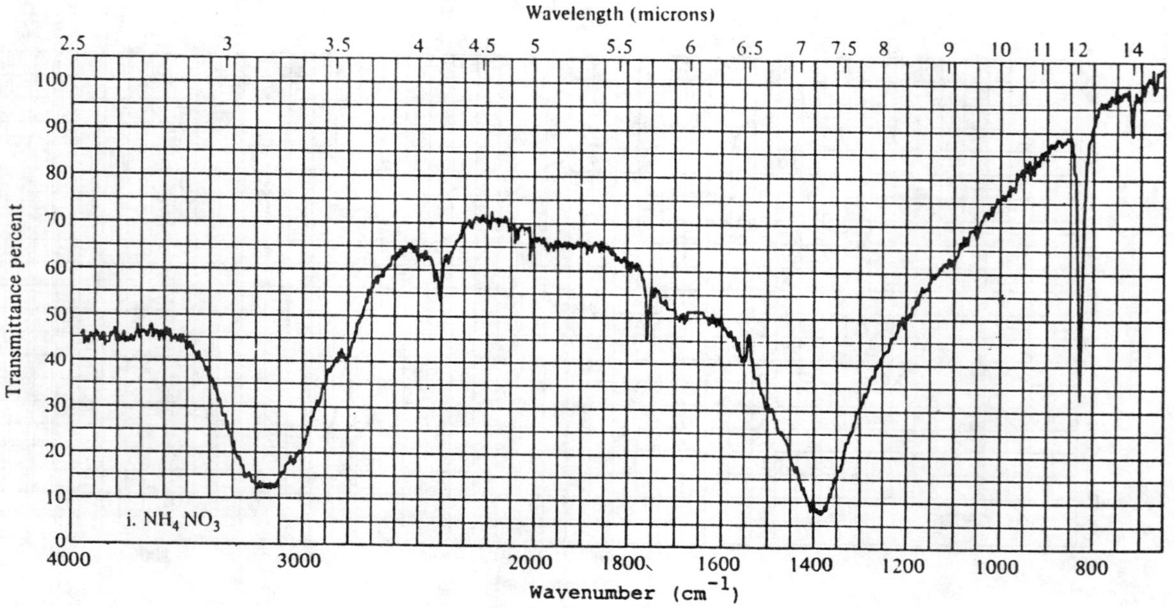
Wavelength (microns)
Transmittance percent
i. $NH_4NO_3$
Wavenumber (cm$^{-1}$)

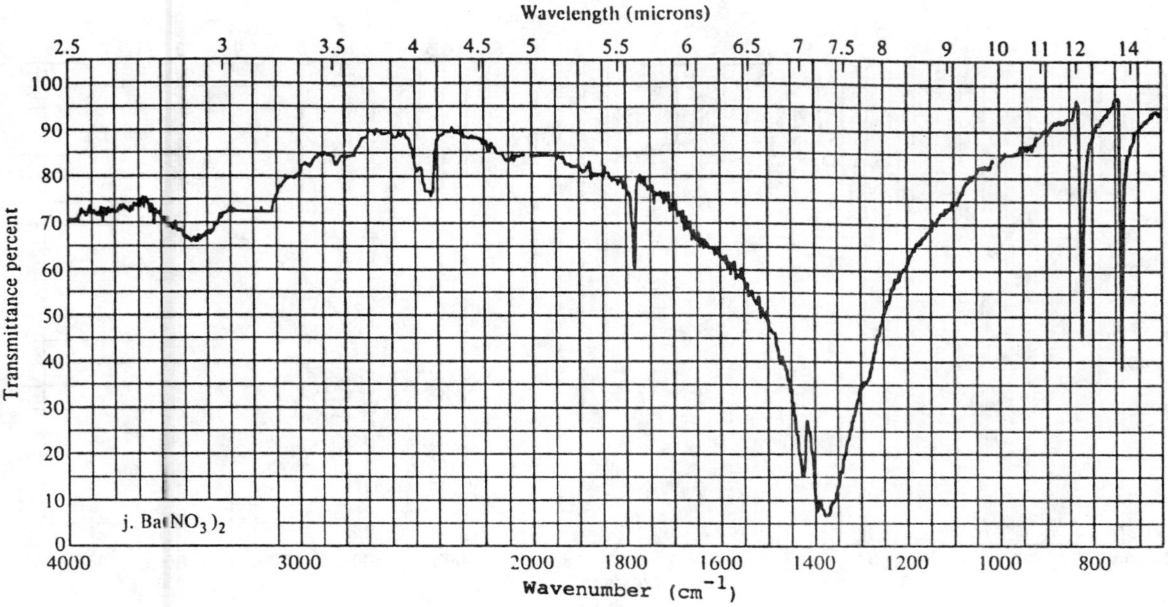
Wavelength (microns)
Transmittance percent
Wavenumber (cm$^{-1}$)
j. Ba(NO$_3$)$_2$

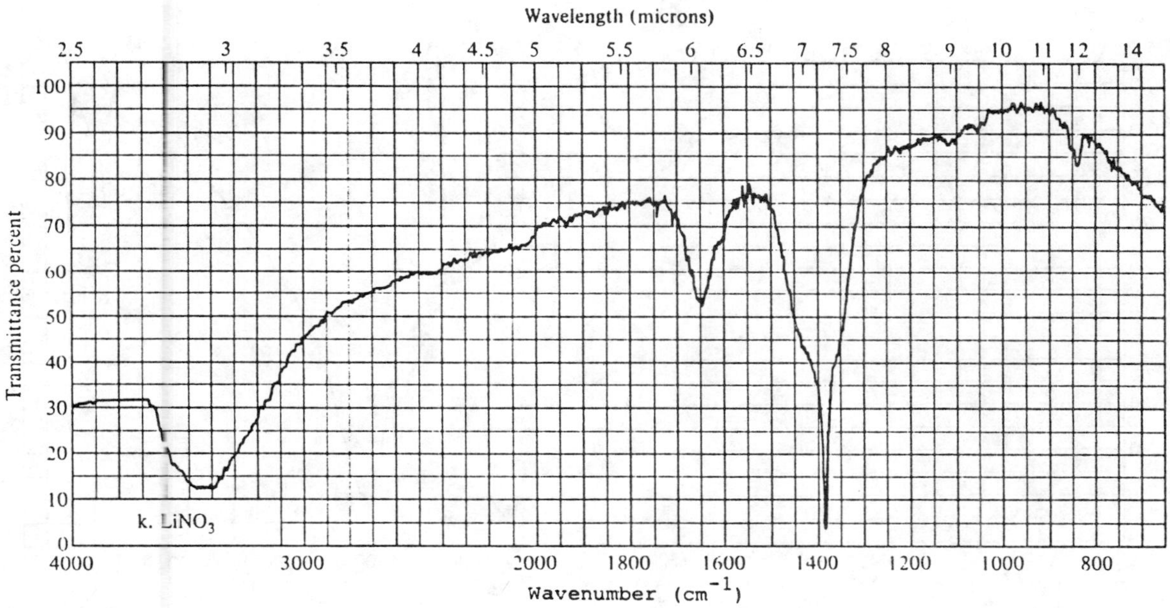
Wavelength (microns)
Transmittance percent
Wavenumber (cm$^{-1}$)
k. LiNO$_3$

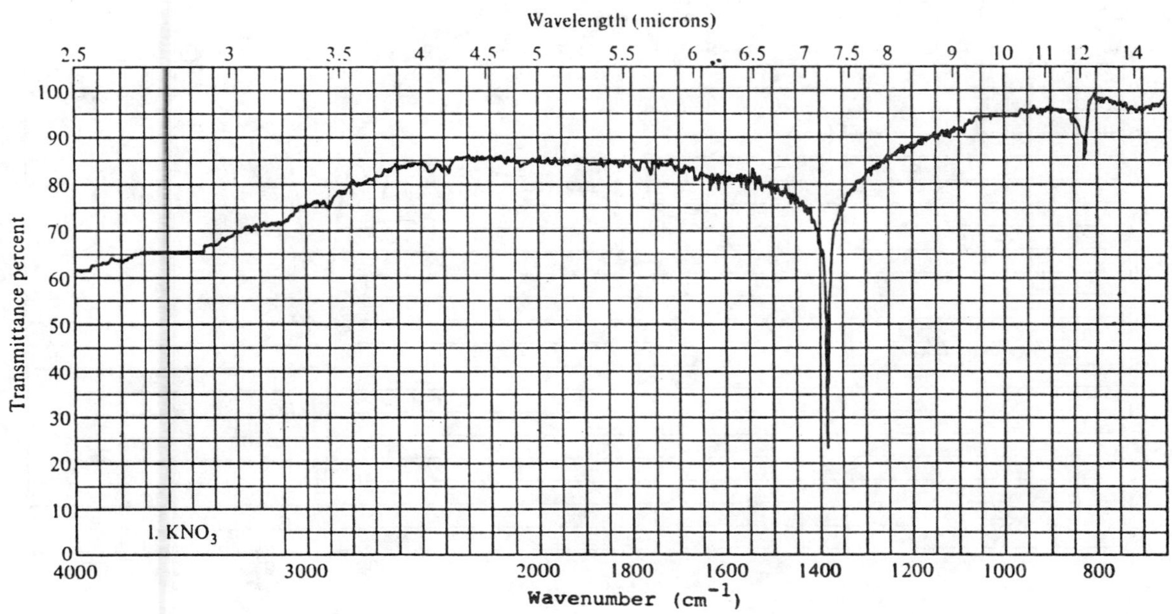
Wavelength (microns)
Transmittance percent
Wavenumber (cm$^{-1}$)
l. KNO$_3$

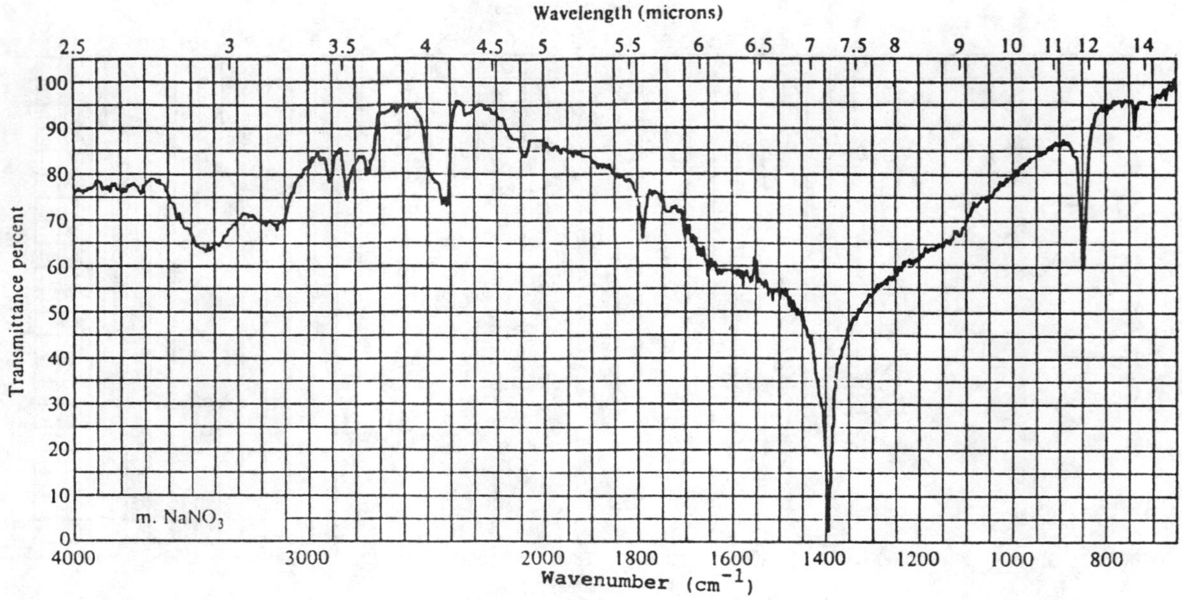
Wavelength (microns)
Transmittance percent
Wavenumber (cm-1)
m. NaNO3

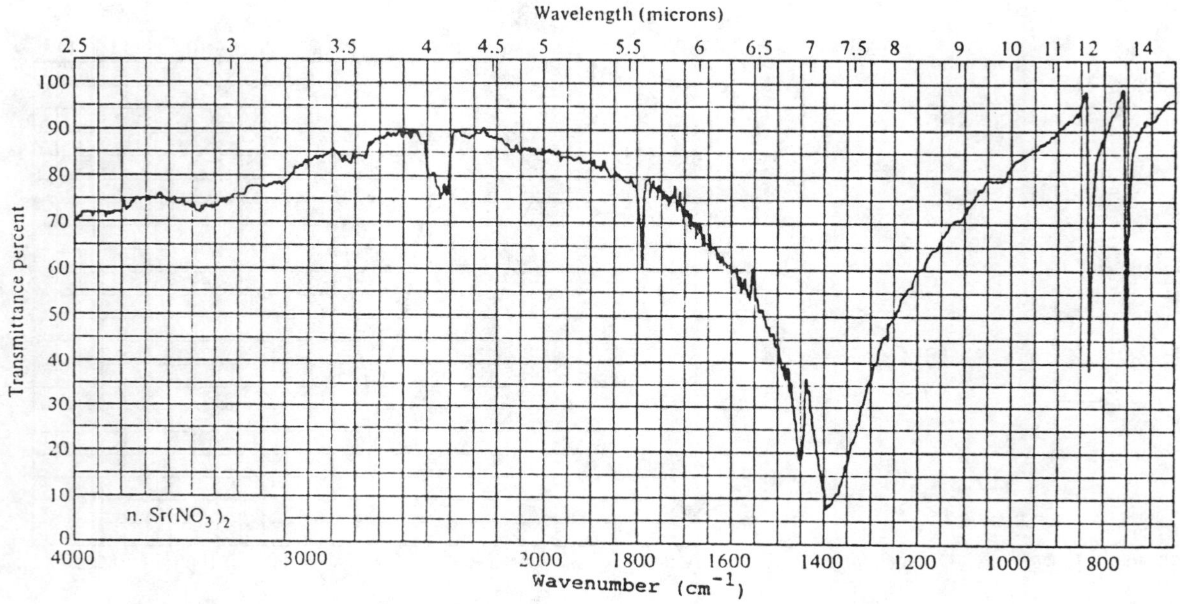
Wavelength (microns)
Transmittance percent
Wavenumber (cm-1)
n. Sr(NO3)2

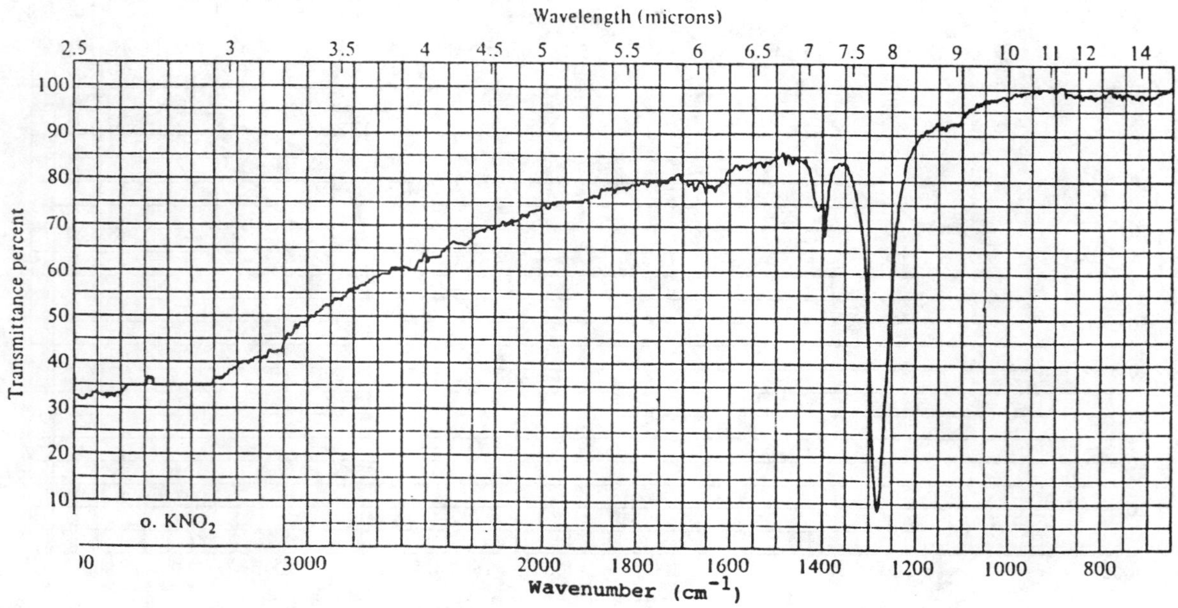
Wavelength (microns)
Transmittance percent
Wavenumber (cm-1)
o. KNO2

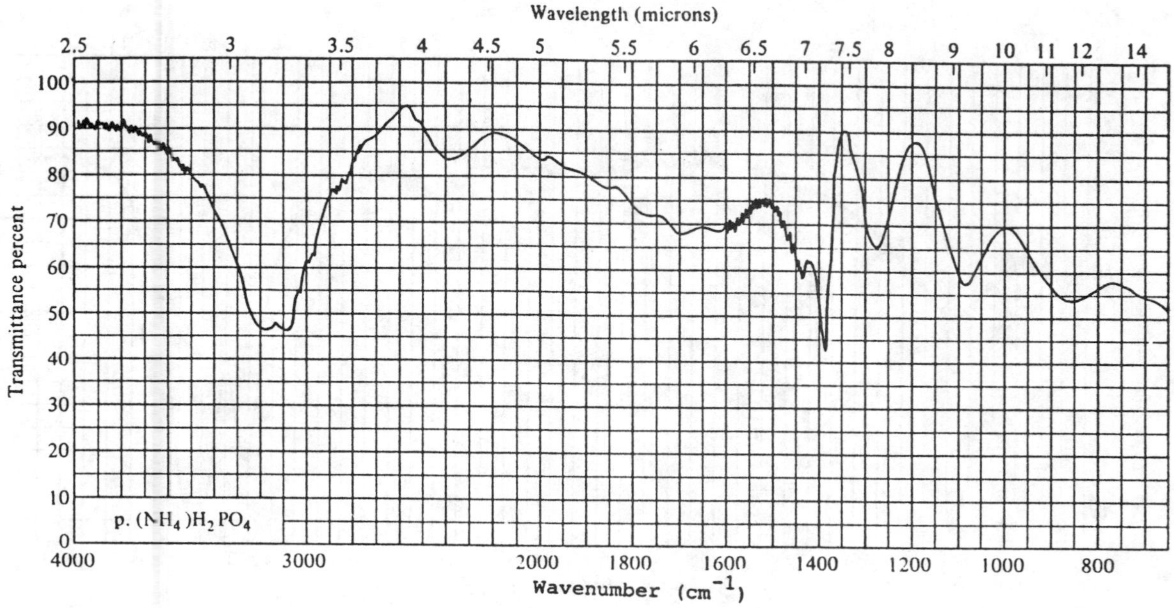
Wavelength (microns)
Transmittance percent
Wavenumber (cm$^{-1}$)
p. $(NH_4)H_2PO_4$

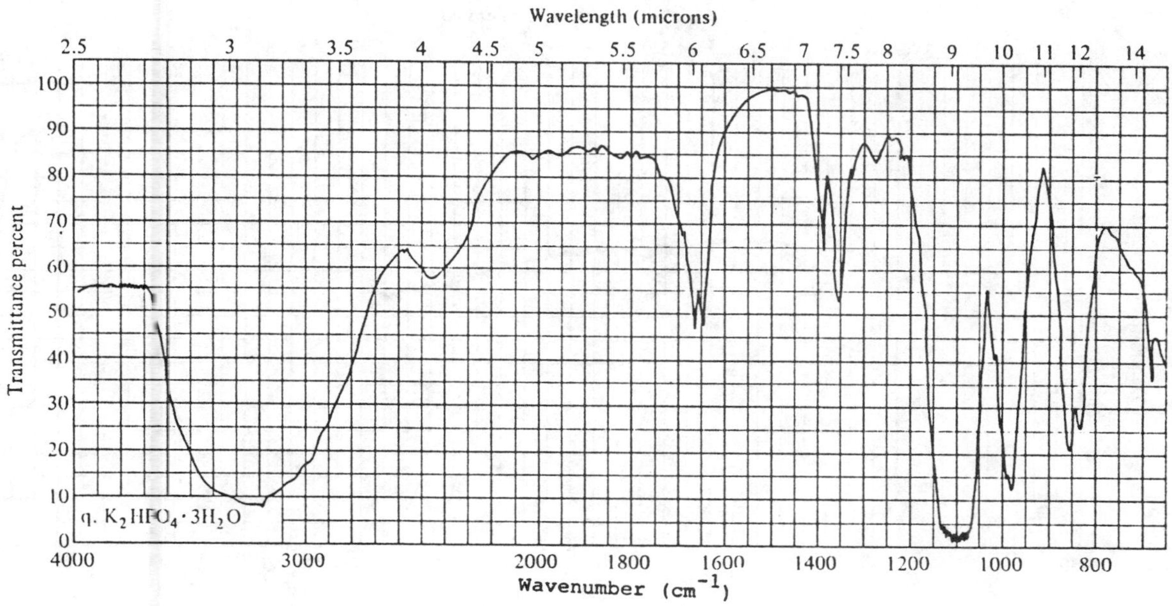
Wavelength (microns)
Transmittance percent
Wavenumber (cm$^{-1}$)
q. $K_2HPO_4 \cdot 3H_2O$

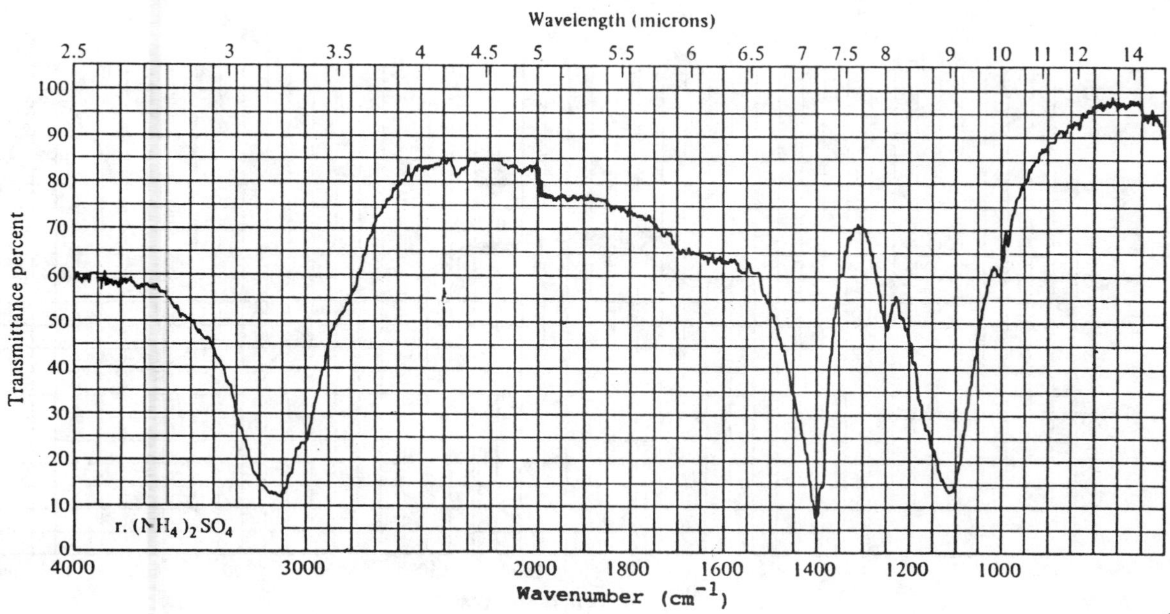
Wavelength (microns)
Transmittance percent
Wavenumber (cm$^{-1}$)
r. $(NH_4)_2SO_4$

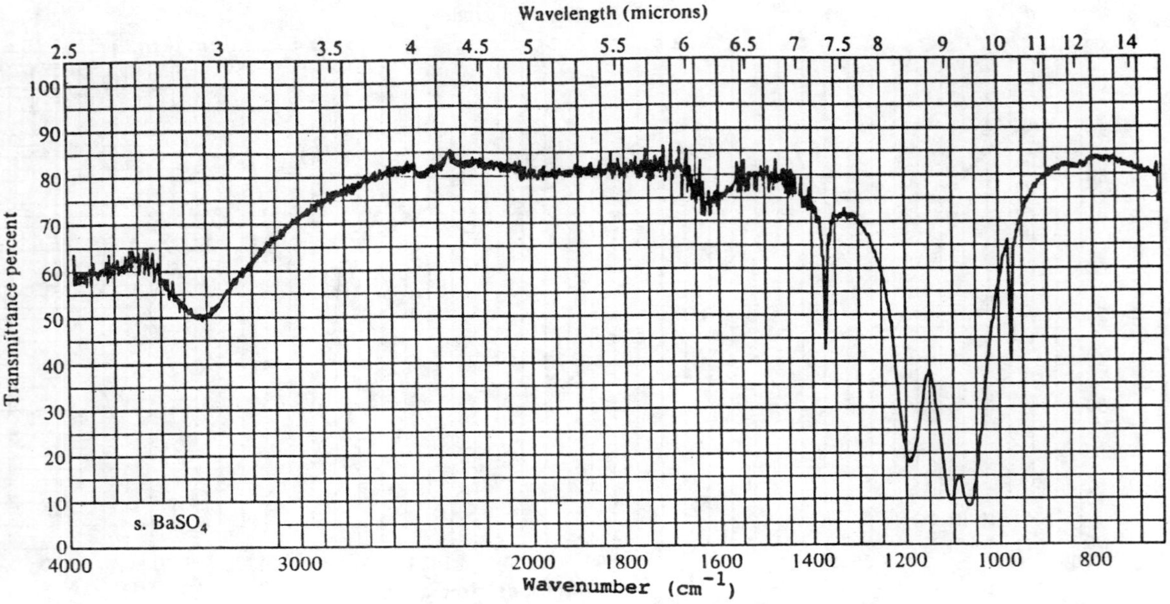
Wavelength (microns)
Transmittance percent
s. $BaSO_4$
Wavenumber ($cm^{-1}$)

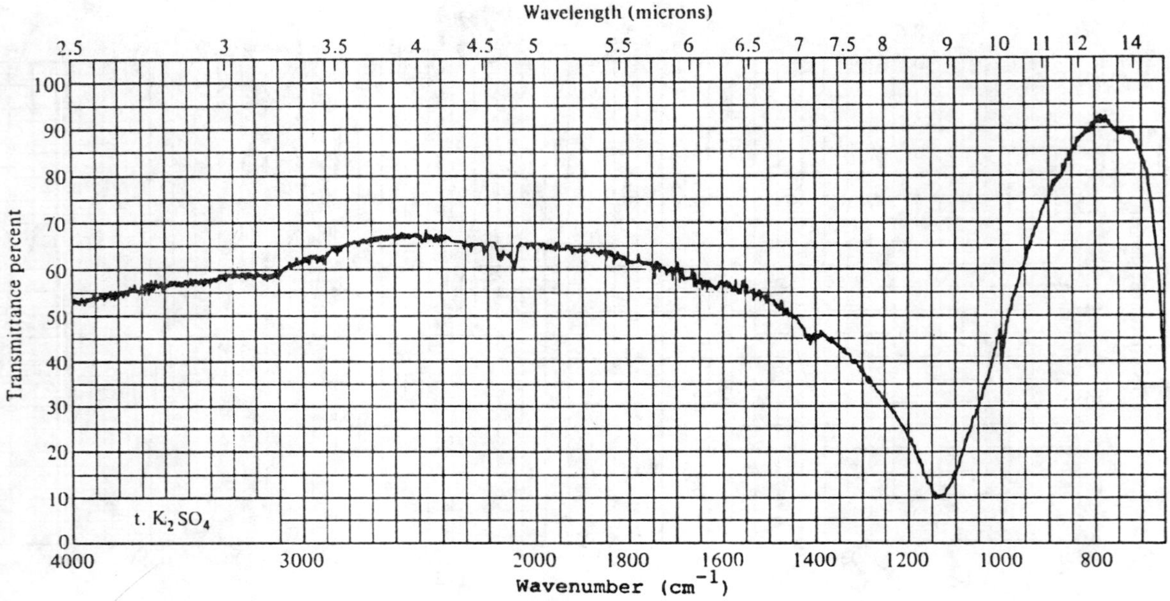
Wavelength (microns)
Transmittance percent
t. $K_2SO_4$
Wavenumber ($cm^{-1}$)

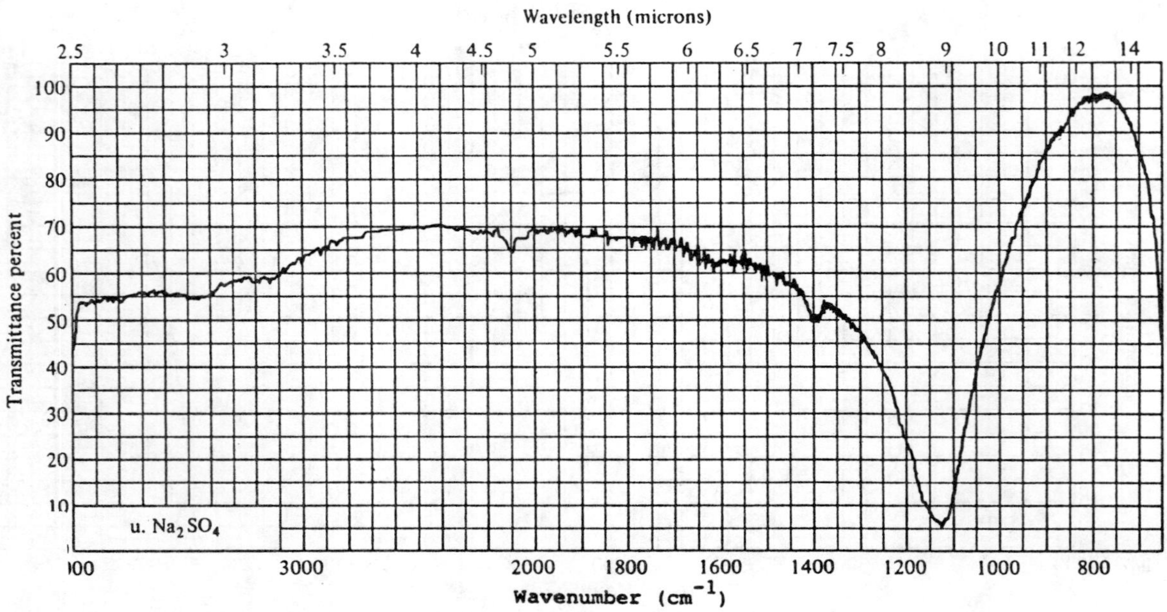
Wavelength (microns)
Transmittance percent
u. $Na_2SO_4$
Wavenumber ($cm^{-1}$)

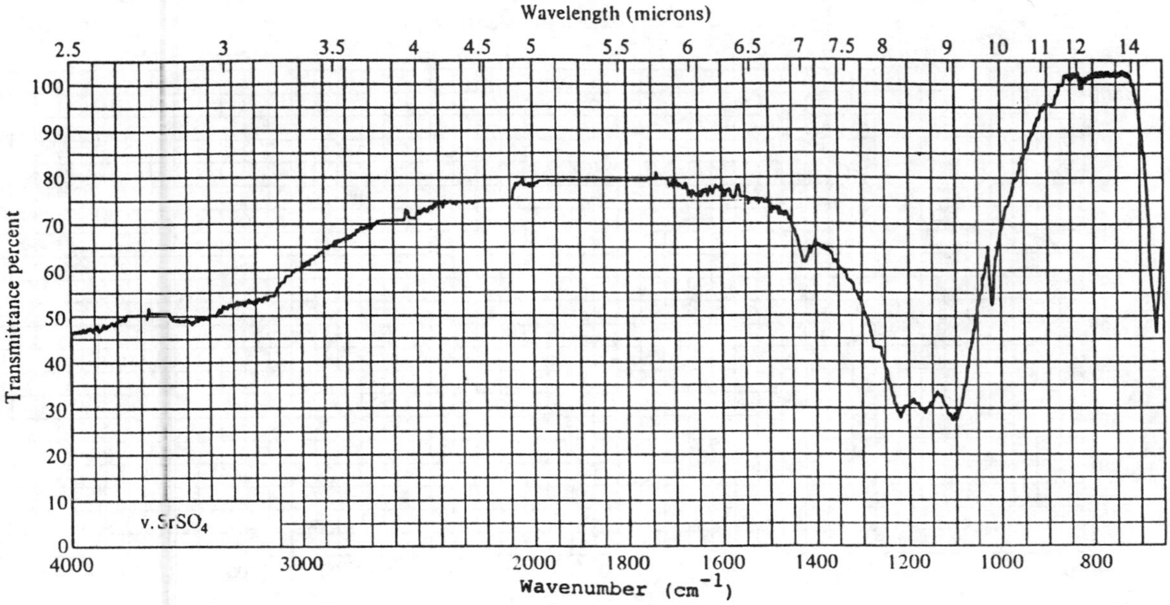
Wavelength (microns)
2.5 3 3.5 4 4.5 5 5.5 6 6.5 7 7.5 8 9 10 11 12 14
Transmittance percent
0 10 20 30 40 50 60 70 80 90 100
v. SrSO4
4000 3000 2000 1800 1600 1400 1200 1000 800
Wavenumber (cm-1)

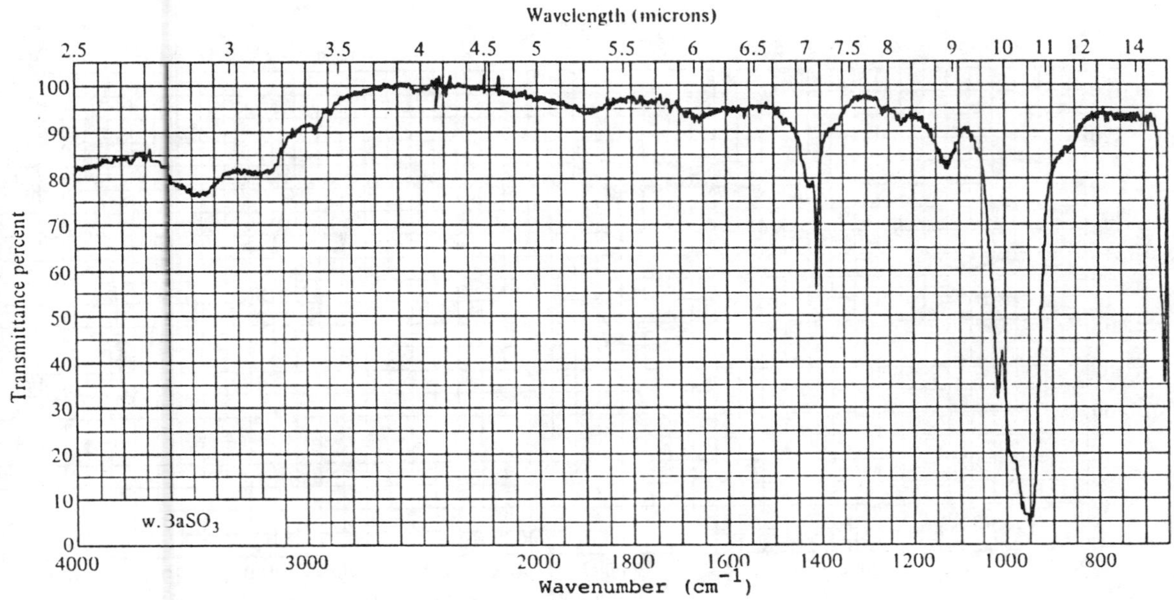
Wavelength (microns)
2.5 3 3.5 4 4.5 5 5.5 6 6.5 7 7.5 8 9 10 11 12 14
Transmittance percent
0 10 20 30 40 50 60 70 80 90 100
w. BaSO3
4000 3000 2000 1800 1600 1400 1200 1000 800
Wavenumber (cm-1)

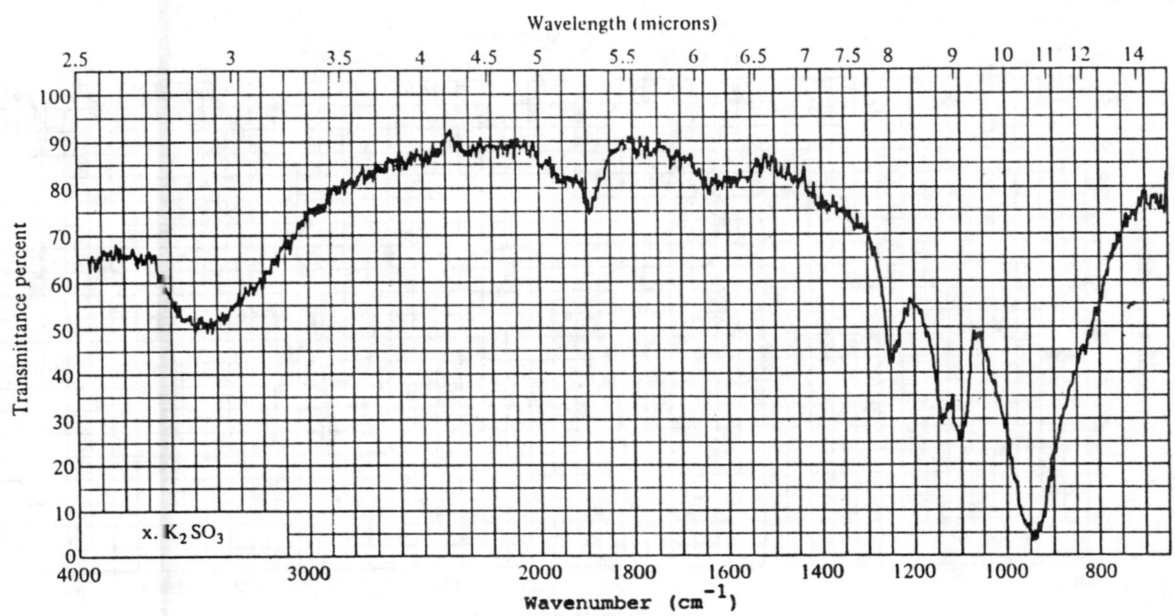
Wavelength (microns)
2.5 3 3.5 4 4.5 5 5.5 6 6.5 7 7.5 8 9 10 11 12 14
Transmittance percent
0 10 20 30 40 50 60 70 80 90 100
x. K2 SO3
4000 3000 2000 1800 1600 1400 1200 1000 800
Wavenumber (cm-1)

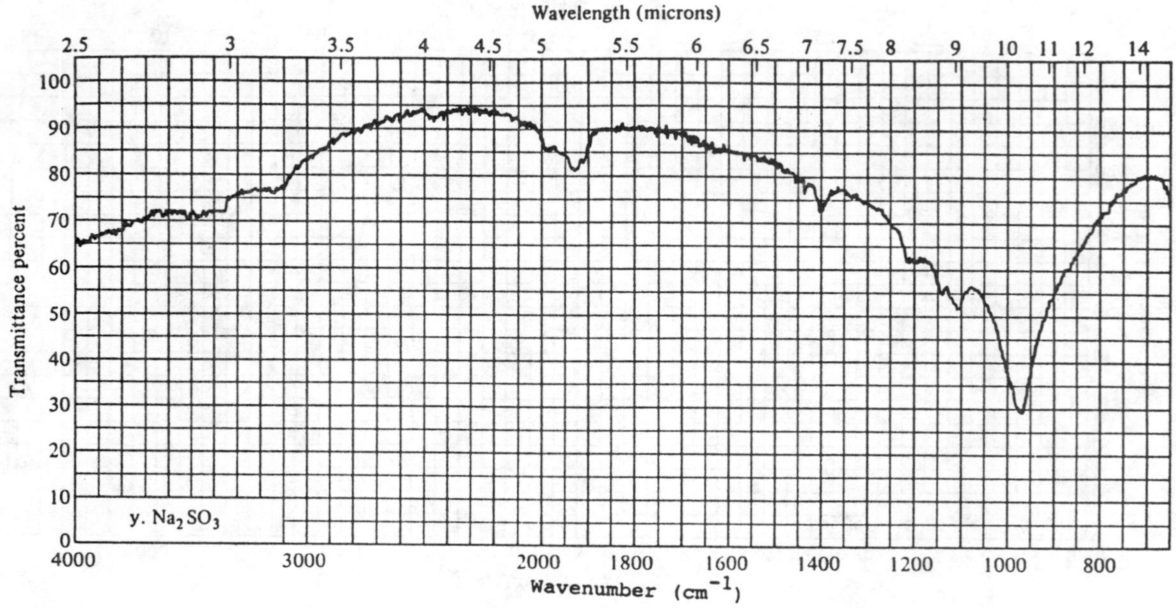
Wavelength (microns)
2.5 3 3.5 4 4.5 5 5.5 6 6.5 7 7.5 8 9 10 11 12 14
Transmittance percent
0 10 20 30 40 50 60 70 80 90 100
y. $Na_2SO_3$
4000 3000 2000 1800 1600 1400 1200 1000 800
Wavenumber ($cm^{-1}$)

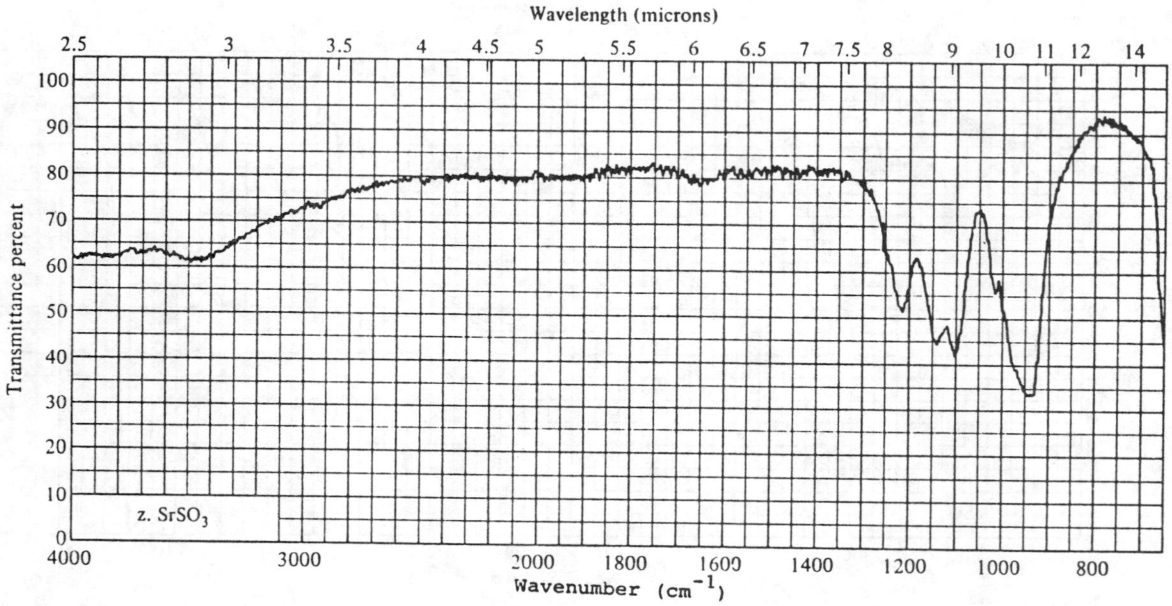
Wavelength (microns)
2.5 3 3.5 4 4.5 5 5.5 6 6.5 7 7.5 8 9 10 11 12 14
Transmittance percent
0 10 20 30 40 50 60 70 80 90 100
z. $SrSO_3$
4000 3000 2000 1800 1600 1400 1200 1000 800
Wavenumber ($cm^{-1}$)

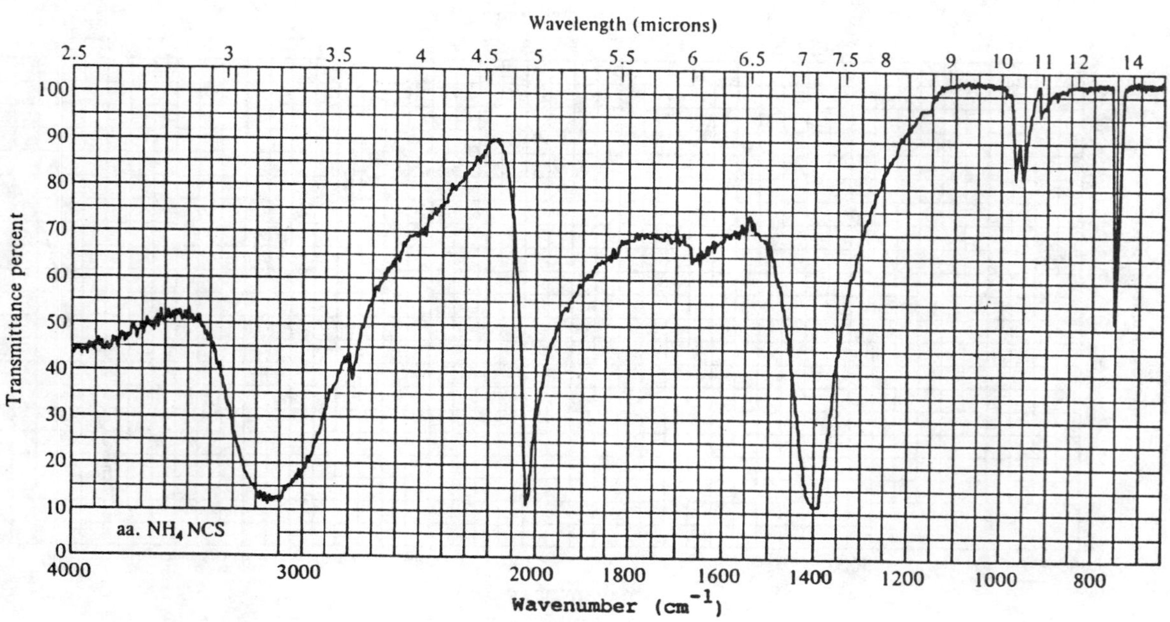
Wavelength (microns)
2.5 3 3.5 4 4.5 5 5.5 6 6.5 7 7.5 8 9 10 11 12 14
Transmittance percent
0 10 20 30 40 50 60 70 80 90 100
aa. $NH_4NCS$
4000 3000 2000 1800 1600 1400 1200 1000 800
Wavenumber ($cm^{-1}$)

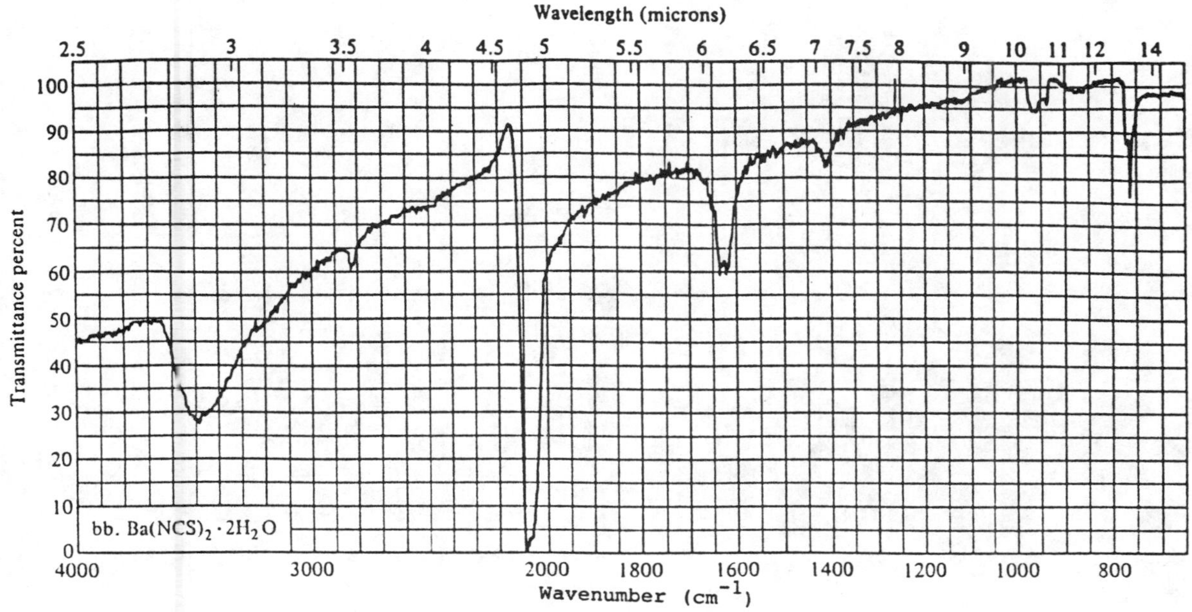
Wavelength (microns)
2.5 3 3.5 4 4.5 5 5.5 6 6.5 7 7.5 8 9 10 11 12 14
Transmittance percent
0 10 20 30 40 50 60 70 80 90 100
bb. $Ba(NCS)_2 \cdot 2H_2O$
4000 3000 2000 1800 1600 1400 1200 1000 800
Wavenumber (cm$^{-1}$)

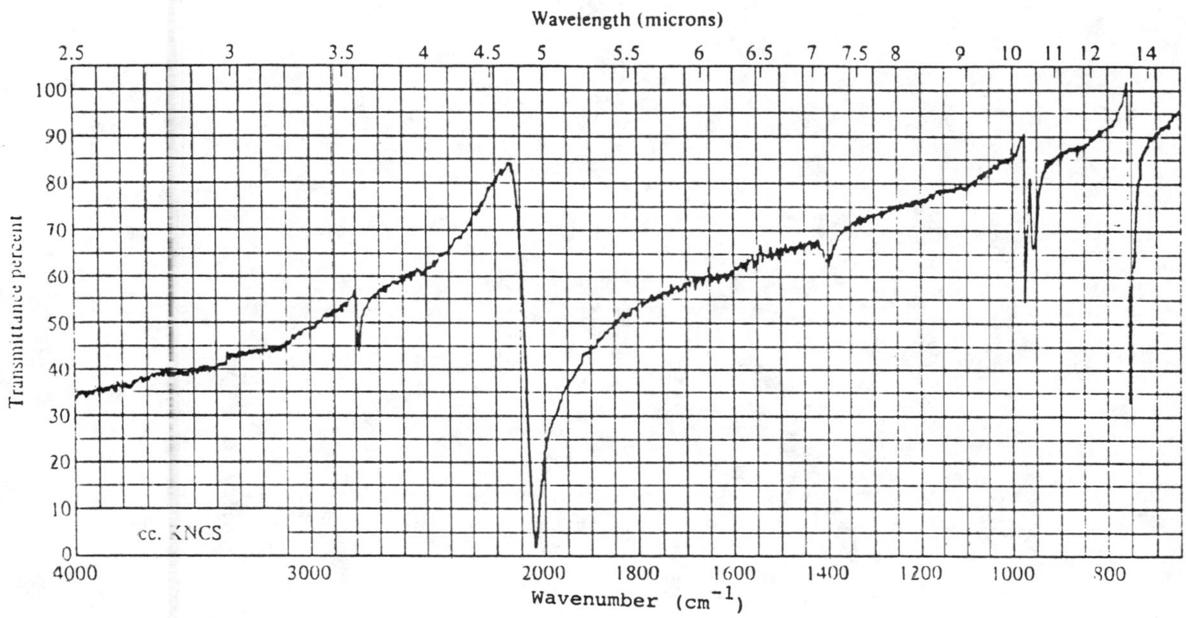
Wavelength (microns)
2.5 3 3.5 4 4.5 5 5.5 6 6.5 7 7.5 8 9 10 11 12 14
Transmittance percent
0 10 20 30 40 50 60 70 80 90 100
cc. KNCS
4000 3000 2000 1800 1600 1400 1200 1000 800
Wavenumber (cm$^{-1}$)

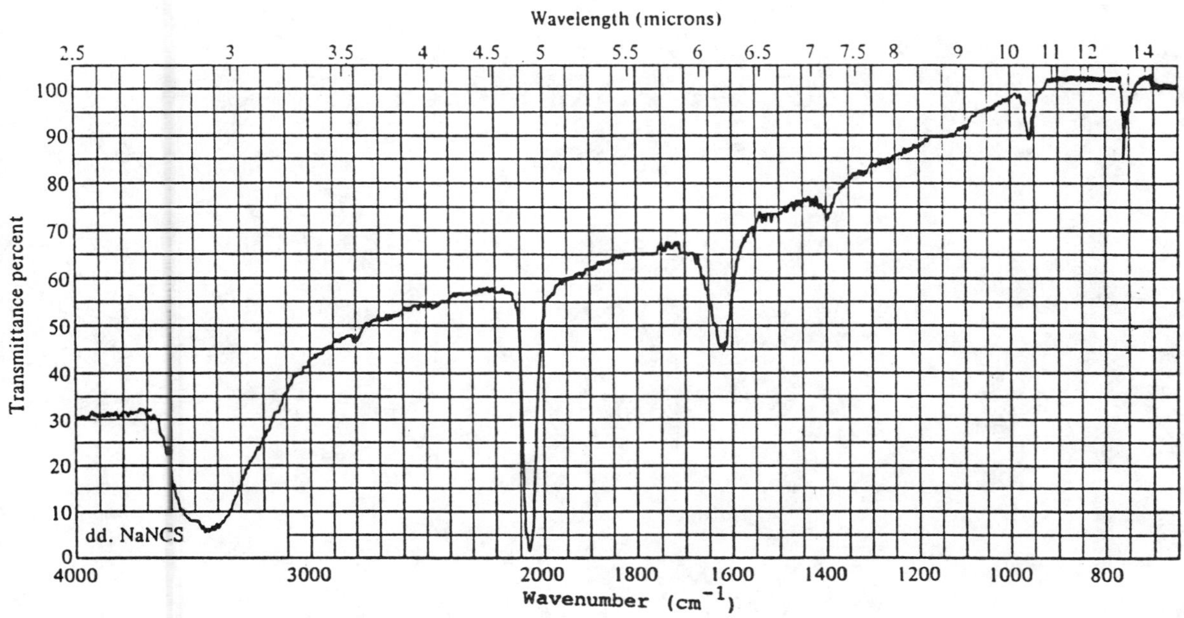
Wavelength (microns)
2.5 3 3.5 4 4.5 5 5.5 6 6.5 7 7.5 8 9 10 11 12 14
Transmittance percent
0 10 20 30 40 50 60 70 80 90 100
dd. NaNCS
4000 3000 2000 1800 1600 1400 1200 1000 800
Wavenumber (cm$^{-1}$)

# EXPERIMENT 9

## REPORT SHEET

Name ______________________________ Date ____________

Section ______________________________

**Data on Flame Tests**

| Ion | Color | Appearance |
|---|---|---|
| $Li^+$ | | |
| $Na^+$ | | |
| $K^+$ | | |
| $NH^{4+}$ | | |
| $Ba^{2+}$ | | |
| $Sr^{2+}$ | | |
| unknown | | |

Identity of unknown salt ______________________________

Attach your infrared spectrum.

# EXPERIMENT 10

# ENTHALPY OF REACTION

## Introduction

Every chemical change is accompanied by an energy change; energy is absorbed or evolved, usually as heat. Energy breaks chemical bonds in reactants, and energy is released as bonds form in products. Whether the combination of these steps absorbs energy or releases energy depends on the relative sizes of the energies associated with bond breaking and bond forming.

The amount of heat involved in a reaction depends not only on what the reaction is, but also on the temperature at which the reaction occurs and whether the reaction occurs under conditions of constant pressure or constant volume. In the laboratory, reactions are conveniently carried out at constant pressure in beakers or flasks that are open to the atmosphere. The amount of heat absorbed or released under this condition is the *enthalpy change*, $\Delta H$, for the reaction, where

$$\Delta H = H_{\text{products}} - H_{\text{reactants}} \tag{10-1}$$

Enthalpy, $H$, can be thought of as the "heat content" of a substance; this heat is stored as potential energy in the form of bond, and other, energies. When atoms rearrange during a reaction, the heat content of the products is usually different from the heat content of the reactants. This difference in heat content appears as heat absorbed or released; it is generally expressed in calories or Joules per mole of a particular material. For example, in this experiment you will work with an acid-base neutralization in aqueous solution:

$$H_3O^+(aq) + OH^-(aq) \rightarrow 2H_2O \tag{10-2}$$

The enthalpy change for this reaction could be given in cal/mole of $H_3O^+$, or in cal/mole of $OH^-$, or in cal/mole of $H_2O$. Similarly, the enthalpy change could be given in Joules/mole of a particular reactant or product, with 1 cal = 4.184 J.

Note that the enthalpy change, $H_{\text{products}} - H_{\text{reactants}}$, is positive if heat is absorbed, that is, if $H_{\text{products}} > H_{\text{reactants}}$; the reaction is *endothermic*. The enthalpy change is negative if heat is released, that is, if $H_{\text{products}} < H_{\text{reactants}}$; the reaction is *exothermic*.

In an exothermic reaction, the formation of new bonds between atoms causes the system to be more stable; "excess" energy is released as heat. The system changes from a higher energy state to a lower energy state. Intuitively, it seems plausible that reactions would have a natural tendency to proceed spontaneously to a state of lower energy; indeed, this negative enthalpy change is part of what causes some reactions to occur spontaneously.

## Technique

*Calorimetry* is the measurement of the heat change for a reaction. This is easier if the reaction is exothermic than if it is endothermic. The heat that is released by the reaction is "trapped" and causes come measurable physical change. This change is then compared to the size of the change caused by adding a known amount of heat to the heat trap.

The device used to measure heat changes is a *calorimeter*. In this experiment, the calorimeter consists of two nested Styrofoam cups with a square of asbestos for a lid; a thermometer and stirrer are inserted through the lid into the cup, as shown in Figure 10-1.

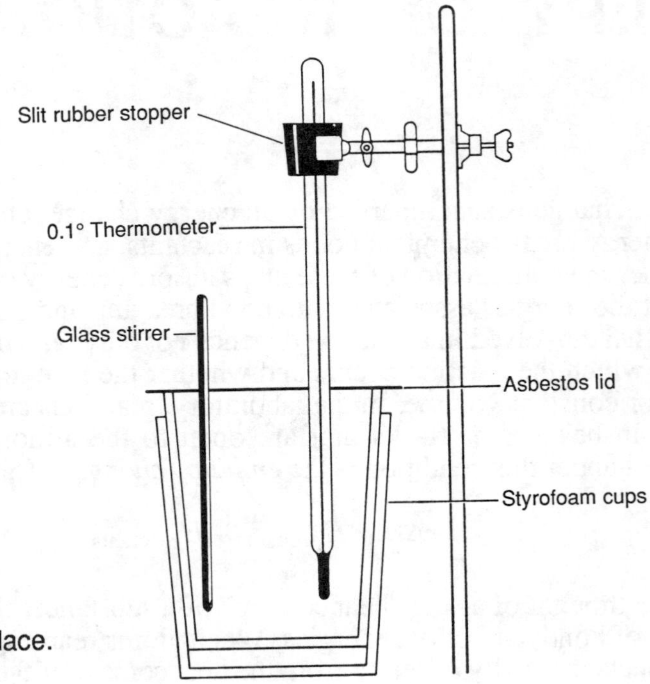


**Figure 10-1**

Calorimeter with lid in place.

As a reaction occurs inside the calorimeter, the calorimeter traps the released heat, which causes the temperature inside the calorimeter to rise. The amount of heat can be related to this temperature change by the equation

$$Q = mC\Delta T \qquad (10\text{-}3)$$

where $Q$ = amount of heat, $m$ = amount of material undergoing a temperature change, $C$ = heat capacity of material undergoing a temperature change, and $\Delta T$ = temperature change = $T_{final} - T_{initial}$. In this experiment, the heat produced by the reaction is absorbed by the calorimeter and by the solution in the calorimeter. Therefore:

$$Q_{total} = Q_{calorimeter} + Q_{solution} \qquad (10\text{-}4)$$

where

$$Q_{calorimeter} = (C_{calorimeter})(\Delta T) \qquad (10\text{-}5)$$

and

$$Q_{solution} = (\text{mass of solution})(C_{solution})(\Delta T) \qquad (10\text{-}6)$$

Heat capacity data for many solutions can be taken from tables in chemical handbooks. However, because every calorimeter is different, $C_{calorimeter}$ must be determined experimentally. This consists of introducing a known amount of heat into the calorimeter and measuring the temperature change that occurs; then you can calculate the heat capacity in cal/C°.

It might seem quite simple to measure $\Delta T$. If no heat escaped from the calorimeter, we would only need to measure $T_{initial}$, mix the reactants, and then measure $T_{final}$. However, a real calorimeter has some heat leakage that must be taken into account in the measurements and calculations. A graphical method can be used to obtain $\Delta T$. A plot of temperature, $T$, vs. time, $t$, for an exothermic reaction is shown in Figure 10-2. The reaction was begun by mixing reactants at time $t = 0$. A large increase in temperature is followed by a steady drop in $T$ because of heat leakage from the calorimeter. The first points on the graph might vary randomly because instantaneous, complete mixing of reactants is impossible; then the temperature drop becomes linear with time. A straight line is drown (perhaps ignoring the first point or two before thermal equilibrium was established) and extrapolated to time $t = 0$. The value of $T$ at $t = 0$ for the *measurements* is $T_{final}$ for the *reaction*, which occurs as rapidly as the reactant solutions can be stirred together. This is the temperature that the calorimeter and its contents *would* have reached *if* complete mixing had been instantaneous and *if* no heat leakage had occurred.

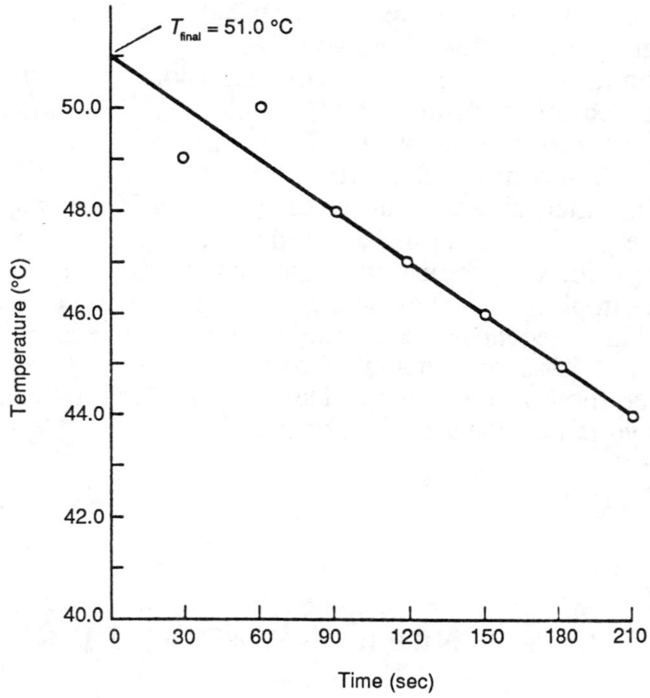


**Figure 10-2**

A plot of temperature vs. time for an exothermic reaction.

## Review

Appendix III

## Equipment Needed

asbestos square
clamp
clock with second hand
glass rod
100-mL graduated cylinders (2)
ring stand
slit rubber stopper
8-oz Styrofoam cups (2)
0.1° thermometer

## Chemicals Needed

0.50M $CuSO_4$; copper sulfate
2.0M HCl; hydrochloric acid
2.0M NaOH; sodium hydroxide
6M $NH_3$; ammonia

## Procedure

### A. Determination of Calorimeter Heat Capacity

Heat about 75 mL of distilled water to 60-70°C. If the thermometer you are using does not register this high, heat the water to about 5° above the highest temperature that your thermometer can record. While you are heating this water, *do not* use the 0.1° thermometer to keep track of its temperature; use the regular thermometer from your desk.

Use a graduated cylinder to measure 50.0 mL of cold (room temperature), distilled water. Put it into the calorimeter and measure its temperature, $T_c$, with the 0.1° thermometer.

Use a graduated cylinder to measure 50.0 mL of the heated water. Allow the water to stand in the cylinder for about 30 seconds. Then stir the water gently and briefly and measure its temperature using the 0.1° thermometer. Put the thermometer in the calorimeter lid. Pour the hot water into the calorimeter; make the transfer as rapid and complete as possible. Record the time of this addition; this is t = 0 for your temperature measurements. Quickly put the calorimeter lid, thermometer, and stirrer in place. Stir the water gently for a few seconds; be careful not to bump the thermometer bulb. Thirty seconds after adding the hot water, record the temperature; continue to record the temperature at 30-second intervals for five minutes.

Dry the calorimeter thoroughly with a clean towel. Repeat the determination. Dry the calorimeter again when you finish the second determination.

### B. Heat of Neutralization

The equation for the reaction is

$$HCl(aq) + NaOH(aq) \rightarrow NaCl(aq) + H_2O \qquad (10\text{-}7)$$

The net equation is

$$H_3O^+(aq) + OH^-(aq) \rightarrow 2H_2O \qquad (10\text{-}8)$$

The heat change per mole of $H_3O^+$ or $OH^-$ that is neutralized, Equation (10-8), is the same for all acids and bases.

You are to combine 50.0 mL of 2.0M HCl and 50.0 mL of 2.0M NaOH. Measure one of these solutions and pour it into the calorimeter. Measure the temperatures of both solutions. (Be sure to rinse and dry the thermometer when transferring it from one solution to the other.) If the temperatures of the two solutions are different, allow them to stand until their temperatures are the same, or record the average of the temperatures of the two solutions as the initial temperature.

Pour the second solution into the calorimeter containing the first solution; record the time. Quickly put the lid, thermometer, and stirrer in place, and stir the mixture gently for a few seconds. Record the temperature at 30-second intervals for five minutes.

To discard the solution in the calorimeter, wash it down the drain. Rinse the calorimeter with tap water and then with distilled water; dry it thoroughly. Repeat the determination. When you finish the second determination, rinse and dry the calorimeter again.

### C. Heat of Formation of $Cu(NH_3)_4^{2+}$

The equation for the reaction is

$$Cu^{2+}(aq) + 4NH_3(aq) \rightarrow Cu(NH_3)_4^{2+}(aq) \qquad (10\text{-}9)$$

An excess of ammonia is used to ensure that the position of equilibrium will be far enough to the right that virtually all of the $Cu^{2+}$ reacts as shown. Your calculations are to be based on the amount of $Cu^{2+}$.

Measure 50.0 mL of 0.50M $CuSO_4$ and 50.0 mL of 6M $NH_3$. Pour one of these solutions into the clean, dry calorimeter. Record the initial temperature of each solution or an average of the temperatures of both solutions. Add the second solution to the first solution in the calorimeter and record the time. Position the lid, thermometer, and stirrer, stir the mixture gently and briefly, and record the temperature at 30-second intervals for five minutes.

To discard the solution in the calorimeter, wash it down the drain. Rinse and dry the calorimeter. Repeat the determination. Thoroughly rinse and dry the calorimeter and thermometer before you return them to the stockroom.

## Results and Calculations

### A. Determination of Calorimeter Heat Capacity

After mixing the cold water and the hot water, the heat gained by the cold water would exactly equal the heat lost by the hot water if there were no heat loss to the calorimeter and other surroundings; the final temperature would be exactly halfway between $T_c$ and $T_h$. However, the final temperature will actually be lower than this, because the hot water will lose more heat than the cold water will gain; the difference between these two heat terms is the heat absorbed by the calorimeter.

Make a graph of $T$ vs. $t$. Draw the best possible straight line through your data points and extrapolate this line to $t = 0$ to determine $T_{final}$ for the heat transfer process within the calorimeter.

Calculate the heat lost by the hot water, $Q_h$:

$$Q_h = V_h \times \frac{1.00 \text{ g}}{\text{mL}} \times \frac{1.00 \text{ cal}}{\text{g C}^\circ} \times (T_{final} - T_h) \qquad (10\text{-}10)$$

In a similar way, calculate the heat gained by the cold water, $Q_c$. Record each of these heat terms as an *amount* of heat *without a sign* that indicates direction of heat flow.

The heat that is the difference between $Q_h$ and $Q_c$ was absorbed by the calorimeter. Divide this heat term, $Q_h - Q_c$, by the temperature change that occurred within the calorimeter, $T_{final} - T_c$, when the hot and cold water were mixed. The result is the heat capacity of the calorimeter in cal/C°.

### B. Heat of Neutralization

Make a graph of $T$ vs. $t$. Extrapolate the straight line to $t = 0$ to determine $T_{final}$ for the reaction.

The heat that is given out during the reaction is absorbed by the solution and by the calorimeter, raising the temperatures of both:

$$Q_{reaction} = Q_{solution} + Q_{calorimeter} \qquad (10\text{-}11)$$

Each of these terms must be calculated separately using the heat capacity of the particular material being considered and the temperature change that it undergoes.

After the neutralization, a solution of NaCl is present. Because heat capacities are defined for mass of material, the volume of this solution must be converted to mass of solution for the calculation of the heat absorbed by the solution:

$$Q_{\text{solution}} = V \text{ x density x } C_{\text{solution}} \text{ x } \Delta T \tag{10-12}$$

The density of a NaCl solution of the concentration you have is 1.06 g/mL. The heat capacity of this solution is 0.931 cal/g C°. Remember to base your calculation on the *entire* amount of solution in the calorimeter.

For the calorimeter

$$Q_{\text{calorimeter}} = C_{\text{calorimeter}} \text{ x } \Delta T \tag{10-13}$$

Equation (10-11) gives the heat of reaction for the amounts of reactants used in the experiment. Use volume and molarity of HCl to calculate the moles of HCl used in the experiment. Then calculate $\Delta H$/mole of acid. Include the proper sign to show the direction of heat flow for the reaction.

### C. Heat of Formation of $Cu(NH_3)_4^{2+}$

Follow the procedure described in Part B for dealing with the data. You need the information that the resulting $Cu(NH_3)_4SO_4$ solution has a density of 1.06 g/mL and a heat capacity of 1.00 cal/g C°. Express $\Delta H$ as cal/mole of $Cu^{2+}$; include the proper sign.

## Question

Assume that after determining the heat capacity of your calorimeter, you use it to study an *endothermic* reaction for which $T_{\text{initial}}$ is 25°C.

(a) Make a sketch showing a graph of $T$ vs. $t$ for such a reaction. On the temperature axis, mark 25°C to make it obvious how $T_{\text{final}}$ would compare to $T_{\text{initial}}$.

(b) What would the direction of heat "leakage" be?

# EXPERIMENT 10

# REPORT SHEET

Name ______________________________ Date ______________

Section ______________________________

## A. Determination of Calorimeter Heat Capacity

| | Trial 1 | Trial 2 |
|---|---|---|
| $T_c$ | | |
| $T_h$ | | |
| $T_{final}$ | | |
| $\Delta T$ for cold water | | |
| $\Delta T$ for hot water | | |
| $Q_c$ (amount, no sign) | | |
| $Q_h$ (amount, no sign) | | |
| $Q_{calorimeter}$ | | |
| $C_{calorimeter}$ | | |
| Average $C_{calorimeter}$ | | |

Show the calculation of $Q_c$ for one trial.

Show the calculation of $Q_h$ for one trial.

Show the calculation of $C_{calorimeter}$ for one trial.

(Attach your two graphs of $T$ vs. $t$. On the back of each graph, give a table showing the data you used to make the graph.)

**B. Heat of Neutralization**

| | Trial 1 | Trial 2 |
|---|---|---|
| $T_{initial}$ | | |
| $T_{final}$ | | |
| $\Delta T$ | | |
| $Q_{calorimeter}$ (amount) | | |
| $Q_{solution}$ (amount) | | |
| $\Delta H$ (with sign) | | |
| moles of HCl | | |
| $\Delta H$/mole of HCl | | |
| Average $\Delta H$/mole HCl | | |

Show the calculation of $Q_{calorimeter}$ for one trial.

Show the calculation of $Q_{solution}$ for one trial.

Show the calculation of $\Delta H$ for one trial.

Show the calculation of moles of HCl.

Show the calculation of $\Delta H$/mole of HCl for one trial.

(Attach your two graphs of $T$ vs. $t$. On the back of each graph, give a table showing the data you used to make the graph.)

## C. Heat of Formation of $Cu(NH_3)_4^{2+}$

| | Trial 1 | Trial 2 |
|---|---|---|
| $T_{initial}$ | | |
| $T_{final}$ | | |
| $\Delta T$ | | |
| $Q_{calorimeter}$ (amount) | | |
| $Q_{solution}$ (amount) | | |
| $\Delta H$ (with sign) | | |
| moles of $Cu^{2+}$ | | |
| $\Delta H$/mole of $Cu^{2+}$ | | |

Average $\Delta H$/mole $Cu^{2+}$ ____________________

Show the calculation of $Q_{calorimeter}$ for one trial.

Show the calculation of $Q_{solution}$ for one trial.

Show the calculation of $\Delta H$ for one trial.

Show the calculation of moles of $Cu^{2+}$.

Show the calculation of $\Delta H$/mole of $Cu^{2+}$ for one trial.

(Attach your two graphs of $T$ vs. $t$. On the back of each graph, give a table showing the data you used to make the graph.)
(On the back of the Report Sheet answer the Questions.)

# EXPERIMENT 11

# GRAVIMETRIC DETERMINATION OF IRON

## Introduction

Chemistry became a modern science when quantitative methods of investigation were developed. Perhaps the most fundamental quantitative measurement in chemistry is the measurement of amounts of materials involved in chemical changes. The nature of many reactions and the composition of many compounds could be understood only after accurate weight (or mass) determinations became possible. We still base much of our knowledge about the composition of substances on measurements of the weights of their components. In this experiment, you will deal with *gravimetric analysis*: analysis by measuring weights, an historical method of quantitative chemical analysis that is still important.

Step by step, the general procedure for the gravimetric determination of the amount of a particular ion in a material is as follows.

1. Dissolve a known weight of the material containing the ion of interest.

2. Add to the solution a substance that will form an "insoluble" compound with the ion of interest. The solubility of this compound must be low enough that the weight of the ion remaining in solution is too low to be measured with the available laboratory instruments.

3. Collect the compound by filtration.

4. Wash the compound to remove soluble contaminants.

5. Dry the compound and weigh it. If the compound has been collected in filter paper, the paper must be burned off so that only the pure, dry compound remains.

6. Calculate the weight of the ion of interest contained in the precipitate.

7. Calculate the percentage by weight of the ion of interest in the original known weight of sample.

An analysis of this type requires the use of three new chemistry laboratory skills: quantitative precipitation, quantitative filtration, and ignition of a precipitate to free it of filter paper and water.

You will carry out a gravimetric analysis for iron with a procedure that can be used to determine the purity of iron salts, to measure the amount of iron that is present in alloys, and to determine the amount of iron in some ores. A precipitate forms when aqueous ammonia is added to a solution containing $Fe^{3+}$. From the simplest point of view, the equation for this reaction is

$$Fe^{3+}(aq) + 3OH^{-}(aq) \rightarrow Fe(OH)_3(s) \qquad (11\text{-}1)$$

Actually, the precipitate is hydrated with an unknown, variable amount of water. Removal of this water by drying the precipitate above 800°C converts the precipitate to $Fe_2O_3$, iron(III) oxide. Knowing the weight of $Fe_2O_3$, you can calculate the weight of iron and the percentage of iron in the original sample.

## Technique

### A. Preparing the Samples

The unknown is a pure or impure salt of $Fe^{2+}$ or $Fe^{3+}$. It is dissolved in acid, because many iron salts are insoluble at pH = 4 or greater. The analysis is based on the presence of $Fe^{3+}$; consequently, any $Fe^{2+}$ in the sample must be oxidized to this state. Nitric acid is used to accomplish this. The reaction is

$$3Fe^{2+}(aq) + NO_3^-(aq) + 4H_3O^+(aq) \rightarrow 3Fe^{3+}(aq) + NO(g) + 6H_2O \qquad (11\text{-}2)$$

The solution is boiled to hasten the oxidation reaction and then to remove NO. The solution is ready for analysis when it is clear and has a yellow color, which is characteristic of $Fe^{3+}$.

### B. Precipitation

Other ions that form insoluble hydroxides in a dilute solution of a weak base must be absent. Examples are $Al^{3+}$ and $Cr^{3+}$.

An excess of $NH_3$ solution is intentionally used in the precipitation to ensure that the precipitation of $Fe^{3+}$ as $Fe(OH)_3$ is as complete as possible.

The precipitate of $Fe(OH)_3$ is not crystalline; instead it has a gelatinous appearance because its individual particles are quite small, below the limits of ordinary microscopic visibility. The smaller the particles are that make up a precipitate, the more difficult it is to collect the precipitate by filtration; extremely small particles might clog the pores of filter paper or might simply pass through the paper instead of being collected in it. However, when the precipitate of $Fe(OH)_3$ forms at a high temperature, its particles are somewhat larger than when it forms at room temperature; consequently, the solution is maintained at nearly boiling during the precipitation. After the precipitation is complete, boiling is continued for one minute so that the particles coagulate to a gelatinous mass that will settle out of suspension. Longer heating must be avoided, for it has just the opposite effect: aggregates break up and the precipitate becomes more difficult to collect by filtration.

### C. Collecting and Washing the Precipitate

The supernatant liquid is decanted through the filter paper keeping as much as possible of the precipitate in the beaker; a large amount of solid in the funnel would slow the filtration.

The precipitate will be contaminated with other ions that were in the solution: $Cl^-$ from the HCl used to dissolve the salt, the anion that was originally combined with the iron, and any impurities that were present in the original material. To ensure that only $Fe_2O_3$ is present after ignition of the filter paper, these adsorbed ions must be removed; they are simply washed away. Water could be used for this, but the aggregated precipitate tends to disperse unless it is in contact with a solution of an electrolyte (a solute that is in solution as ions). Therefore, a dilute solution of ammonium nitrate is used to wash away the unwanted ions. Any ammonium nitrate that is retained by the precipitate will volatilize during the ignition.

Not all the contaminating ions that might be present are known, but $Cl^-$ is known to be among them. Therefore a test for this ion allows you to determine whether the precipitate has been washed sufficiently. A few drops of the filtrate are collected in a small test tube containing $AgNO_3$ solution. If $Cl^-$ is present, a white precipitate of AgCl forms. If this happens, the test solution should be discarded and the precipitate should be washed until the filtrate shows no precipitate with $AgNO_3$ solution.

After the precipitate in the beaker has been washed, it is transferred into the funnel with the aid of a stream of water from a wash bottle and allowed to drain as thoroughly as possible.

### D. Ignition of the Precipitate

**This procedure must be carried out in a hood.** As filter paper chars it emits vapors that have a noticeable odor and that should not be inhaled.

The crucible is placed in a slanted position on a clay triangle on a ring with its lid slightly ajar. Gentle heat is used first to dry the filter paper; intense heat could cause spattering and some precipitate could be thrown out of the crucible.

When the filter paper is dry, it is charred at as low a temperature as possible. The crucible must be partially uncovered to allow contact with air. In the absence of air, $Fe_2O_3$ could be reduced to $Fe_3O_4$. The filter paper must not be allowed to burst into flame, because gas currents might carry particles of the precipitate out of the crucible. If flames or even sparks appear, the crucible should be covered immediately to smother them.

When the filter paper is entirely charred, the heat is increased to complete the combustion of the carbon that remains and to drive all water out of the precipitate. When the ignition is complete, the crucible is cooled and weighed.

## Review

Equipment and Techniques A, E, H
Appendix II

## Equipment Needed

- ashless filter paper
- beakers
- Bunsen burner
- ceramic mat or wire gauze
- clay triangle
- funnels (2)
- glass stirring rods (2)
- medicine dropper
- Meeker (or other high-temperature) burner
- porcelain crucibles with lids (2)
- ring
- ring stand
- small test tubes
- wash bottle

## Chemicals Needed

0.1M $AgNO_3$; silver nitrate
6M $HCl$; hydrochloric acid
conc. $HNO_3$; nitric acid
conc. $NH_3$; ammonia
$NH_4NO_3$; ammonium nitrate
unknown

## Procedure

### A. Preparing the Samples

In an 800-mL beaker, prepare 600 mL of a 1% (by weight) solution of $NH_4NO_3$ (ammonium nitrate), which you will use to wash the precipitates. (*Before* you come to the laboratory, calculate and record in your notebook the amounts of $NH_4NO_3$ and water to use.)

On the analytical balance weigh out two 0.7-0.9 g samples of your unknown into two clean, dry 250-mL (or 400-mL) beakers. (The beakers do not have to be dry for your analytical procedure, but they must be completely dry if you weigh them and then weigh your samples into them. If the weight of these beakers exceeds the weighing capacity of your analytical balance, use this alternative procedure; weigh the sample into a small beaker and then use a stream of water from your wash bottle to wash *all* the sample into a larger beaker.) Be sure to mark the beakers in some way that will enable you to know which beaker contains which sample; record these markings in your notebook beside the sample weights.

Put 40 mL of distilled water and 10 mL of 6M HCl into each beaker. Heat one of these mixtures to nearly boiling. Add 2 mL of concentrated $HNO_3$ to this solution dropwise from a medicine dropper. (**Caution! Concentrated nitric acid is extremely corrosive**. If you get it on your skin, wash it off immediately under lots of cold, running water; it reacts very quickly with proteins in your skin leaving a yellow stain. If you spill any on the outside of the bottle or on the table top, sponge it off and wash out the sponge.) Boil the mixture gently until it is clear and yellow; this takes at least three to five minutes.

### B. Precipitation

Put 10 mL of distilled water into a small beaker and take it to the hood where the concentrated $NH_3$ solution is kept. Pour 10 mL of conc. $NH_3$ into the water and stir it with a glass rod. **Cover the beaker with a watch glass before you take it out of the hood**. Keep this solution covered with the watch glass except when you are pouring it into the $Fe^{3+}$ solution. Add about 15 mL of this solution in a slow stream, with constant, gentle stirring, to the $Fe^{3+}$ solution. Allow the precipitate to settle slightly and then run a few drops of the $NH_3$ solution down the inside wall of the beaker; watch closely to see if any more precipitate forms. If it does, add $NH_3$ dropwise until there is no evidence of precipitate formation; then add about two mL excess of $NH_3$. If precipitation appears to be complete, simply add the excess two mL of $NH_3$. There will be a noticeable odor of $NH_3$ in the steam above the mixture after an excess of $NH_3$ has been added. Any unused $NH_3$ solution should **immediately** be washed down the drain with lots of water.

### C. Collecting and Washing the Precipitate

*Gently* boil the mixture for *one minute*. Then turn off the burner and allow the precipitate to settle. Decant as much as possible of the supernatant liquid through filter paper in a funnel; transfer as little as possible of the precipitate. Use a stirring rod to direct the solution, to prevent splashing, and to prevent dripping from the beaker spout. Collect the filtrate in a beaker or flask; to discard it when you finish the filtration, wash it down the drain. Then add about 30 mL of hot $NH_4NO_3$ solution to the precipitate in the beaker; stir the mixture thoroughly and then allow the precipitate to settle. Decant as much as possible of the liquid into the funnel. Repeat this washing two more times. When the last of the third wash liquid passes through the funnel, collect a few drops of it in a small test tube; add one drop of $AgNO_3$ solution to this sample of the filtrate. (Discard the mixture in the test tube as your instructor directs.) If $Cl^-$ is still present, wash the precipitate in the beaker with another 30 mL of $NH_4NO_3$ solution. When the last of this wash liquid passes through the funnel, collect a few drops of it and test it for $Cl^-$. Repeat this washing and testing for $Cl^-$ as many times as necessary. When washing is complete, transfer the precipitate to the funnel. Use a stream of distilled water from a wash bottle to remove particles of precipitate from the beaker wall and stirring rod. This transfer must be complete.

During the filtration of the first sample, carry out the oxidation and precipitation of the second sample. Be careful not to interchange the two samples after you begin filtering the second sample.

Once you start a filtration, you must complete it. If the precipitate is allowed to dry, its surface will crack and form channels through which the wash solution will run without effectively washing the precipitate.

Drain each precipitate as thoroughly as possible in the funnel. If your schedule permits, allow the precipitates and filter paper to dry in your desk (where dirt cannot get into them) until the next laboratory period.

### D. Ignition of the Precipitate

Wash and dry a crucible and lid. Support them on a clay triangle on a ring and apply the full heat of a Meeker (or other high-temperature burner) to them for at least 15 minutes. Allow the crucible and lid to cool on the triangle for a minute and then to cool to room temperature on a ceramic mat or wire gauze; this should take about 15-20 minutes. While this crucible is cooling, wash, dry, and heat a second crucible and lid. Weigh each crucible (with lid) on the analytical balance. If, during the weighing, the weight readout changes continuously, the crucible and lid are not cool; immediately remove them from the balance, return to your desk while they cool, and wait five minutes before you weigh them again. When you record the weights, also record the identifying numbers and letters that are on each crucible and lid, so that you will know which sample you put into which crucible. Heat, cool, and weigh the crucibles as many times as necessary until you obtain two consecutive weighings that differ by no more than 0.0008 g. Use the final weight in your calculations.

Fold the open edge of the filter paper over to enclose the precipitate; place the paper, open-end down, in a crucible that has been heated to constant weight. Position the crucible lid so that the crucible is nearly covered. **In the hood**, heat the crucible slowly to dry the paper; hold the burner in your hand, so that you can immediately remove it if the paper begins to spark or flame. When the paper is completely dry and charred, reposition the crucible lid so that air can freely enter the crucible. Continue heating the crucible until there is no longer any visible carbon. Be sure that the lid, as well as the crucible, is free of carbon. After no more carbon is visible, continue heating the open crucible at the highest possible temperature for at least ten minutes to remove all traces of carbon and water. Be careful to keep the gases of the flame from entering the crucible, for they could react with the $Fe_2O_3$. Cool the crucible for a minute on the triangle; then cover the crucible and cool it to room temperature on a ceramic mat or wire gauze. Repeat this process with the second crucible and precipitate. Weigh each crucible, lid, and contents on the analytical balance.

Repeat the high-temperature heating, cooling, and weighing for each crucible until two consecutive weighings of a crucible differ by no more than 0.0008 g. These repeated heatings (*after* the filter paper has been burned off) do not need to be done in the hood. This agreement between consecutive weighings of a crucible will indicate that volatile and combustible materials (water, $NH_4NO_3$ from the wash solution, and filter paper) have been removed, leaving pure $Fe_2O_3$. Use the final weight in your calculations.

## Results

For each sample, calculate

(a) weight of $Fe_2O_3$

(b) weight of Fe in this weight of $Fe_2O_3$

(c) percentage that this weight of Fe is of the original sample weight.

Describe any departures that you made from the procedure that is given, any unexpected problems you encountered, or any blunders you made that could affect the accuracy of your results for either trial.

## Questions

1. What effect (too high, too low, no change) would each of the following have on your calculated percentage of iron in comparison to the correct percentage of iron for your unknown? Explain each answer.

   (a) The unknown contains a mixture of $Fe^{2+}$ and $Fe^{3+}$

   (b) Some unknown sample is lost after it is weighed.

   (c) The $Fe^{3+}$ is not completely precipitated.

   (d) The precipitate is not thoroughly washed.

   (e) The filter paper is not completely burned off.

   (f) Some $Fe_2O_3$ is reduced to $Fe_3O_4$.

2. Using the procedure described in this experiment, a 0.8014 g sample of wet, hydrated $Fe(NO_3)_3$ yields a residue of $Fe_2O_3$ that weighs 0.1483 g after ignition. Calculate the percentages of Fe, $Fe(NO_3)_3$, and $H_2O$ in the sample.

3. If you use the quantities of reagents described in this experiment, the concentration of $OH^-$ remaining in the solution should not be less than $10^{-5}$M.

   (a) What concentration of $Fe^{3+}$ can remain in solution with this concentration of $OH^-$, rather than be precipitated as $Fe(OH)_3$? For $Fe(OH)_3$, $K_{sp} = 2.5 \times 10^{-39}$.

   (b) What weight of $Fe^{3+}$ would remain in the volume of solution (approximately 70 mL) that you had in this experiment?

   (c) What weight of $Fe_2O_3$ would contain the weight of $Fe^{3+}$ remaining in solution? If this $Fe^{3+}$ precipitated entirely, could you detect this weight of $Fe_2O_3$? Can this be considered an essentially complete precipitation of $Fe^{3+}$?

## EXPERIMENT 11

## REPORT SHEET

Name ______________________________ Date ____________

Section ______________________________

Unknown identification code ____________________

| | Trial 1 | Trial 2 |
|---|---|---|
| Weight of unknown sample | | |
| Weight of crucible and lid<br>After first heating | | |
| After second heating | | |
| After third heating | | |
| After fourth heating | | |
| Weight of crucible, lid, and $Fe_2O_3$<br>After first heating | | |
| After second heating | | |
| After third heating | | |
| After fourth heating | | |
| Weight of $Fe_2O_3$ | | |
| Weight of Fe in this $Fe_2O_3$ | | |
| Weight of Fe in unknown sample | | |
| Percentage Fe in unknown sample | | |
| Average percentage Fe | | |

On the back of the Report Sheet show the calculation of weight and percentage of Fe in the unknown for one trial.

(Attach a separate sheet with the answers to the Questions.)

# EXPERIMENT 12

# MOLAR WEIGHT BY VAPOR DENSITY

## Introduction

The *ideal gas law* states the quantitative relationship among the pressure exerted on or by a gas sample, the volume the gas occupies, its temperature, and the number of moles of gas, *n*, in the sample:

$$PV = nRT \tag{12-1}$$

An important aspect of this relationship is that it applies to all gases; it does not depend on the size, shape, mass, or composition of the particles of the gas. Consequently, it is possible to use measured pressure, volume, and temperature data to "count" the number of moles (or the number of molecules) of gas in any gas sample without knowing the identity of the gas. If one more piece of data is known, it is also possible to calculate another important physical property of the particular gas in the sample; if the mass of the gas sample is known, the weight of one mole of the gas can be calculated. The number of moles of a substance in any weighed sample can be calculated by dividing the mass of the sample by the weight of one mole of the substance:

$$\text{moles} = \frac{\text{grams}}{\text{grams/mole}} \quad \text{or } n = \frac{\text{g}}{M} \tag{12-2}$$

This expression for *n* can be put into the ideal gas law:

$$PV = \left(\frac{g}{M}\right)RT \tag{12-3}$$

The equation now contains the four measurable, variable properties of a gas sample that can be used to calculate its molar weight—pressure, volume, temperature, and mass. Rearranging the terms in Equation (12-3) to solve for molar weight gives

$$M = \frac{gRT}{PV} \tag{12-4}$$

In this experiment, you will work with a compound that is a liquid at room temperature but has a boiling point below that of water. You will measure the mass of this gas that occupies a particular volume at 100°C and atmospheric pressure. Then you will use your measured values for these properties to calculate the molar weight of the compound.

Remember that the equation you use to calculate molar weight is called the *ideal* gas law. It is an exact description of the behavior a gas only if the gas is ideal; this would mean there are no attractive forces between particles of the gas and the particles are volumeless points. Of course neither of these is true of a real gas. However, the behavior of a real gas is very nearly described by the ideal gas law if the total volume occupied by the gas is much greater than the volume taken up by gas particles and if the temperature is high enough that particles have enough energy to resist intermolecular attractions. With the volume and temperature conditions you will use, the behavior of your unknown is sufficiently close to ideal that you should by able to determine a molar weight within a few per cent of the correct value.

After you have used your experimental data to calculate the molar weight of your unknown, your instructor will give you the elemental percentage composition of the compound. From these data, you can calculate the empirical formula to determine the molecular formula of the compound and its correct molar weight. Chemists use this procedure to help them identify unknown substances. The main difference between your work here and the work of a practicing chemist is the sophistication of the equipment that is used, not in the basic procedure.

The equation $PV = nRT$ describes ideal behavior. It has been modified to describe real gases that behave nonideally. This modification is called the *van der Waals equation*:

$$\left(P + \frac{an^2}{V^2}\right)(V - nb) = nRT \tag{12-5}$$

There are two terms in this equation that take into account the ways in which real gases differ from ideal gases. The van der Waals constant $a$ is a measure of the *attractive forces* between particles of the gas; the greater the attractive forces, the greater the numerical value of this constant for the gas. These attractions cause the measured pressure $P$ to be lower than predicted by the ideal gas law for a particular volume of gas. The measured pressure plus the correction factor containing $a$ equals the ideal pressure. The van der Waals constant $b$ is the *volume occupied by the particles* themselves in one mole of gas; the compressible volume (that is, the free space into which the particles can be pushed) is reduced by this amount. In Equation (12-5), $V$ is the measured volume, which includes the actual particle volume, and $(V - nb)$ is the compressible, or ideal, volume. After you have calculated the molar weight of your unknown, your instructor will also give you the density of the liquid. In a liquid, particles are about as close to one another as possible, and so the liquid density can be used to calculate an approximate value of $b$. Then you will use this calculated value of $b$, the correct molar weight for the unknown, and your measured values of pressure, volume, temperature, and mass to calculate a numerical value for $a$.

## Technique

A small amount of liquid is put into a weighed flask. The flask is closed with a stopper through which a short piece of capillary-bore glass tubing extends. The flask is heated in a water bath to about 100°C. As air in the flask is heated and as the liquid vaporizes, air and vapor push out of the flask. Vapor will escape from the flask only until enough vapor remains to exert a pressure that equals the pressure being exerted on the vapor by the atmosphere. The flask is then allowed to cool to condense the vapor. As the flask cools and vapor condenses, the pressure inside the flask drops; air pushes back into the flask until the pressure inside the flask again equals atmospheric pressure at room temperature. The flask with the condensed liquid is then weighed to determine the mass of the compound that was present as vapor when the flask was in the water bath.

The volume of the flask is determined by weighing the flask when it is empty and then weighing it filled with water. Since the density of water is 1.00 g/mL, the measured mass of water is numerically equal to the volume of water, which, in turn, equals the volume of the flask and of the vapor.

## Review

Equipment and Techniques H
Appendix II

## Equipment Needed

- aluminum foil
- balances
- barometer
- large beaker
- burner
- clamp
- 250-mL Florence flask
- 10-mL graduated cylinder
- #4 Neoprene stopper with capillary-bore tubing
- ring
- ring stand
- thermometer
- tissues
- wire
- wire gauze

## Chemicals Needed

Unknown liquids with boiling points below 90°C (elemental analysis and density at room temperature to be provided by instructor)

## Procedure

Obtain a #4, one-hole, Neoprene stopper that has a piece of capillary-bore glass tubing through it. Place the bottom of the stopper on a small piece of aluminum foil. Wrap the foil snugly and smoothly up and around the sides and over the top of the stopper. Remove excess foil that extends beyond the wrapped top of the stopper. Use a piece of wire to poke a hole through the aluminum foil; push the wire all the way up through the glass tubing. Put the stopper into a clean, dry 250-mL Florence flask.

Weigh the flask and stopper on the analytical balance. Put 2 mL of the unknown liquid into the flask. Insert the stopper very snugly.

**Caution! Your unknown is flammable**. Do not pour it near a flame and do not use more of it than instructed. If a fire does start, immediately turn off your burner and use a towel to smother the flames.

**Caution! Avoid inhaling the vapor from your unknown**. Do not stand over the flask while the sample vaporizes.

Immerse the flask as completely as possible in a water bath by clamping it in place as shown in Figure (12-1). Heat the water to boiling and then lower the burner flame to maintain gentle boiling. You might be able to see the liquid boil and you will probably see droplets of liquid or a stream of vapor rush from the flask. After the water is boiling and after you can no longer tell there is liquid in the flask or that vapor is being emitted from the flask, leave the flask in the bath for another minute to ensure that all the vapor is at the same temperature. Note that you are to leave the flask in the bath for just one more minute after both of these conditions exist. Measure the temperature of the bath close to the middle of the flask, not with the thermometer touching the bottom of the beaker.

Remove the flask from the bath by removing the clamp, still holding the hot flask, from the ring stand. Remove the clamp when the flask has cooled enough that you can handle it easily. Dry the outside of the flask. If the water left a deposit on the neck of the flask, wipe it off. If water has splashed into folds in the aluminum foil, absorb it with a tissue. Set the flask on your desk in a safe, clean place and allow it to cool until it is at room temperature, about ten minutes. Handle the flask with a test tube holder or tongs so that you will not leave fingerprints on it or warm it above room temperature.

Weigh the cooled flask on the analytical balance. If the weight readout increases continuously, the flask probably has not cooled to room temperature and air is still pushing into the flask. If this happens, remove the flask from the balance, return to your laboratory desk, and reweigh the flask in two to three minutes. If the weight readout decreases continuously, the stopper is probably loose and vapor is escaping from the flask. If this happens, record the highest weight reading that you see.

Add another one mL of your unknown to the flask and repeat the experiment. It is not necessary to remove the remainder of the first sample of liquid; no matter how much liquid is present

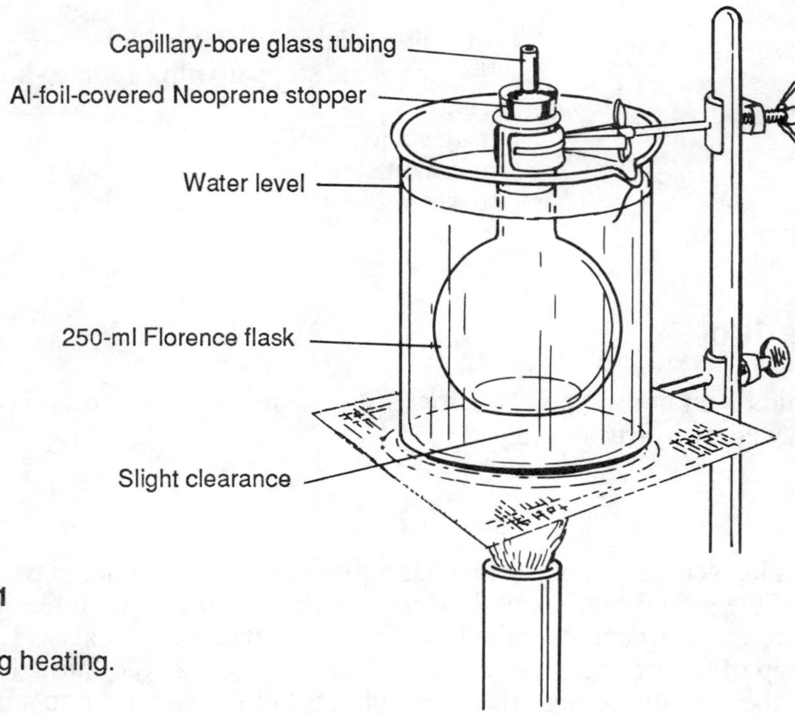


**Figure 12-1**

Flask during heating.

originally, vaporization will occur until enough vapor remains to exert a pressure that is equal to atmospheric pressure.

Carry out a third determination. If your three values for weight of liquid (not weight of flask and liquid) differ by no more than 10%, report all three values and their average; that is,

$$\frac{\text{highest value} - \text{lowest value}}{\text{average value}}(100) = 10\% \text{ or less} \tag{12-6}$$

If the three values are not within 10%, but two consecutive values are within 5%, report all the values but calculate the average of these two values. Repeat the experiment until you have three values that differ by no more than 10% or two consecutive values that differ by no more than 5%.

After you have completed your determinations of the mass of vapor in the flask, measure the volume of this vapor. Pour the remaining liquid out of the flask into the specified waste container. Dry the outside of the flask, put the stopper in place, and weigh the stoppered flask on the laboratory balance. Then fill the flask with water and put the stopper in place again. If necessary, dry the outside of the flask. Weigh the water-filled flask. The volume of water in the flask and stopper equals the volume previously occupied by the vapor of your unknown. Then pour out the water and fill and weigh the stoppered flask again. As usual, use an average of your results.

Read the laboratory barometer to determine the atmospheric pressure.

## Results and Calculations

Calculate the molar weight of your unknown liquid using Equation (12-4) and your measured values of volume, mass, temperature, and atmospheric pressure. Report the average of your trials. Briefly describe any difficulties you had with the experiment that might affect your results. List what you think your main sources of error were and how they would affect your final result.

Determine the empirical formula of your unknown from the elemental analysis provided by your instructor. Then use this formula and your experimental value of molar weight to determine the molecular formula and correct molar weight of your unknown.

Use the liquid density of your unknown (given to you in g/mL) and the correct molar weight to calculate an approximate value for the van der Waals constant *b* in the units liters/mole.

Next, consider Equation (12-5), the van der Waals equation for real gases. You have a numerical value for each variable in this equation. You have just calculated a value for the constant *b*. You can now calculate a value for the constant *a* for your unknown. To do this, you could rearrange the equation to obtain an algebraic expression for *a* in terms of the other terms in the equation. Or you could put your numerical values into the equation and then complete the arithmetic to obtain a value for *a*. The latter procedure is probably easier. Use your experimental values of volume, temperature, and atmospheric pressure. Use the correct molar weight and an average of your measured weights to calculate moles. As you work through this problem, retain and pay close attention to all units.

## Questions

1. In this experiment, there are some avoidable sources of error. State how the calculated molar weight of your unknown would compare to its correct molar weight (high, low, no change) as a result of each of these errors. Explain each answer.

   (a) The flask is not completely submerged in the boiling water bath.

   (b) Vapor is still being emitted from the flask when it is removed from the bath

   (c) The temperature of the flask and vapor is still well above room temperature when the flask and vapor are weighed.

2. What does a large value for the van der Waals constant *a* tell you about a substance? What would a value of zero for this constant mean?

## EXPERIMENT 12

## REPORT SHEET

Name ______________________________ Date ____________

Section ______________________________

Unknown identification code____________________

**Volume of flask**

Weight of dry flask ____________________

Weight of water and flask ________________ ____________________

Weight of water ____________________ ____________________

Volume of flask ____________________

**Molar weight determination**

Weight of dry flask and stopper ______________

Flask, stopper, and condensed liquid __________ __________ __________

Weight of condensed liquid __________ __________ __________

Temperature ____________

Atmospheric pressure ____________

Molar weight __________ __________ __________

Average molar weight ____________

Show a sample calculation of molar weight.

State difficulties and your main sources of error.

**Determination of formula and correct molar weight**

Elemental analysis_______________________________

Empirical formula _______________________________

Experimental molar weight _______________________

Molecular formula _______________________________

Correct molar weight ____________________________

Show calculation of empirical and molecular formulas.

Attach a separate sheet with the calculations of the van der Waals constants *b* and *a* (in that order).

(Attach a separate sheet with the answers to the Questions.)

# EXPERIMENT 13

# VAPOR PRESSURE AND HEAT OF VAPORIZATION

## Introduction

Any liquid evaporates to at least a slight extent; that is, some molecules leave the surface of the liquid and become vapor. As molecules in the vapor phase move about randomly, some strike the surface of the liquid and again become part of the liquid phase; vapor condenses. If a liquid is in a closed container, vapor cannot escape and eventually a state of equilibrium is reached in which vaporization and condensation occur at the same rate; there is no change in the number of molecules in either the liquid phase or vapor phase. The vapor, like any gas, exerts pressure. When the liquid and vapor phases of a substance are in equilibrium, this pressure is called the *vapor pressure of the liquid*; it is the maximum pressure that can be exerted by the vapor phase of a substance at a particular temperature. It is possible to tell when this equilibrium between phases has been reached, because the volume of liquid no longer decreases and the pressure exerted by vapor no longer increases. In this experiment, you will determine when equilibrium has been established between the liquid and vapor phases of a substance by measuring the pressure exerted by the vapor.

The size of a liquid's vapor pressure reflects the ability of molecules to escape from the surface of the liquid. In the liquid phase, molecules exert attractive forces on one another and energy is required for molecules to overcome these forces—to leave the liquid and to become part of the gas phase. Temperature is a measure of average molecular kinetic energy, and as temperature increases more molecules possess the energy required for escape. Consequently, vapor pressure increases as temperature increases. In this experiment, you will measure the vapor pressure of one liquid at several temperatures.

After you have made measurements showing that the vapor pressure of a liquid is related to temperature, you will use these measurements to calculate the energy needed to overcome intermolecular attractions in the liquid. The *molar heat of vaporization*, $\Delta H_v$, is the heat necessary to change one mole of liquid into one mole of vapor at a constant temperature. The mathematical relationship between vapor pressure and molar heat of vaporization is given by the Clausius-Clapeyron equation:

$$\log P_v = \frac{-\Delta H_v}{2.303\,R}\left(\frac{1}{T}\right) + C \qquad (13\text{-}1)$$

Equation (13-1) has the form of the equation for a straight line, $y = mx + b$. Here, $y = \log P_v$ and $x = 1/T$. When the logarithm of vapor pressure is plotted against the reciprocal of Kelvin temperature, a straight line results. The $y$ intercept $b$ corresponds to the constant $C$ in Equation (13-1); you will not be concerned with this constant. The slope of the line equals $-\Delta H_v/2.303R$, where $R$ is the gas constant, 1.987 cal/mole•K or 8.314 J/mole•K. After the slope of the line is calculated from the graph, the value of $\Delta H_v$ can be calculated. This determination of the heat of vaporization is an example of how scientists determine a quantity that is not easy to measure ($\Delta H_v$) by using measurements that are easy to make ($P_v$ and $T$) and then applying an equation that relates these.

## Technique: Measuring Pressure With a Manometer

The pressure of a confined gas sample can be measured with a manometer. This instrument consists of a U-shaped glass tube partially filled with mercury. When both arms of the U are open to the atmosphere, the pressure will be the same in both arms. When one arm of the manometer is attached to a closed flask containing air at atmospheric pressure, the mercury levels will remain the same in both arms, since the pressure on the mercury in the two arms is still the same (atmospheric pressure). When a liquid is put into the flask and the liquid vaporizes, the pressure inside the flask increases above atmospheric pressure; the mercury is pushed down in the arm that is connected to the flask (see Figure 13-1). The difference in the mercury levels, $\Delta h$, is the amount by which the pressure in the flask exceeds atmospheric pressure:

$$P_{\text{flask}} = P_{\text{atm}} + \Delta h \tag{13-2}$$

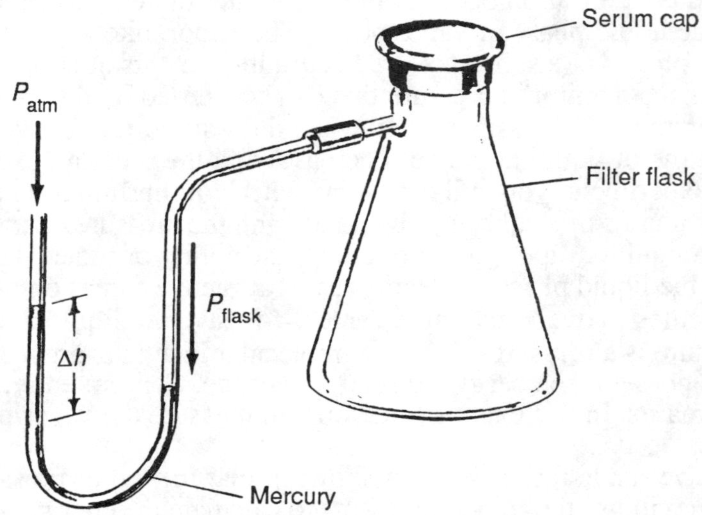


**Figure 13-1**

Using a manometer to measure vapor pressure.

The total pressure in the flask after the liquid vaporizes is the sum of the air pressure in the flask and the vapor pressure of the liquid:

$$P_{\text{flask}} = P_{\text{air}} + P_{\text{v}} \tag{13-3}$$

Since $P_{\text{air}} = P_{\text{atm}}$, the total pressure is the same as $P_{\text{atm}} + P_{\text{v}}$, and the difference $\Delta h$ is simply the vapor pressure of the liquid. You can measure $\Delta h$ in either mm Hg or cm Hg; just be sure that you know which unit you are reading and that you record the proper unit.

## Review

Experiment 2
Appendix III

## Equipment Needed

- barometer
- beakers
- capillary tubes
- clamps
- 125-mL filter flask
- ice
- manometer
- ring
- ring stand
- rubber band
- rubber tubing
- serum cap
- slit rubber stopper
- syringe
- 20-mL test tube
- thermometer
- wire gauze

## Chemicals Needed

unknown liquids

## Procedure

The filter flask must be clean and dry. Stopper it with a serum cap. Place the smaller, center portion of the serum cap inside the neck of the flask. A larger portion of the cap extends upward; pull it down around the neck of the flask on the outside of the flask; pull this outside portion of the cap down as low as possible so that it fits tightly. Clamp the flask to a ring stand. Use a short piece of rubber tubing to attach the flask to the manometer; do this *after* the serum cap is in place. When you put the rubber tubing onto the manometer, grasp the glass tubing as close to the end as possible; push the rubber tubing on only far enough to ensure that it cannot fall off. The mercury levels should be the same in both arms of the manometer; if they are not, remove and reconnect the rubber tubing.

Obtain an unknown liquid; immediately record its identification code. Check out a syringe. Draw about one mL of your liquid into the syringe. Turn the syringe upside down and push the plunger far enough to expel any air bubbles. Then push the needle through the center of the serum cap and inject the liquid into the flask. As you do this, keep a firm grip on the plunger of the syringe so that it cannot be pushed upward by pressure from the vaporizing liquid. Remove the needle from the cap. As you do this, be sure that the needle and syringe do not separate from one another; if they do, air and vapor can escape through the needle. After about three minutes, read the mercury levels in the two arms of the manometer. Repeat this reading at one-minute intervals until the levels no longer change. When the levels have stabilized, their difference $\Delta h$ is the vapor pressure of the liquid. If you find that the difference $\Delta h$ is decreasing after several minutes, there is a leak in the system which must be found and eliminated; you might need your instructor's assistance. After completing the measurement, disconnect the tubing from the flask, pour the liquid into the specified waste container, and thoroughly dry the flask with a stream of compressed air. It is absolutely essential to dry the flask. If vapor remains in the flask, it will limit the amount of additional liquid that can vaporize; this would cause your measured values of vapor pressure to be low.

Repeat this measurement of the liquid's vapor pressure at room temperature. If the two values are within 5% of one another, you have satisfactory precision. If they differ by more than 5%, repeat the measurement until you do have two values that are within 5% of one another; use the average of these two values for your graph.

Measure the liquid's vapor pressure at 0°C. Immerse the stoppered flask as deeply as possible in a beaker of water and ice; be sure to use enough ice to maintain a temperature of 0°C all the way to the bottom of the flask. Let the flask cool for about ten minutes and then attach the flask to the manometer. Wait five minutes; if the mercury levels remain the same in both arms, inject the liquid and measure its vapor pressure. If the levels change during this five minutes, the air in the flask has not yet cooled to 0°C and its pressure is still dropping as it cools further; disconnect the flask and reattach it before injecting the liquid. After you are finished, pour out the liquid and dry the flask. You are not required to repeat this measurement.

Measure the vapor pressure of your liquid at a temperature of about 10-15°C. Immerse the stoppered flask in a beaker of cold tap water or in water that has been cooled by ice. Add more ice or cold water as needed to maintain a constant temperature. Be sure to record the actual temperature of this bath. After you are sure that the flask and air have cooled to the bath temperature, add the liquid and measure its vapor pressure. You are not required to repeat this measurement.

A fourth set of temperature and vapor pressure data can be obtained by applying the definition of boiling point: the temperature at which a liquid's vapor pressure equals the pressure of the atmosphere on the liquid. Use the technique you learned in Experiment 2 to measure the boiling point. Read the barometer to determine the atmospheric pressure. The vapor pressure of the liquid equals this measured barometric pressure at the measured boiling point.

When you complete your work, discard the serum cap.

## Results

Make a graph on which you plot vapor pressure of your liquid on the *y* axis and temperature in °C on the *x* axis. On the back of the graph, show a data table with the values of $P_v$ and $T$ that you used. The line through the experimental points should be a continuous one; in response to one of the questions, discuss what the general shape of this plot indicates about the relationship between vapor pressure and temperature.

Make a graph on which you plot log (to the base 10) $P_v$ on the *y* axis and $1/T$ ($T$ in Kelvin) on the *x* axis. You must convert each $1/T$ to a decimal; do not use fractions. On the back of the graph, show a data table with the values of $P_v$, log $P_v$, $T$ (in K), and $1/T$ for each measurement. Draw the best possible *straight* line through the points.

Determine the slope of the straight line for the second graph. On the graph mark the points (*on* the line) used in calculating the slope. From the slope, calculate the value of $\Delta H_v$ for the liquid by rearranging the relationship from the Clausius-Clapeyron equation:

$$\text{slope} = \frac{-\Delta H_v}{2.303\,R} \qquad (13\text{-}4)$$

Be sure to include the proper units in each calculation.

## Questions

1. How will each of the following affect the measured vapor pressure in comparison to the actual vapor pressure for your liquid (increase, decrease, no change)? Explain each answer.

   (a) A 250-mL flask is used instead of a 125-mL flask.

   (b) A long length of slightly porous tubing is used instead of nonporous tubing.

   (c) A large air bubble is injected into the flask along with the liquid.

   (d) The flask is immersed only halfway into the 0°C bath.

2. Comment on the shape of the plot of $P_v$ vs. $T$.

3. Liquid A has vapor pressure = 102 mm Hg at room temperature and liquid B has vapor pressure = 268 mm Hg at the same temperature. In which liquid are there stronger intermolecular attractions? Which liquid would have the higher normal boiling point? Higher heat of vaporization?

# EXPERIMENT 13

## REPORT SHEET

Name ______________________________ Date ______________

Section ______________________________

Unknown identification code ____________________

Measurement of vapor pressure

At room temperature, __________ °C

| | Hg level, open arm | Hg level, flask arm | Vapor pressure |
|---|---|---|---|
| Trial 1 | | | |
| Trial 2 | | | |
| Trial 3 | | | |
| Trial 4 | | | |
| Average vapor pressure | | | |

At other temperatures

| $T$, °C | Hg level, open arm | Hg level, flask arm | Vapor pressure |
|---|---|---|---|
| | | | |
| | | | |

Boiling Point of Liquid

Boiling temperature ____________________

Barometric pressure ____________________

Vapor pressure at b.p. ____________________

Attach your graph of $P_v$ vs. $T$.

Calculation of Heat of Vaporization

Slope of straight line for log $P_v$ vs. $1/T$ ______________________

Show calculation of slope.

$\Delta H_v$ for liquid ______________________

Show calculation of $\Delta H_v$ from slope.

Attach your graph of log $P_v$ vs. $1/T$.

(Attach a separate sheet with the answers to the Questions.)

# EXPERIMENT 14

# MOLECULAR INTERACTIONS IN SOLUTION

## Introduction

There are several kinds of solutions including solid in solid, solid in liquid, liquid in liquid, gas in liquid, and gas in gas. In first-year chemistry you work mainly with liquid solutions of solids in liquids. This experiment, however, deals with liquid–liquid solutions. You will study how vapor pressure over a liquid–liquid solution reflects the interaction of solute and solvent particles.

Consider a solution of two liquid components A and B. In the simplest case of such a solution, the interactions between A molecules and B molecules in the solution are like the interactions between A molecules in component A's pure state and between B molecules in component B's pure state. That is, the attraction between A and B molecules in solution is about the same, in kind and strength, as the attractions between molecules in each of the pure liquids.

Consider the vapor pressure of such a solution. The vapor pressure of a liquid reflects the ability of molecules to escape from the surface of the liquid into the vapor phase. If A—A and A—B interactions in solution are alike, the only factor affecting the escape ability of A molecules in solution is that there are fewer of them per unit of surface area, since some space is occupied by B molecules. Consequently, there are fewer chances per second for A molecules to leave the surface of the solution. Obviously then, the opportunity for vaporization of A is related to the fraction of the molecules that are A. This relationship is formulated mathematically by Raoult's Law:

$$P_A = P_A{}^{\circ} X_A \tag{14-1}$$

This law states that the vapor pressure of A in solution is equal to the vapor pressure of pure A, $P_A{}^{\circ}$, multiplied by the mole fraction of A, $X_A$.

If B is volatile (has measurable vapor pressure), the same argument can be made about B's escape ability and:

$$P_B = P_B{}^{\circ} X_B = P_B{}^{\circ}(1\text{-}X_A) \tag{14-2}$$

The second equality follows from the fact that the sum of the mole fractions of all components of a solution must be one.

A solution is said to be ideal if the escape ability of each component is proportional to the mole fraction of that component in the solution, that is, if Equations (14-1) and (14-2) accurately describe the extents of vaporization of A and B. The total vapor pressure above an ideal solution is the sum of the individual vapor pressures of the components of the solution (Dalton's Law):

$$P_{\text{total}} = P_A + P_B = P_A^{\circ} X_A + P_B^{\circ}(1\text{-}X_A) \tag{14-3}$$

Equation (14-3) can be used to calculate the vapor pressure over a solution of A and B if the vapor pressure of each pure liquid is known and the mole fraction of each component of the solution is known.

Equation (14-3) is also the basis for a graphical method of determining the vapor pressure over a two-component solution. This is shown in Figure 14-1. A plot of the vapor pressure of one component, $P_A$, of a two-component solution vs. mole fraction of that component, $X_A$, is a straight line of positive slope. A plot of the vapor pressure of the second component, $P_B$, vs. mole fraction of the first component is a straight line of negative slope. Values of total vapor pressure of the solution for all compositions, $P_{total}$, are given by the straight line that has vapor pressures of pure B and pure A as its end points. This graphical representation of Equation (14-3) gives a simple criterion that can be used to determine whether liquids form an ideal solution. Solutions of various compositions can be prepared and their vapor pressures measured. If the solutions are ideal, a plot of measured $P_{total}$ vs. mole fraction of one component will be a straight line, as in Figure 14-1.

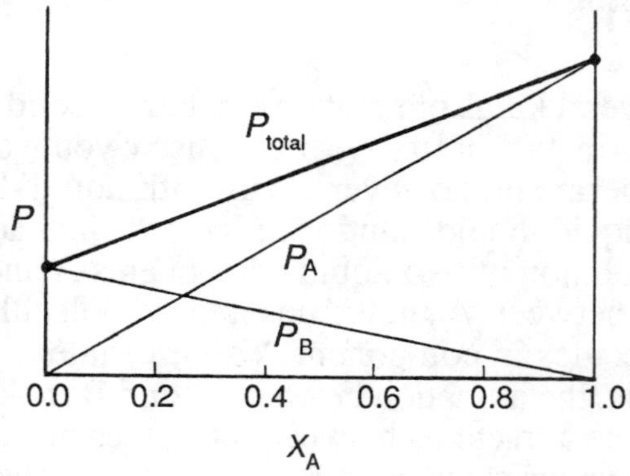


**Figure 14-1**

Vapor pressure vs. composition for an ideal solution.

One factor on a molecular level that tends to produce interactions in solution that are like interactions in pure components is a similarity in the shape and polarity of component molecules. Consequently, the greater the similarity between A and B molecules, the more likely it is that their solution will be ideal.

Many liquid-liquid solutions do not have the vapor pressure that would be predicted from Raoult's Law. Deviations from Raoult's Law occur when molecular interactions in a solution are significantly different from those in the pure liquids. Suppose that two liquids C and D form a solution and in this solution C molecules are attracted to D molecules more strongly than C molecules are attracted to one another in pure C. The escape ability of C is not directly proportional to the mole fraction of C, for its escape ability is further decreased by the increased attraction holding it in the solution. That is, $P_C$ is less than $P_C°X_C$. Similarly, the vapor pressure of D could be decreased by a C—D attraction that is stronger than the D—D attraction in pure D; that is, $P_D$ could be less than $P_D°X_D$. Consequently, the total vapor pressure of the solution is less than predicted by Raoult's Law. The solution exhibits *negative deviation* from Raoult's Law. Figure 14-2 is a graph of vapor pressure vs. composition for such a solution. The dotted lines show ideal vapor pressures and the solid lines show actual vapor pressures.

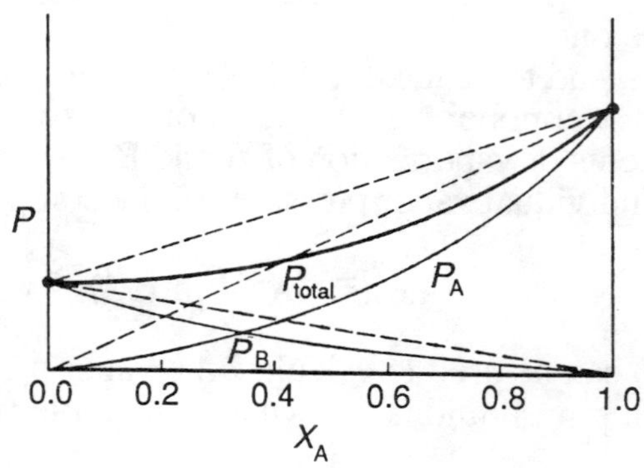


**Figure 14-2**

Vapor pressure vs. composition for a solution with negative deviation from Raoult's Law.

A third possible solution behavior is that molecules of each of the components more strongly attract their own kind of molecule than they attract a different kind of molecule. Consequently, molecules are held in the solution by weaker attractive forces than they are held in the pure materials; the escape ability of each component is greater in the solution than in the pure liquid. The vapor pressure of each component is greater than predicted by Raoult's Law and the total vapor pressure of the solution is greater than predicted by Raoult's Law. Figure 14-3 is a plot of vapor pressure vs. composition for such a solution. There is a *positive deviation* from Raoult's Law.

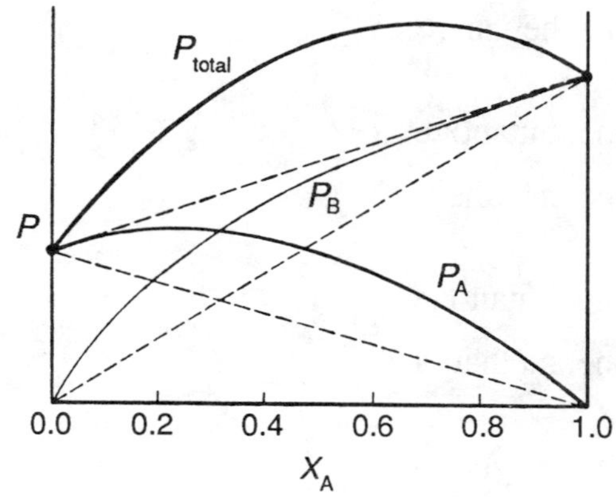


**Figure 14-3**

Vapor pressure vs. composition for a solution with positive deviation from Raoult's Law.

## Technique

Your instructor will assign to you (or allow you to choose) one of the sets of liquids listed on the next page.

You will work with one pair of liquids and prepare three solutions of different compositions. You will measure the vapor pressure of each solution and make a graph of vapor pressure vs. mole fraction of one component. (Of course this procedure enables you to measure only the total vapor pressure of the solution, not the vapor pressures of the individual components.) If the plot is a straight line (within about 10% to allow for experimental error), the solution behavior is considered ideal. If the plot is a curve, the solution behavior deviates in a negative or positive direction from Raoult's Law.

Then you will work with a second pair of liquids, one of which is also in the first pair. The two pairs of liquids will show different behaviors in relationship to Raoult's Law.

After you have determined the kind of deviation shown by each solution, you are to attempt to account for the observed behavior on the basis of molecular shapes, polarity, and kinds of attractive forces between molecules.

## Liquid Pairs for Studying Raoult's Law

1. acetone, butanone
   acetone, dichloromethane

2. acetone, butanone
   acetone, heptane

3. acetone, butanone
   acetone, hexane

4. acetone, butanone
   acetone, methanol

5. acetone, ethyl acetate
   acetone, dichloromethane

6. acetone, ethyl acetate
   acetone, heptane

7. acetone, ethyl acetate
   acetone, hexane

8. acetone, ethyl acetate
   acetone, methanol

9. butanone, acetone
   butanone, dichloromethane

10. butanone, acetone
    butanone, heptane

11. butanone, acetone
    butanone, hexane

12. butanone, acetone
    butanone, methanol

13. ethyl acetate, acetone
    ethyl acetate, dichloromethane

14. ethyl acetate, acetone
    ethyl acetate, heptane

15. ethyl acetate, acetone
    ethyl acetate, hexane

16. ethyl acetate, acetone
    ethyl acetate, methanol

17. heptane, hexane
    heptane, acetone

18. heptane, hexane
    heptane, butanone

19. heptane, hexane
    heptane, dichloromethane

20. heptane, hexane
    heptane, ethyl acetate

21. hexane, heptane
    hexane, acetone

22. hexane, heptane
    hexane, butanone

23. hexane, heptane
    hexane, dichloromethane

24. hexane, heptane
    hexane, ethyl acetate

## Review

Equipment and Techniques G
Experiment 13
Appendix III

## Equipment Needed

small beakers
burets (two)
clamps
125-mL filter flask
manometer
ring stands
serum caps (2)
syringe

## Chemicals Needed

acetone; $C_3H_6O$
butanone; $C_4H_8O$
dichloromethane; $CH_2Cl_2$
ethyl acetate; $C_4H_8O_2$
heptane; $C_7H_{16}$
hexane; $C_6H_{14}$
methanol; $CH_4O$

## Procedure

**Caution!** Some of these liquids can be harmful if their vapors are inhaled in large amounts for a prolonged time; when transferring them, keep them away from your face and hold one container as close as possible to the other to minimize the amount of vapor in the air. When you complete your work with a liquid, dispose of it by pouring it into the waste container that is provided.

Clean two burets. Then rinse each one with an appropriate liquid to remove all traces of water. If you are working with acetone, butanone, ethyl acetate, or methanol, use this liquid in three 3-mL portions to rinse the buret. Dichloromethane, heptane, and hexane do not mix with water; if you are working with one of these liquids, remove water from the buret by rinsing it with three 3-mL portions of acetone and then rinse it with three 3-mL portions of your liquid to remove the acetone. Pour into the buret only enough of the liquid to fill the tip and to provide the amount of it that you will need to measure out for the mixtures. You will need to measure out 12 mL of the liquid that is in both of your assigned pairs of liquids. You will need to measure out 6 mL of the liquid that is in just one pair. Be sure that you keep track of which buret contains which liquid.

For each buret, record the initial reading. From the burets measure into a small beaker (clean and dry) one mL of the first liquid and three mL of the second liquid. You do not have to take exactly 1.00 mL and 3.00 mL; just to be sure to record the volume that you actually do use. Wait about a minute to record the final reading for each buret so that liquid on the wall of the buret will have time to flow into the bulk of the liquid; this is especially important when you are measuring such small volumes. Be sure that you use at least 1.00 mL of a liquid so that every measurement will have three figures. Cover the beaker with a watch glass to prevent liquid loss by evaporation; gently swirl the beaker until the liquids are thoroughly mixed. Measure the room temperature vapor pressure of the solution using the equipment and technique described in Experiment 13; use only one mL of the solution for the measurement, not the entire solution that you prepared. After the measurement, pour the remaining liquid out of the flask into the waste container that is provided; use a stream of compressed air to dry the flask and remove all vapor from it. Completely dry the syringe by using its plunger to pump air through it. Pour the unused solution into the specified waste container.

In a clean, dry beaker prepare a solution containing two mL of each liquid. Measure the vapor pressure of this solution.

In a clean, dry beaker prepare a solution containing three mL of the first liquid and one mL of the second liquid. Measure the vapor pressure of this solution.

Drain from the buret the liquid that you are not using in your second pair of liquids; collect this liquid in a clean, dry beaker. Measure the vapor pressure of this liquid.

When you complete your work with the first pair of liquids, discard the serum cap. Use a new serum cap for your second pair of liquids. Discard it when you complete your work.

Rinse the drained buret with three 3-mL portions of your new liquid and then put liquid into the buret. Measure the vapor pressures of three solutions containing this liquid and the liquid that is in both of your pairs of liquids; use the liquids in volumes of one mL and three mL, two mL and two mL, and three mL and one mL. Then drain the unused liquid from each buret and measure the vapor pressure of each pure liquid. Remember to dispose of the liquid in the specified waste container.

## Data and Results

Use the names of your liquids to look them up in a chemistry handbook; record the room temperature density of each liquid. Use the density and measured volumes to calculate the mass of each liquid in each solution. Then calculate the number of moles of each liquid in each solution. On the Report Sheet the material that is identified by the letter A is the liquid that is in both of your pairs of liquids; calculate the mole fraction of this liquid for each solution. Record the vapor pressure of each solution and of each pure liquid.

For each pair of liquids plot vapor pressure ($y$ axis) vs. mole fraction of the liquid that is designated as A ($x$ axis). Be sure that the graph nearly fills the page. Use a scale on the $y$ axis that allows as much expansion as possible; for example, if the measured vapor pressures range from 120 to 220 torr, you might use a scale that goes from 100 at the bottom of the page to 240 at the top of the page. Be sure that each graph is completely labelled to identify the liquids.

Each graph should have five points plotted on it: vapor pressure of pure B ($X_A = 0$), of three solutions, and of pure A ($X_A = 1.00$). After you put these five points on the graph, use a ruler to draw a dashed straight line connecting the two end points ($X_A = 0$ and $X_A = 1.00$); this straight line corresponds to vapor pressure values for ideal solution behavior. The remaining three points might be scattered slightly above and below this line; if so, the solution behavior is ideal, or nearly so. The three points might all be below this line; this is negative deviation from Raoult's Law. The three points might all be above this line; this is positive deviation from Raoult's Law. For either of the latter two cases, draw a smooth curve through the points (as shown in Figures 14-2 and 14-3).

For each pair of liquids use your graph (the line through experimental points) to determine the solution's vapor pressure at $X_A = 0.50$. Use the line representing ideal behavior to calculate the vapor pressure the solution would have if it were ideal. Calculate the percentage deviation of the actual vapor pressure from the ideal vapor pressure:

$$\left(\frac{\text{actual V.P.} - \text{ideal V.P.}}{\text{ideal V.P.}}\right)100 = \%\ \text{deviation} \qquad (14\text{-}4)$$

Be sure to include the proper sign on this deviation. A value less than about 10% corresponds to nearly ideal behavior.

Draw the electron dot structure for a molecule of each of your liquids. You will probably have to use an organic chemistry textbook to look up the structure; *do not guess* at the structure. Describe the shape of each molecule and state whether it is polar. For each pure liquid state the predominant type of intermolecular attraction. For each pair of liquids state what type of intermolecular attraction there can be in solution. Try to account for a negative or positive deviation from Raoult's Law on the basis of solution intermolecular attractions that are stronger or weaker than pure liquid intermolecular attractions.

## Questions

1. Explain why the mole fraction of acetone in a solution of 3.00 mL acetone and 1.00 mL of butanone is different from the mole fraction of acetone in a solution of 3.00 mL acetone and 1.00 mL of dichloromethane.

2. How would the graph that you prepared have differed if you had graphed $P_{total}$ vs. $X_B$ or $X_C$ instead of $P_{total}$ vs. $X_A$?

# EXPERIMENT 14

## REPORT SHEET

Name ______________________________ Date ____________

Section ______________________________

Identity of A ____________ Density ____________ Molar weight ________

Identity of B ____________ Density ____________ Molar weight ________

| Solutions of A and B | #1 | #2 | #3 |
|---|---|---|---|
| Volume of A | | | |
| Moles of A | | | |
| Volume of B | | | |
| Moles of B | | | |
| Mole Fraction A | | | |

Vapor pressure measurements

| | Hg level, open arm | Hg level, flask arm | Vapor pressure |
|---|---|---|---|
| Pure A | | | |
| Solution #1 | | | |
| Solution #2 | | | |
| Solution #3 | | | |
| Pure B | | | |

Data from graph (Attach the graph to this Report Sheet.)

Experimental vapor pressure for mole fraction A = 0.50 ____________

Ideal vapor pressure for mole fraction A = 0.50 ____________

% deviation of experimental from ideal vapor pressure ____________

On the back of this Report Sheet show the calculation of moles of A, moles of B, and mole fraction of A. Also on the back of this Report Sheet give the information about A and B and about the solution that is requested in Data and Results.

Identity of A ______________________ Density __________ Molar weight __________

Identity of C ______________________ Density __________ Molar weight __________

| Solutions of A and C | #1 | #2 | #3 |
|---|---|---|---|
| Volume of A | ________ | ________ | ________ |
| Moles of A | ________ | ________ | ________ |
| Volume of C | ________ | ________ | ________ |
| Moles of C | ________ | ________ | ________ |
| Mole Fraction A | ________ | ________ | ________ |

Vapor pressure measurements

| | Hg level, open arm | Hg level, flask arm | Vapor pressure |
|---|---|---|---|
| Pure A | ________ | ________ | ________ |
| Solution #1 | ________ | ________ | ________ |
| Solution #2 | ________ | ________ | ________ |
| Solution #3 | ________ | ________ | ________ |
| Pure C | ________ | ________ | ________ |

Data from graph (Attach the graph to this Report Sheet.)

Experimental vapor pressure for mole fraction A = 0.50 ________

Ideal vapor pressure for mole fraction A = 0.50 ________

% deviation of experimental from ideal vapor pressure ________

On the back of this Report Sheet show the calculation of moles of A, moles of C, and mole fraction of A. Also on the back of this Report Sheet give the information about A and C and about the solution that is requested in Data and Results.

(Attach a separate sheet with the answers to the Questions.)

# EXPERIMENT 15

# MOLAR WEIGHT BY FREEZING POINT DEPRESSION

## Introduction

In this experiment you will study the nature of solutions by measuring one of their colligative properties—freezing point depression. Previously, you observed this phenomenon in Experiment 2 when you saw that a mixture of similar amounts of two solids has a melting point that is lower than the melting point of either pure solid. Since melting point and freezing point are the same temperature considered from opposite directions, you dealt with freezing point depression when you measured the melting point of a mixture.

The subject of this experiment is also related to Experiment 14. The difference is that here one of the two components of the solution is a solid having negligible vapor pressure. Hence $P_{vapor}$ of the solution is determined by the solvent. The reason for the lowering of freezing point of a solvent when a nonvolatile solute is added to it is discernible from Figure 15-1. The solid lines represent vapor pressure vs. temperature for pure solid (frozen) solvent and liquid solvent. The lines intersect at point A, meaning that at point A solid and liquid solvent have the same vapor pressure. Any liquid that vaporizes at this temperature and pressure could just as likely recondense as solid, and *vice versa*; that is, solid, liquid, and vapor are in equilibrium at point A. The temperature at this point, $T_p$, is the freezing point of pure solvent.

Consider the dashed line in Figure 15-1; it represents the vapor pressure of a solution of specific composition prepared by dissolving a nonvolatile solute in a volatile solvent. The vapor

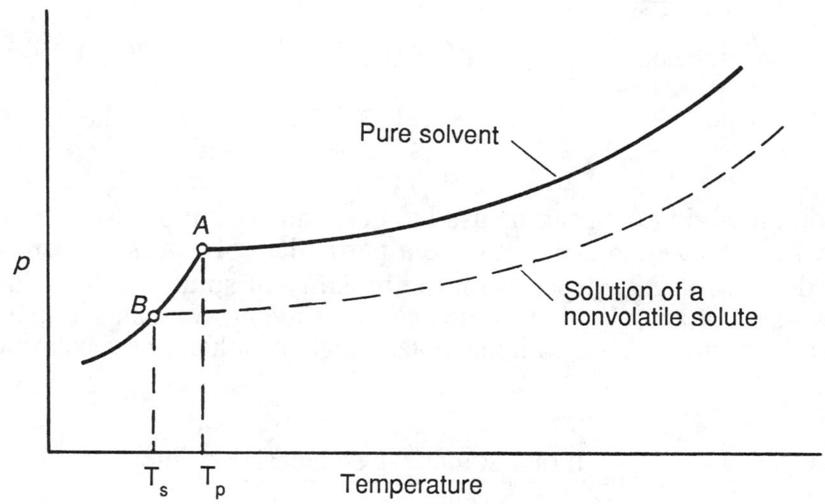


**Figure 15-1**

Vapor pressure as a function of temperature for pure solvent and solution.

pressure of this solution is always lower than the vapor pressure of pure solvent. The intersection of the dashed line with the vapor pressure curve of solid solvent, point B, is at a lower temperature than for pure solvent. This temperature, $T_s$, is the freezing point of solvent in the solution. When the temperature of the solution is $T_s$, solvent in the solution is in equilibrium with pure solvent; crystals of solvent form. As solvent freezes, the composition of the solution changes; concentration of solvent decreases and concentration of solute increases. This changes the position of the solution vapor pressure curve and lowers the temperature at which freezing of solvent occurs; the freezing point decreases continuously as solute concentration increases. For our purpose here, we define the freezing point of solvent in the solution as the temperature at which the first crystals of solvent appear.

Lowering of freezing point of solvent by a nonvolatile, molecular solute in a dilute solution depends on two factors, as shown in Equation (15-1):

$$\Delta T_f = T_p - T_s = K_f m \tag{15-1}$$

The two factors are the identity of the solvent, which determines the value of the freezing point depression constant $K_f$, and the concentration of solute particles, as expressed by molality $m$ (moles of solute/kg of solvent).

Table 15-1 lists the freezing points and molal freezing point depression constants for some solvents. By combining the information in Table 15-1 and Equation (15-1), you can see that a one-molal concentration of solute particles in cyclohexane results in a freezing point of -13.8°C for the solution, compared to 6.6°C for pure cyclohexane.

Table 15-1

| Solvent | Freezing point, °C | $K_f$, C° kg/mole |
|---|---|---|
| acetic acid | 16.6 | 3.90 |
| cyclohexane | 6.6 | 20.4 |
| naphthalene | 80.2 | 6.80 |
| water | 0.0 | 1.86 |

Freezing point lowering data can be used to determine the molar weight of a dissolved solid. From Equation (15-1) you can see that once a particular solvent is chosen, $K_f$ is fixed and the freezing point depression $\Delta T_f$ depends on the molality of solute in the solution: $m = \Delta T_f/K_f$. Using known weights of an unknown solute and of a solvent, you can set up and solve an expression for the molality of the solution with the molar weight of solute as the unknown, Equation (15-2):

$$m = \frac{\text{moles solute}}{\text{kg solvent}} = \frac{\text{g solute/MW solute}}{\text{kg solvent}} \tag{15-2}$$

## Technique

Freezing points can be measured with the equipment shown in Figure 15-2.

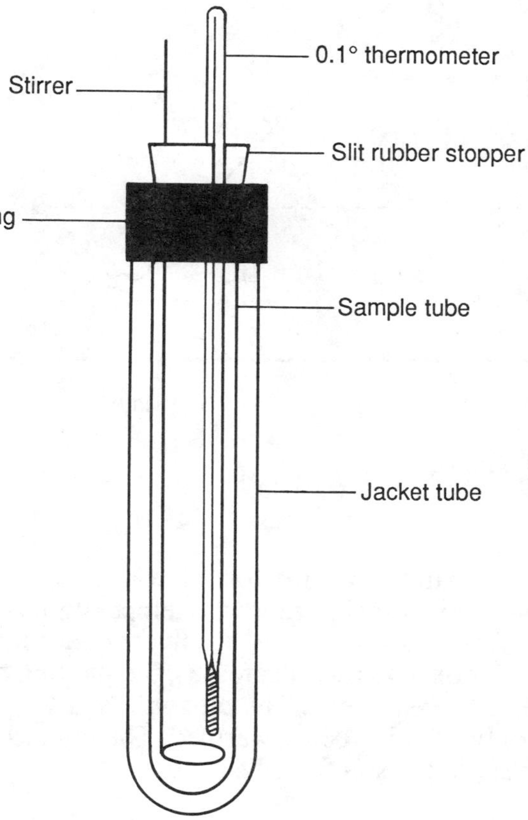


**Figure 15-2**

Freezing point apparatus.

The larger, outer test tube contains a smaller test tube; this inner tube contains the material for which the freezing point is to be determined. The air jacket between the two tubes allows for slow and uniform cooling or warming in the inner test tube.

The freezing point of pure liquid is measured on a cooling cycle. Liquid is put into the inner tube and the equipment is placed in a container of water and ice. The liquid is stirred continuously to maintain thermal homogeneity and its temperature is recorded at regular time intervals. Initially, the temperature will drop as the liquid cools. Then the temperature will remain constant as the liquid freezes. However, the liquid might supercool; that is, the temperature might drop below the freezing point before freezing begins. If this happens, the temperature will rise to the freezing point as freezing occurs; this temperature rise occurs because of the release of heat as freezing occurs, heat equal in amount to the heat of fusion. A plot of temperature vs. time for the cooling, supercooling, and freezing of a liquid will look like Figure 15-3. The horizontal part of the plot, corresponding to a constant temperature, is the freezing point, $T_p$.

The freezing point of a solution is most easily measured on a warming cycle. A known weight of solute is dissolved in a known volume of solvent in the inner tube. The inner tube containing the solution is placed directly into an iced-water bath; it is cooled and continuously stirred until the solution is semi-frozen and has a "slushy" appearance. Then the inner tube is wiped dry and placed in the outer tube, warmed by the surrounding air, and stirred continuously. The temperature will gradually rise as the mixture of liquid and solid is warmed, all the solid melts, and liquid is

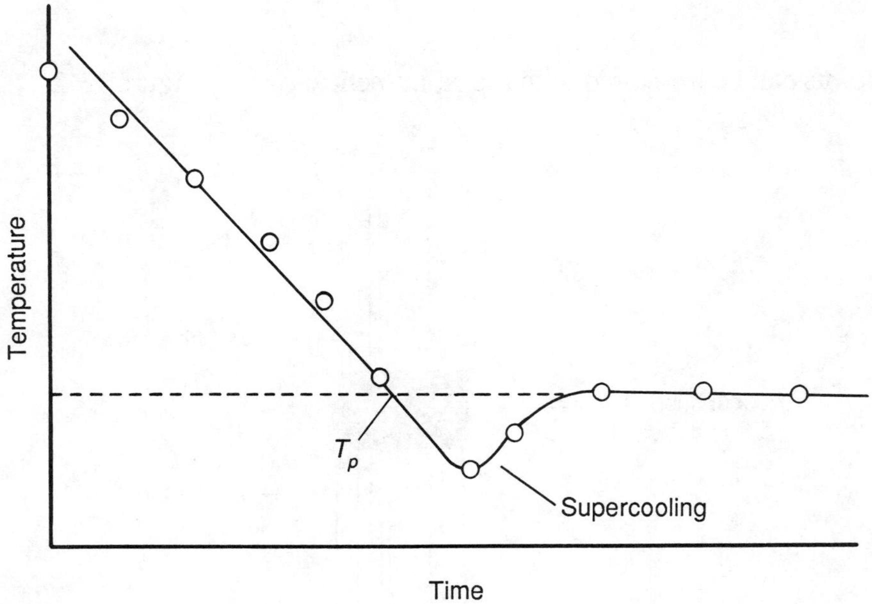


**Figure 15-3**

Cooling curve of a pure liquid.

warmed. A plot of temperature vs. time for the warming and melting will look like Figure 15-4. The plot does not have a horizontal portion; the temperature is not constant during melting or freezing of solvent in a solution. As solvent melts, solvent concentration increases and solute concentration decreases; as solute concentration decreases, the melting point rises. The plot will be a straight line of positive slope as solvent melts. The plot will be a straight line of greater slope after all solvent has melted and only liquid is being warmed. The intersection of these two straight lines is the freezing point of solvent in the solution, $T_s$.

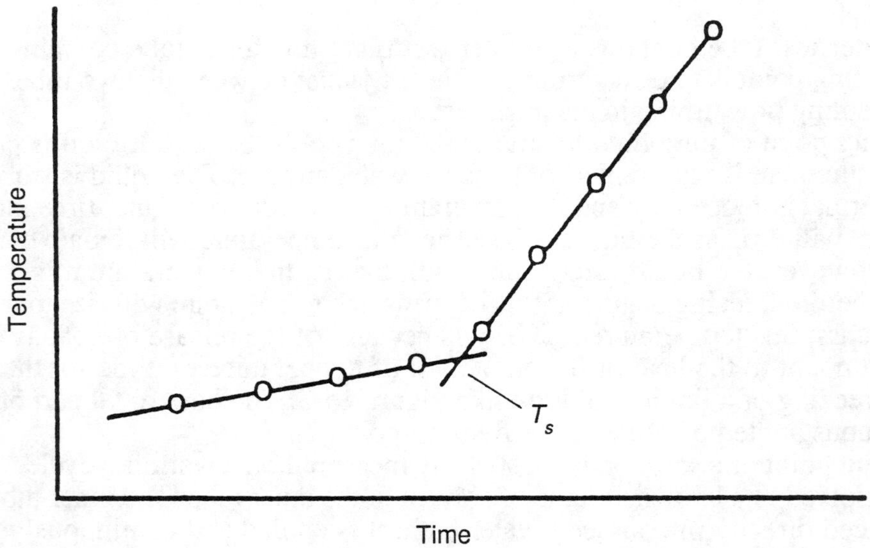


**Figure 15-4**

Warming curve of a solution.

## Review

Equipment and Techniques G
Appendix III

## Equipment Needed

large beaker or
 pneumatic trough for ice bath
small beaker
clamp
clock or watch with second hand
freezing point apparatus
graph paper (provided by student)
ice
ring stand
0.1° thermometer

## Chemicals Needed

$C_6H_{12}$; cyclohexane
unknowns

## Procedure

### A. Freezing Point of Cyclohexane

Prepare an ice-and-water bath in a large beaker or pneumatic trough.

From a buret that is already set up in the laboratory, measure about 20 mL of cyclohexane into the clean, dry inner test tube. You do not have to take exactly 20 mL; just be sure to record to the nearest 0.01 mL the volume that you actually take. Place this test tube inside the larger test tube; close it with the stopper fitted with a stirrer and the 0.1° thermometer. Be sure that the thermometer bulb is completely covered by liquid.

Clamp the tube onto a ring stand with the bottom of the tube slightly above the bottom of the ice-and-water bath. Be sure there is ice above the level of the liquid inside the test tube. It is not necessary that you be able to see the liquid in the test tube at all times. Do not hold the test tube in your hand, for you might create warm spots on it rather than allowing it to have a uniform temperature.

Begin stirring the cyclohexane; record its temperature every 30 seconds. The stirring must be continuous, but it should be slow and gentle; be especially careful not to bump the thermometer bulb. Before each reading tap the thermometer lightly with your finger to ensure that the mercury does not stick to the wall of the thermometer. It is possible that the mercury column might separate; examine the mercury column very closely before you begin to use the thermometer and again after you conduct each part of the experiment to see whether this has happened. If this separation does occur, consult your instructor; you cannot make accurate temperature readings with the thermometer in this condition.

Continue recording the temperature of the cyclohexane until you can see that the liquid is partially frozen and until the temperature has remained constant for at least four consecutive readings. To see the cyclohexane, you might need to push some ice away from the tube.

After you complete the first trial, remove the inner tube from the outer tube and melt all the frozen cyclohexane by warming the tube in your hand; gently stir the cyclohexane during this warming. Return the inner tube to the outer tube and repeat the determination of the freezing point. The two trials should agree within 0.1°; if they do, use the average of the two trials, or your result from the trial that you think you did more accurately, as the freezing point of cyclohexane, $T_p$. If the two trials do not agree within 0.1°, repeat the determination until two trials do agree within 0.1°.

If you compare your result with those of other students or with the value given in the Introduction, you might find the values for the freezing point of pure cyclohexane differ by a degree or more. Do not let this upset you. The thermometer that you are using might not be perfectly accurate. However, since you are concerned with the *difference* between two temperatures—the freezing point of pure liquid and the freezing point of the liquid as solvent in a solution—both of your measurements will have the same error and you can accurately determine a difference between the two temperatures. Because of a possible inaccuracy in your thermometer, it is essential that you use the same thermometer to make measurements for cyclohexane and for the solution.

**B. Molar Weight of an Unknown Solid.**

Obtain an unknown solid; record its identification code. Its molar weight is about 100-200 g. Weigh out 0.3-0.4 g of the unknown in the following manner. On the analytical balance weigh the proper amount of the unknown into a clean, dry small beaker; record the weight of beaker and unknown. Go back to your laboratory desk and pour the unknown into the cyclohexane in the freezing point apparatus. A little of the unknown might stick to the beaker, but that is all right; just be sure not to spill any of the unknown. Reweigh the beaker; the difference between its weight at this point and its weight with the unknown sample is the weight of the unknown that you put into the cyclohexane. Gently stir the cyclohexane until all the solid dissolves.

You will measure the *melting* point of solvent in the solution, rather than its *freezing* point. Begin by placing the test tube that contains the solution directly into the iced-water bath without the outer test tube. *Immediately* begin stirring the solution; maintain constant, gentle stirring as the solution cools and solvent freezes. Be sure that you stir all the way to the bottom of the test tube and all the way to the top of the solution so that cooling and freezing are uniform. When there is frozen solvent throughout the mixture and the mixture is "slushy", remove the test tube from the bath, dry the outside of it, place it in the outer test tube, and clamp the outer test tube to a ring stand; try to do this quickly and without warming the solution very much. (**Do not let solvent freeze completely solid.** If you do, you cannot stir the mixture, you will not have a uniform temperature as melting occurs from the warmer outside of the tube, and you might break the thermometer if you do try to stir the mixture.) Begin recording the mixture's temperature. Record the temperature every 30 seconds while maintaining gentle stirring. After you can see that all the frozen solvent has melted, continue recording the temperature for at least three more minutes.

Make a graph of temperature vs. time for your readings. There should be a definite linear portion of the plot after all solvent melted and liquid was warming. Ideally, there will also be a linear portion of the plot as solvent was melting. If this portion does not have obvious linearity, you will need to cool the partially frozen mixture to a lower temperature when you repeat the measurement. You will probably need temperature readings for about three minutes before all solvent melts. Do not expect your plot to be as neat as the one shown in Figure 15-4. In the region of intersection of the two linear portions of the plot you will probably have several points that do not fit on either straight-line portion; disregard these and base your straight lines on the points farther from this region. Be sure each line is based on at least three points.

After repeating the measurement, make another graph of your data. Is there clear linearity during melting? If so, determine the point of intersection of the two straight-line portions of the plot. This point of intersection is $T_s$, the initial melting point (freezing point) of solvent in the solution. If there is not clear linearity, begin the next trial by cooling the "slush" to a lower temperature. Repeat the measurement until trials agree within 0.2°.

When you complete your work, pour the cyclohexane solution into the waste container that is provided. Scrub the inner test tube with detergent solution and a brush until water runs freely down its inside wall. Rinse it with tap water and distilled water; dry it with a clean towel.

## Results

### A. Freezing Point of Cyclohexane

Make a graph of temperature vs. time for two trials. Report the freezing point, $T_P$, as the average for the two trials if they give different results, or use the value you think is more accurate.

### B. Molar Weight of an Unknown Solid

Make a graph of temperature vs. time for each of your two best trials. Report the freezing point, $T_S$, as the average for the two trials if they give different results.

Look up the density of cyclohexane in a chemistry handbook or encyclopedia. Use this density and your measured volume of cyclohexane to calculate the mass of cyclohexane you used.

Use your measured freezing point of pure cyclohexane, your measured freezing point of cyclohexane in the solution, and cyclohexane's freezing point depression constant to calculate the molality of solute in the solution.

Set up an expression for calculating molality in terms of the weights of solute and solvent and the molar weight of the solute. Equate this expression for molality to the value of molality that you calculated from the freezing point lowering. Solve the equation to determine the numerical value for the molar weight of the unknown solute.

## Question

1. How would the molar weight that is calculated for the unknown solute be affected (high, low, no change) by each of the following blunders? Explain each answer.

   (a) Some cyclohexane is accidentally spilled before the unknown solid is added.

   (b) Not all of the solid dissolves in the cyclohexane.

   (c) Some of the solution is spilled.

# EXPERIMENT 15

# REPORT SHEET

Name ______________________________ Date ______________

Section ______________________________

Unknown identification code ____________________

**Data**

Weight of solute ______________

Volume of cyclohexane ______________

Density of cyclohexane ______________

Weight of cyclohexane ______________

| | Trial 1 | Trial 2 | Average |
|---|---|---|---|
| Freezing point of cyclohexane, $T_p$ | ________ | ________ | ________ |
| Freezing point in solution, $T_s$ | ________ | ________ | ________ |

Attach your graphs of temperature vs. time for cyclohexane and solution.

**Results**

$\Delta T_f = T_p - T_s$ ____________________

$m = \Delta T_f / K_f$ ____________________

Show the expression for *m* on the basis of weights of solution components.

Molar weight of unknown ____________________

Show calculation of molar weight from molality and weights of solution components.

(On the back of the Report Sheet answer the Question.)

# Experiment 16

# ACID–BASE TITRATION

## Introduction

Much of chemistry and biology is concerned with the behavior of acids and bases. Acids and bases are participants in many reactions in nature, and many reactions require a particular level of acidity or basicity. For example, the concentration of acid in the stomach must be quite high, in comparison to the rest of the body, to break down proteins during digestion.

This experiment is an introduction to the determination of the amount of acid or base in an aqueous solution, acid—base titration. A *titration* is the addition of a carefully measured volume of a solution of one substance to a solution of another substance until a reaction between the two substances is just complete. The titration in this experiment is based on a fundamental chemical property of acids and bases: their neutralization reaction producing water and a salt:

$$HA + BOH \rightarrow H_2O + BA \tag{16-1}$$

for which the net ionic equation is

$$H_3O^+(aq) + OH^-(aq) \rightarrow 2H_2O \tag{16-2}$$

If the amount of acid in one solution is known, then the amount of base in a second solution can be determined by measuring how much of it neutralizes the acid. Similarly, if the amount of base in one solution is known, the amount of acid in another solution can be determined by a titration.

Equation (16-1) can be used as a general equation to represent acid—base neutralization reactions. However, the general formulas in it do not accurately represent all acids and bases, for some acids contain more than one reactive hydrogen atom per formula unit and some bases contain (or produce) more than one hydroxide ion per formula unit; examples are $H_2SO_4$, $H_3PO_4$, $Mg(OH)_2$, and $Fe(OH)_3$. Equation (16-2) is a more general equation; it shows that a neutralization involves equimolar amounts of hydrogen (in aqueous solution as hydronium ions) from the acid and hydroxide ions from the base:

$$\text{moles } H_3O^+ \text{ from acid} = \text{moles } OH^- \text{ from base} \tag{16-3}$$

This is true regardless of the identities of the acid and base.

The term *equivalent weight* emphasizes the essential chemical behavior of acids and bases. One equivalent weight of an acid is the mass of the acid that contains one mole of ionizable hydrogen atoms. One equivalent weight of a base is the mass of the base that produces one mole of hydroxide ions. For some compounds, equivalent weight equals molar weight; examples are HCl, $HC_2H_3O_2$, NaOH, and $NH_3$. For other compunds, equivalent weight is a particular fraction of the molar weight. For example, one mole of $Ca(OH)_2$ contains two moles of hydroxide ions and so one equivalent = 1/2 mole = 37 g. For $H_3PO_4$, one equivalent = 1/3 mole = 33 g. In each case

$$\text{equiv. wt.} = \frac{\text{molar weight}}{\text{number of } H^+ \text{ or } OH^-\text{/formula unit}} \tag{16-4}$$

In an acid–base titration the *end point* is reached when equimolar amounts of hydrogen from acid and hydroxide from base have been combined in solution, Equations (16-2) and (16-3). We can also say that at the end point equal numbers of equivalent weights (or equivalents) of acid and base have been combined. If the identity of one reactant is known, the number of equivalents of that reactant can be calculated. The titration requires the same number of equivalents of the other reactant, and so the number of equivalents of it is known also.

## Technique

You will carry out acid—base titrations using the acid—base indicator phenolphthalein. It is colorless in acidic solutions and pink in basic solutions; the color ranges from pale pink to dark purplish pink, depending on the concentrations of indicator and base. If you titrate to the appearance of a light pink color, the amount of excess base will be numerically insignificant; hence, this lightest possible pink color indicates the end point.

The water you use to prepare solutions is boiled to expel dissolved $CO_2$, for $CO_2$ reacts with water giving an acidic solution:

$$CO_2 + 2H_2O \rightarrow H_3O^+ + HCO_3^- \qquad (16\text{-}5)$$

Neutralizing this acidity would use up some of the base that you use in the titration, thereby decreasing the accuracy of your titrations.

Your first activity will be to *standardize* a solution of NaOH; this means that you determine its concentration as accurately as possible. You do this by using the NaOH to neutralize an accurately weighed sample of $KHC_8H_4O_4$, a monoprotic acid that has an equivalent weight of 204.22 g. You can calculate the number of equivalents of acid used in the titration:

$$\text{no. of equivalents} = \frac{\text{g acid}}{\text{equiv. wt. of acid}} \qquad (16\text{-}6)$$

This is the same as the number of moles of hydrogen provided by the acid. It is also the number of equivalents of base used in the titration; this is the number of moles of hydroxide ions from base that the acid neutralizes. The NaOH solution is dispensed from a buret and its volume is carefully measured; you use the numbers of moles of hydroxide ions and volume of NaOH solution to calculate molarity (moles/liter) of NaOH solution.

Then you will use your standardized NaOH solution to determine the equivalent weight of an unknown solid acid. You do this by using your standardized NaOH solution to neutralize a carefully weighed sample of the acid. Using molarity and volume of NaOH solution, you can calculate moles of hydroxide ions required to neutralize the acid. This equals moles of hydrogen provided by the acid, Equation (16-3). The equivalent weight of the acid is the mass of the acid that contains one mole of ionizable hydrogen. Therefore, you can calculate the equivalent weight from your data:

$$\text{equivalent weight} = \frac{\text{g acid}}{\text{mole } H_3O^+} \qquad (16\text{-}7)$$

## Review

Equipment and Techniques F, G, H
Experiment 1

## Equipment Needed

analytical balance
beakers
buret
burner
clamp
Erlenmeyer flasks
graduated cylinders
large flask or bottle
ring
ring stand
rubber stopper
wire gauze

## Chemicals Needed

$KHC_8H_4O_4$; potassium hydrogen phthalate
3M NaOH; sodium hydroxide
phenolphthalein indicator solution
unknown solid acid

## Procedure

### A. Preparation of NaOH Solution

Boil about 500 mL of distilled water in a large beaker for about five minutes. Allow the water to cool sufficiently that you can pick up the beaker by wrapping a towel around it. Use a graduated cylinder to measure 290 mL of this water into a large flask or bottle. To this water add 10.0 mL of 3M NaOH. (**Caution! Concentrated base solutions are very corrosive.** If you spill 3M NaOH on yourself, **immediately** wash it off under lots of cold, running water for several minutes. If you spill any on the outside of the bottle or on the table top, sponge it off and wash out the sponge.) Stopper the flask, but be careful not to allow the solution to contact the stopper; the flask should be stoppered as much as possible throughout the experiment to prevent absorption of $CO_2$ from the air. Swirl the flask to mix the NaOH and water homogeneously. Put the rest of the boiled distilled water into your wash bottle; use it to dissolve your acid samples and to wash the flask wall during the titrations.

### B. Standardization of NaOH Solution

Clean a buret and fill it with the NaOH solution that you prepared. Don't forget to rinse the buret with the solution before filling it.

On the analytical balance, weigh out 0.2-0.4 g of $KHC_8H_4O_4$ and put it into an Erlenmeyer flask. To do this, first weigh a clean, dry crucible or small beaker on the analytical balance. Then add 0.2-0.4 g of the acid to this container. Go back to your laboratory desk and pour the acid into the flask; use a stream of water from a wash bottle to complete the transfer. Dissolve the acid in about 20 mL of boiled distilled water. Add two drops of phenolphthalein indicator solution to this acid solution.

Record the initial buret reading. If there is a drop of solution on the tip of the buret, remove it by touching the tip of the buret to the inside wall of a waste beaker. Place the flask (containing the acid) under the buret and place a piece of white paper under the flask so that you can easily see any color in the solution. Position the buret so that the top of it is just slightly below the rim of the flask.

Begin the titration by adding several mL of base solution in a rapid stream; constantly swirl the flask. Then slightly close the buret's stopcock so that individual drops are released rapidly. At the site of addition, there will be a pink color that disappears quickly as the acid and base solutions mix. When this color seems to linger, decrease the rate of addition of base so that the color disappears between the addition of one drop and the addition of the next drop. Occasionally wash solution off the inside wall of the flask with a stream of water from a wash bottle. Near the end point you can add a partial drop by barely opening the stopcock so that a drop forms on the tip of the buret but does not fall; close the stopcock and use a stream of water from a wash bottle to wash this droplet into the acid solution. Add base until the lightest possible pink color persists throughout the solution for at least 20 seconds. If the solution has a very dark color, you probably passed the end point; be sure to record this information to help you decide later about the relative accuracy of the titrations. Record the final buret reading about a minute after completing the titration; this time lag allows solution on the inside wall of the buret to flow into the bulk of the solution.

Before you begin the next titration of a weighed $KHC_8H_4O_4$ sample, use base volume and acid weight from the first titration as a conversion factor to calculate the approximate volume of base that this weight of acid will require. Add base rapidly until you approach this volume and then add base slowly. Keep in mind that you probably over-titrated the first sample; therefore, this calculated approximate volume will probably be high.

Carry out at least three titrations. Repeat the titrations until two values for the molarity of the base differ by no more than 1%, that is:

$$\frac{\text{higher value} - \text{lower value}}{\text{average value}} \times 100 = 1\% \text{ or less} \qquad (16\text{-}8)$$

If you know that you over-titrated a particular acid sample, or made some other mistake that would affect the accuracy of your results, you should record this in your notebook and state it on the Report Sheet, and you should not include that trial in your averaging.

### C. Titration of Unknown Acid

Obtain an unknown solid acid and record its identification code. On the analytical balance, weigh out 0.1-0.2 g of the acid and put it into an Erlenmeyer flask. Dissolve the acid in about 20 mL of boiled distilled water; add two drops of phenolphthalein solution. Titrate the acid to the pale pink end point with your standardized NaOH solution.

If this first titration requires less than 10.00 mL of base solution, you will have only three figures in your calculations. For the next titrations, increase the acid sample size so that the titration will require at least 10.00 mL of base, but less than the capacity of the buret.

If this first titration requires more base than the capacity of the buret, refill the buret, record the buret reading, and continue the titration. For the next titrations, decrease the acid sample size so that the titration will require less base solution than the capacity of the buret, but at least 10.00 mL.

Carry out at least two titrations that require a volume of base that is at least 10.00 mL but less than the capacity of the buret. Repeat the titrations until two consecutive values for the equivalent weight of the acid differ by no more than 1%.

All the solutions from your titrations may be disposed of by washing them down the drain. Dispose of leftover solid acid as your instructor directs.

## Results

### B. Standardization of NaOH Solution

Use weight of acid sample and equivalent weight of acid to calculate the number of equivalents of acid. This equals moles of hydrogen from the acid, which equals moles of hydroxide ions from base. Use moles of hydroxide ions and volume of NaOH solution required for the titration to calculate molarity of NaOH solution. Report the average of all trials or of your two best trials as the molarity of the base.

### C. Titration of Unknown Acid

Use molarity (average from preceding) of NaOH solution and volume of NaOH solution required for the titration to calculate moles of hydroxide ions. This equals moles of hydrogen from the acid, which equals number of equivalents of the acid. Use this value and the weight of the acid sample to calculate the acid's equivalent weight, Equation (16-7). Report an average for all your trials or for your two best trials.

## Questions

1. How would your calculated molarity of the NaOH solution be affected (too high, too low, no change) by each of the following errors? Explain each answer.

   (a) The buret is not rinsed with solution before it is filled.

   (b) Some $KHC_8H_4O_4$ is spilled out of the flask after it is weighed.

   (c) Some solution splashes out of the flask during titration.

   (d) You go past the end point in the titration.

2. How would your calculated value for the equivalent weight of the acid be affected (too high, too low, no change) by each of the following errors? Explain each answer.

   (a) Some acid is spilled out of the flask after it is weighed.

   (b) Some solution splashes out of the flask during titration.

   (c) You go past the end point in the titration.

# EXPERIMENT 16

# REPORT SHEET

Name________________________________________ Date ______________

Section _______________________________________

**B. Standardization of NaOH Solution**

| | Trial 1 | Trial 2 | Trial 3 | Trial 4 |
|---|---|---|---|---|
| 1. wt. of $KHC_8H_4O_4$ | | | | |
| 2. equiv. of $KHC_8H_4O_4$ (moles of $H_3O^+$) | | | | |
| 3. equiv. of NaOH (moles of $OH^-$) | | | | |
| 4. initial buret reading | | | | |
| 5. final buret reading | | | | |
| 6. vol. of NaOH | | | | |
| 7. molarity of NaOH | | | | |
| 8. ave. M of NaOH | | | | |

Show the calculation of 2, 3, and 7 for one trial.

## C. Titration of Unknown Acid

Unknown identification code ____________________

| | Trial 1 | Trial 2 | Trial 3 | Trial 4 |
|---|---|---|---|---|
| 1. wt. of acid | | | | |
| 2. initial buret reading | | | | |
| 3. final buret reading | | | | |
| 4. volume of NaOH | | | | |
| 5. molarity of NaOH | | | | |
| 6. moles of $OH^-$ (equiv. of NaOH) | | | | |
| 7. moles of $H_3O^+$ (equiv. of acid) | | | | |
| 8. equiv. wt. of acid | | | | |

9. ave. equiv. wt. of acid ____________

Show the calculation of 6, 7, and 8 for one trial.

(Attach a separate sheet with the answers to the Questions.)

# EXPERIMENT 17

# REDOX REACTIONS

## Introduction

Redox reactions are chemical changes that involve the transfer of electrons. For example, in the reaction of magnesium and chlorine, magnesium atoms lose electrons; they are oxidized by chlorine, which is the oxidizing agent. Chlorine atoms gain electrons; they are reduced by magnesium, which is the reducing agent.

$$Mg(s) + Cl_2(g) \rightarrow MgCl_2(s) \qquad (17\text{-}1)$$

Magnesium's loss of electrons is a characteristic chemical behavior of metals. However, among the metals there is considerable variation in the ease with which this electron loss occurs. Some of the Alkali Metals (Group I of the Periodic Table) are oxidized so easily and rapidly that they burst into flame in air as they react with oxygen; in contrast, objects made of gold can remain untarnished for centuries. One way to compare the reactivity of metals is to compare their ability to reduce hydrogen from the +1 oxidation state to the elemental state. The most reactive metals can reduce the hydrogen in water:

$$M + H_2O \rightarrow H_2 + OH^- + M^{x+} \qquad (17\text{-}2)$$

Any metal that reacts with water will react even more vigorously with hydronium ions from an acid:

$$M + H_3O^+ \rightarrow H_2 + H_2O + M^{x+} \qquad (17\text{-}3)$$

Metals of intermediate reactivity will not be oxidized by water, but will react with hydronium ions of an acidic solution. Less reactive metals are not affected by either water or hydronium ions but might be oxidized by stronger oxidizing agents such as nitric acid. In this experiment, you will use water, dilute hydrochloric acid, and dilute nitric acid to compare the reactivities of some familiar metals.

Chlorine and other highly electronegative nonmetals are good oxidizing agents. Also, many compounds containing atoms in high oxidation states are good oxidizing agents, for example, $MnO_4^-$, which contains Mn in a +7 oxidation state. You will compare the effectiveness of this and several other ions to act as oxidizing agents in aqueous solution. The main evidence of a reaction will be a color change or formation of an insoluble product, either solid or gas. Keep in mind that "color change" might mean the color of a reactant disappears, a product that forms has color, or both a reactant and a product have color but their colors are different.

The reactions for the reduction of the oxidizing agents you will use include these changes:

$$Cr_2O_7^{2-} \rightarrow Cr^{3+} \text{ (color change)} \qquad (17\text{-}4)$$

$$Fe^{3+} \rightarrow Fe^{2+} \text{ (color change)} \qquad (17\text{-}5)$$

$$MnO_4^- \rightarrow MnO_2(s) \text{ or } Mn^{2+} \text{ (color change)} \qquad (17\text{-}6)$$

$$OCl^- \rightarrow Cl^- \qquad (17\text{-}7)$$

The reactions for the oxidation of the reducing agents you will use include these changes:

$$Cl^- \rightarrow Cl_2(g) \qquad (17\text{-}8)$$

$$Br^- \rightarrow Br_2 \text{ (color change)} \qquad (17\text{-}9)$$

$$I^- \rightarrow I_2 \text{ (color change)} \qquad (17\text{-}10)$$

$$Fe^{2+} \rightarrow Fe^{3+} \text{ (color change)} \qquad (17\text{-}11)$$

## Equipment Needed

beakers
burner
emery cloth
medicine droppers
ring
ring stand
4-mL test tubes
20-mL test tubes
wire gauze

## Chemicals Needed

Metals

Small pieces of Al, Ca, Cu, Fe, Mg, Pb, Sn, and Zn

Acids

6M HCl; hydrochloric acid
3M $HNO_3$; nitric acid
3M $H_2SO_4$; sulfuric acid

Oxidizing Agents

0.5M $FeCl_3$; iron(III) chloride
0.5M $K_2Cr_2O_7$; potassium dichromate
0.5M $KMnO_4$; potassium permanganate
0.5M NaOCl; sodium hypochlorite

Reducing Agents

0.1M $FeSO_4$; iron(II) sulfate
0.1M NaBr; sodium bromide
0.1M NaCl; sodium chloride
0.1M KI; potassium iodide

Other

litmus paper
phenolphthalein

## Procedure

### A. Metals as Reducing Agents

Into each of eight 20-mL test tubes, put 5 mL of distilled water and a drop of phenolphthalein solution. Obtain a small piece (or several granules) of each of the eight metals. If the surface of the metal is not shiny, polish it with emery cloth; however, do not attempt to polish calcium if it appears corroded. Drop each piece of metal into a different test tube containing water and phenolphthalein. Note whether there is any evolution of gas (hydrogen); if so, is it vigorous or slight? Note whether the basic color (pink) of phenolphthalein appears showing that the metal reacted with water producing $OH^-$.

Next, heat in a beaker of boiling water each test tube containing a metal that did not react with water at room temperature or for which there was only a slight reaction. Look for evidence of reaction as the mixture is heated. If the water in the test tube is hot enough to boil, you might mistake this boiling for the evolution of hydrogen. Even if the water is not hot enough to boil, there might be slight bubbling as dissolved air comes out of solution while the water is heated. If hydrogen is being evolved, bubbles of it will form *only* on the surface of the metal. This evolution of hydrogen will *continue* after the test tube is removed from the hot water bath.

To each test tube containing a metal that did not react with water, even when heated, add 5 mL of 6M HCl. Look for evolution of hydrogen. If there is no reaction, or only a slight reaction, heat the tube in a boiling water bath. Again, be careful not to mistake boiling for evolution of hydrogen.

If there is any metal that did not react with hydrochloric acid, take a different piece of the metal and put it into a clean test tube with 5 mL of 3M $HNO_3$. When $NO_3^-$ acts as an oxidizing agent, it is most apt to be reduced to $NO_2$ (a red-brown gas) or to NO (a colorless gas). Decide whether either of these gases forms; if necessary, heat the tube in a boiling water bath. If there is a reaction with nitric acid, immediately fill the tube with water to dilute the acid and slow down or stop the reaction, for both $NO_2$ and NO are very irritating and should not be inhaled.

When you complete this part of the experiment, your piece of any metal that is so unreactive it did not react with water or with either acid may be discarded in the regular waste basket. All your solutions and any unused acid may be washed down the drain.

### B. Comparison of Oxidizing Agents

Put about one mL of each of the four reducing agents into separate 4-mL test tubes; put about one mL of distilled water into a fifth tube to use for color comparisons. To each of the five tubes, add two drops of $K_2Cr_2O_7$ solution and mix thoroughly. Note any evidence of reaction. The dichromate ion is often a better oxidizing agent in an acidic solution than it is in a neutral or basic solution; to each solution for which there was no reaction (and also the water) add two drops of 3M $H_2SO_4$ and note whether or not a reaction occurs. Use litmus paper to determine whether each reaction mixture to which you did not add acid is acidic or basic.

For each of the three remaining oxidizing agents, prepare fresh tubes of the reducing agents (and water) and look for a reaction. For $KMnO_4$ (*only*), add two drops of 3M $H_2SO_4$ if there was not a reaction without it and look for evidence of reaction. Decide whether insoluble $MnO_2$ or soluble $Mn^{2+}$ forms. Use litmus paper to determine whether each reaction mixture to which you did not add acid is acidic or basic. **Do not add acid to solutions containing $OCl^-$.**

Any unused $K_2Cr_2O_7$ should not be washed down the drain; dispose of it as directed by your instructor. All other unused reagents and your reaction mixtures may be washed down the drain.

## Results

### A. Metals as Reducing Agents

1. State what metals reacted with water and what the evidence of reaction was. Describe the vigor of the reaction. Write a balanced equation for each reaction.
2. State what metals (that did not react with water) reacted with hydrochloric acid (that is, with hydronium ions). Describe each reaction. Write a balanced equation for each reaction.
3. State what metals (that did not react with water or hydrochloric acid) reacted with nitric acid (that is, with nitrate ions). Describe each reaction. Write a balanced equation for each reaction.
4. State what metals did not react with water or with either acid.
5. Use your observations to list the eight metals in order from the most reactive to the least reactive.

### B. Comparison of Oxidizing Agents

Prepare a chart with oxidizing agents listed horizontally and reducing agents listed vertically. Give the formula of the reacting ion, not the compound. State "yes" or "no" to indicate whether or not each pair of materials reacted. State whether acid was required.

Below the chart list the oxidizing agents in order from strongest to weakest.

Write a balanced net ionic equation for each reaction that occurred. After balancing the electron transfer (or oxidation number change) between oxidizing and reducing agents, use $H_2O$ and $H_3O^+$ or $OH^-$ (depending on whether the solution was acidic or basic to litmus) to complete the balancing of atoms.

## Questions

1. Use a table of standard electrode potentials to list the eight metals in order from most reactive to least reactive. Does the order differ from your list that is based on your observations in this experiment? If so, try to account for the differences.
2. Compare the stabilities of the halide ions $Cl^-$, $Br^-$, and $I^-$; that is, compare their ease of or resistance to oxidation.
3. Each of the following formulas is acceptable in terms of cation–anion charge balance. However, some of these compounds do not exist. Use your results to identify the compounds that could not exist in aqueous solution, that is, to identify hypothetical compounds containing ions that react with one another and, therefore, could not exist together.

   $FeCl_3$, $FeBr_3$, $FeI_3$, $FeCr_2O_7$, $Fe(MnO_4)_2$, $Fe(OCl)_2$

# EXPERIMENT 18

# ELECTROCHEMISTRY: VOLTAIC CELLS

## Introduction

This experiment deals with cells in which spontaneous oxidation–reduction reactions can be used to produce electricity. The reactants in the oxidation–reduction reaction are separated physically, so there cannot be a direct electron transfer from one reactant to the other as there would be if the reactants were in physical contact. Instead, the two reactants are joined by a wire through which electrons travel from one reactant to the other. This flow of electrons through a wire constitutes an electric current. You should recognize one important difference between voltaic and electrolytic cells: in the *voltaic cell*, electrons flow unassisted from one reactant to the other because the reaction is thermodynamically spontaneous; in an electrolytic cell, a non-spontaneous reaction is forced to occur by the application of an outside power supply (a battery) which "pumps" electrons through the electrolytic cell.

Figure 18-1 shows a voltaic cell in which $Cu^{2+}$ is reduced by Zn metal, the cell reaction being:

$$Zn + Cu^{2+} \rightarrow Cu + Zn^{2+}$$

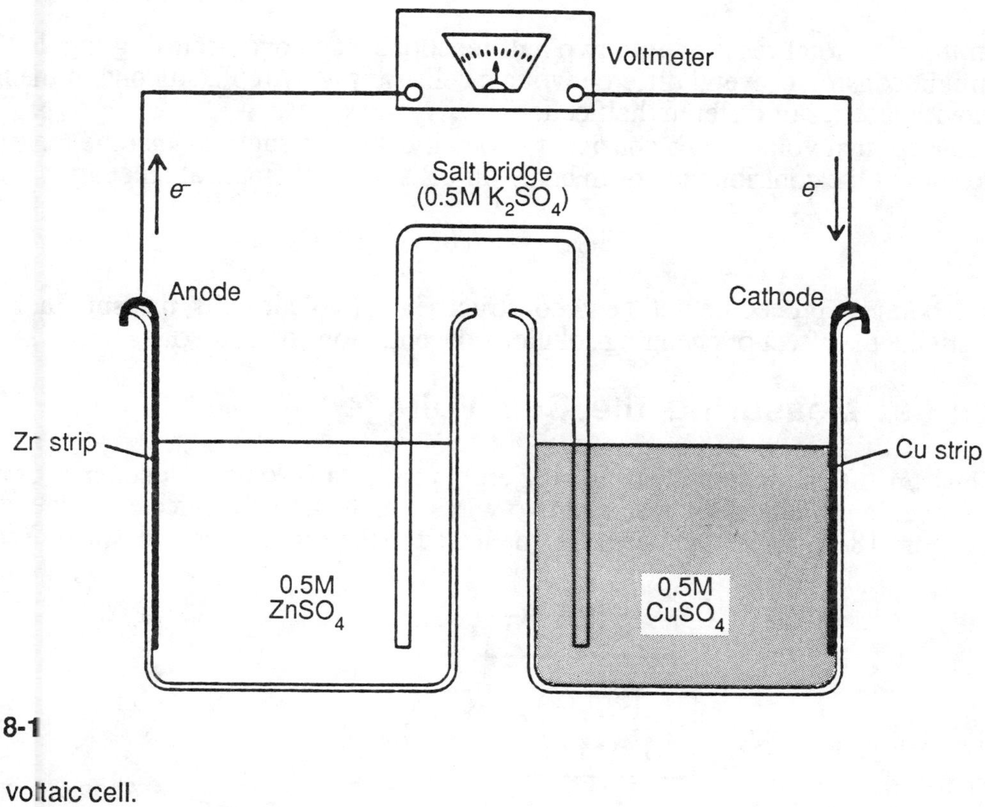


**Figure 18-1**

A Cu-Zn voltaic cell.

The cell consists of:

1. a Zn strip immersed in 0.5M $ZnSO_4$ solution
2. a Cu strip immersed in 0.5M $CuSO_4$ solution
3. a salt bridge which is constructed by filling a U-shaped tube with 0.5 M $K_2SO_4$, plugging the ends of the tube with cotton, inverting the tube, and immersing one arm of the tube in each beaker
4. a conducting wire through which electrons flow from the Zn electrode to the Cu electrode.

A voltmeter is inserted in the circuit in Figure 18-1 to measure the *cell voltage*. The value of the voltage or the *electromotive force* (emf), can be regarded as a quantitative measure of the tendency for electrons to flow in the cell, that is, the tendency for the electrons to be transferred from one reactant (Zn) to the other ($Cu^{2+}$).

The overall reaction in a voltaic cell can be divided into two half-cell reactions, an oxidation half and a reduction half, each one occurring at a different electrode:

Oxidation: $$Zn \rightarrow Zn^{2+} + 2e^- \text{ (at the anode)} \quad (18\text{-}1)$$

Reduction: $$Cu^{2+} + 2e^- \rightarrow Cu \text{ (at the cathode)} \quad (18\text{-}2)$$

Overall Reaction: $$Cu^{2+} + Zn \rightarrow Cu + Zn^{2+} \quad (18\text{-}3)$$

Separating the total reaction into two half reactions, each one occurring in a half cell, implies that we might construct several different voltaic cells simply by replacing one of the half cells, say, $Zn/Zn^{2+}$, with a series of different half cells.

The concept that voltaic cells consist of two half cells also suggests that the measured cell voltage is the sum of the contributions from both half cells. In mathematical language:

$$E_{total} = E_{oxidation} + E_{reduction} \quad (18\text{-}4)$$

Your laboratory procedure will be to construct several voltaic cells, measure their voltages, and then investigate the effect of changing solution concentrations on cell voltage.

## Technique: Measuring the Cell Voltage

The instrument employed for measuring emf is called a *voltmeter*. It consists of a scale with a deflecting needle mounted in a case and two wires which can be connected to the electrodes of a cell (see Figure 18-2). When the wires are connected to the proper electrodes, the needle will move

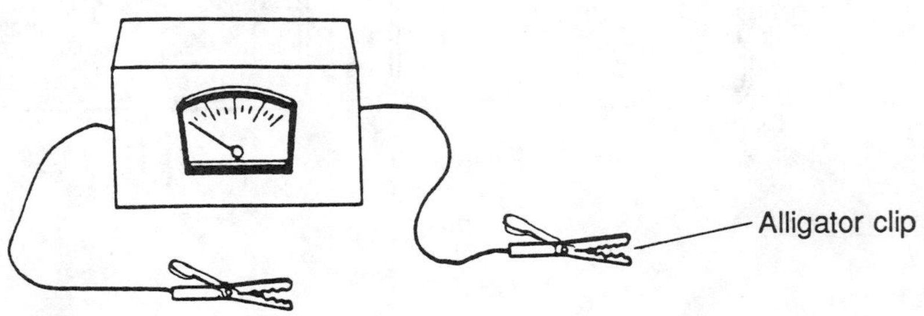


**Figure 18-2**

A voltmeter.

from zero and the voltage generated by the cell can then be read. If the wires are connected in reverse, the needle will not deflect. Hence, when you begin the experiment you must first determine how to connect the voltmeter to a cell. In the reaction between $Cu^{2+}$ and Zn you know that the Zn electrode is the anode and the Cu electrode is the cathode. Determine and record which wire of the voltmeter must be connected to the anode and which to the cathode before a voltage will register. For the remaining cells, you can use this information to determine which electrode is the cathode and which electrode is the anode, and, consequently, in which direction electrons flow through the external cell circuit.

## Equipment Needed

alligator clips
100-mL beaker
250-mL beaker
porous cup
sandpaper
voltmeter (1000 ohms/volt resistance)

## Chemicals Needed

Cu strip
Fe strip
Mg strip
Pb strip
Sn strip
Zn strip
0.5M $CuSO_4$; copper sulfate
0.5M $FeSO_4$; iron(II) sulfate
dil. HCl; hydrochloric acid
0.5M $MgSO_4$; magnesium sulfate
conc. $NH_3(aq)$; aqueous ammonia
0.1M $Na_2S$; sodium sulfide
0.5M $Pb(NO_3)_2$; lead nitrate
0.5M $SnCl_2$; tin(II) chloride
0.5M $ZnSO_4$; zinc sulfate

## Procedure

To avoid the time-consuming process of making U-tube salt bridges, you can construct a cell as illustrated in Figure 18-3. The porous cup allows ions to flow back and forth through its walls to equalize the electrical charges in the two halves of the cell. The cup fulfills the same function as the salt bridge but is easier to handle in setting up the cells.

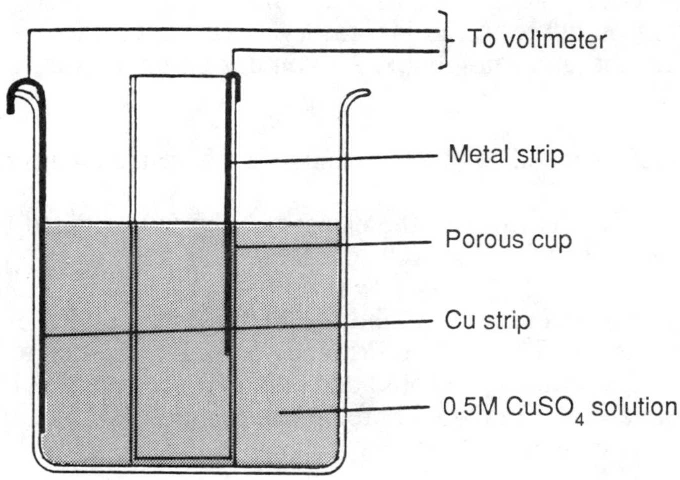


**Figure 18-3**

Constructing a voltaic cell with a porous cup.

## Cell Voltages

First, pour about 150 mL of 0.5M $CuSO_4$ into a 250-mL beaker. Bend a sandpapered Cu strip and hang it over the side of the beaker so that it is partly immersed in the solution. Clean a Zn strip by briefly immersing it in about 75 mL dil. HCl in a 100-mL beaker. Do not let the strip remain for very long in the HCl. Otherwise, it will react liberating $H_2$. (The HCl cleans the Zn by dissolving the surface layer exposing fresh metal.) Immediately rinse the Zn strip thoroughly in distilled water. Place the porous cup in the beaker. Add 0.5M $ZnSO_4$ to the porous cup until the liquid level is the same as it is in the beaker. Bend the Zn strip you just cleaned, and hang it over the side of the porous cup.

Use alligator clips to connect the terminals of the voltmeter to the metal electrodes. Record the initial voltage of the cell. Be sure to determine which wire must be connected to the anode and which to the cathode.

Disconnect the voltmeter and remove the porous cup from the beaker. Rinse the $ZnSO_4$ solution down the drain with plenty of water. Clean and dry the cup.

Repeat this procedure replacing the Zn and 0.5M $ZnSO_4$ with the following list of metal strips and solutions of their salts:

Fe and 0.5M $FeSO_4$
Mg and 0.5M $MgSO_4$
Pb and 0.5M $Pb(NO_3)_2$
Sn and 0.5M $SnCl_2$

All metal strips should be sandpapered before you use them. This cleans off any coating of metal oxide on the surface so that the pure metal will be in contact with the solution. For each cell identify the anode and the cathode. All solutions may be rinsed down the drain with plenty of water. Replace the metal strips in the proper containers on the supply shelf after rinsing them thoroughly with distilled water.

## Concentration Effects on Cell Voltage

All the solutions that you have used in this experiment have been 0.5M. Next you will determine (at least qualitatively) what effect changing the concentrations of salt solutions has on cell voltage.

You are to design experiments that will reveal the effects of varying the solution concentrations. Before you come to the laboratory, your experiments should be thought out and written down in your laboratory notebook. As you design your experiments, keep the following points in mind:

1. Use a Zn-Cu cell so that the results of different students can be easily compared.

2. If your time is limited, the effect may be investigated for only the cathode compartment solution instead of for both solutions.

3. Vary the concentration slightly from 0.5M to see what effect this has. Then try larger variations such as 0.05M, 0.005M, and so on. If you wish to reduce the metal concentration to a very low value, you could add an equal volume of 0.1M $Na_2S$ to the 0.005M solution of the metal salt. This would precipitate the metal sulfide and reduce the metal ion concentration to less than $10^{-30}$ M.

4. Try to determine if the cell voltage varies in some regular way with concentration. For example, is the variation linear? That is, does a tenfold decrease in concentration lower the voltage to one tenth of its former value? And so on.

### Complexation Effects on Cell Voltage

The blue color of the 0.5M $CuSO_4$ solution is actually due to the presence of a complex ion of copper $[Cu(H_2O)_6]^{2+}$. Design and carry out an experiment that will show what effect (if any) is produced by the presence of copper in the form of the ammonia complex $[Cu(NH_3)_4]^{2+}$ instead of in the form of the hydrated ion. The hydrated ion can be converted to the ammoniated ion by adding concentrated $NH_3$ solution to a $CuSO_4$ solution until an intense deep blue color results.

For both of the above experiments that you design for yourself, your report should (1.) describe the experimental procedures you use, (2.) present the data you collect, and (3.) discuss the conclusions you can draw from your data.

## Results and Calculations

1. For each cell for which you measured voltage, write the anode half reaction and the cathode half reaction. In each case this involves determining which electrode is the anode and which electrode is the cathode.

2. For each cell add the two half reactions to obtain the overall spontaneous cell reaction.

3. The potential (emf) of each cell is the sum of the contributions from each half cell. The two half-cell contributions are difficult to separate; however, if we *define* that the $Cu^{2+}/Cu$ couple has $E° = 0.000$ volts, we can assign the entire measured cell potential (emf) to the other couple. In order for the reaction determined by experiment to be spontaneous to have a positive voltage, the entire measured positive voltage must be assigned to the half reaction occurring at the non-Cu electrode. Write the half reaction that occurs at each of the non-Cu electrodes and the voltage assigned.

4. Some of the half reactions to which you just assigned voltages are oxidations and some are reductions. In order to compare voltages on an "equal" basis, it is customary to write all reactions as reductions. Do this for the half reactions in (3.) and assign to each reduction half reaction a voltage with the proper algebraic sign.

5. Arrange the list of metals in order of their decreasing ability to donate electrons to $Cu^{2+}$; that is, list the metal with greatest tendency to donate electrons first.

6. From your textbook or from a handbook obtain accepted values of standard half-cell potentials for the couples used in this experiment. Next to the appropriate half reaction written as a reduction, record this value of E°. For each half reaction, record the difference between this value and the value you assigned for E in this experiment; that is, record $E° - E_{assigned}$.

7. What do the differences between the tabulated standard potentials and the values you assigned from your measurements tell you about the E° of the $Cu^{2+}$ couple? Explain your answer.

8. Discuss your data concerning cell voltage and its concentration dependence.

9. What effect does the complexation of $Cu^{2+}$ by $NH_3$ to produce $[Cu(NH_3)_4]^{2+}$ have on the half-cell potential of $Cu^{2+}$?

# EXPERIMENT 18

# REPORT SHEET

Name ______________________________ Date ______________

Section ______________________________

**Experimental Data**

| Cell | Measured Voltage |
|---|---|
| I. Zn /$Zn^{2+}$ ‖ $Cu^{2+}$ /Cu | |
| II. Fe /$Fe^{2+}$ ‖ $Cu^{2+}$ /Cu | |
| III. Mg /$Mg^{2+}$ ‖ $Cu^{2+}$ /Cu | |
| IV. Pb /$Pb^{2-}$ ‖ $Cu^{2+}$ /Cu | |
| V. Sn /$Sn^{2-}$ ‖ $Cu^{2+}$ /Cu | |

**Concentration Effects on Cell Voltage**

**Complexation Effects on Cell Voltage**

1. Half Reactions:

| | Anode | Cathode |
|---|---|---|
| I. | $Zn \rightarrow Zn^{2+} + 2e^-$ | $Cu^{2+} + 2e^- \rightarrow Cu$ |
| II. | | |
| III. | | |
| IV. | | |
| V. | | |

2. Overall Cell Reactions:

I. $Zn + Cu^{2+} \rightarrow Zn^{2+} + Cu$

II.

III.

IV.

V.

3. Reaction at Non-Cu Half Cell

| | Reaction at Non-Cu Half Cell | Assigned Voltage |
|---|---|---|
| I. | $Zn \rightarrow Zn^{2+} + 2e$ | 1.10 volts |
| II. | | |
| III. | | |
| IV. | | |
| V. | | |

4. Reduction Half Reaction

| | Reduction Half Reaction | Assigned Voltage |
|---|---|---|
| I. | $Zn^{2+} + 2e^- \rightarrow Zn$ | -1.10 volts |
| II. | | |
| III. | | |
| IV. | | |
| V. | | |

5. List the metals in order of decreasing tendency to reduce $Cu^{2+}$

6. Half-Cell Reaction

| Half-Cell Reaction | Tabulated E° | $E_{assigned}$ | Difference |
|---|---|---|---|
| $Cu^{2+} + 2e^- \rightarrow Cu$ | ________ | ________ | ________ |
| $Zn^{2+} + 2e^- \rightarrow Zn$ | ________ | ________ | ________ |
| $Fe^{2+} + 2e^- \rightarrow Fe$ | ________ | ________ | ________ |
| $Mg^{2+} + 2e^- \rightarrow Mg$ | ________ | ________ | ________ |
| $Pb^{2+} + 2e^- \rightarrow Pb$ | ________ | ________ | ________ |
| $Sn^{2+} + 2e^- \rightarrow Sn$ | ________ | ________ | ________ |

7. What is the meaning of the difference between the assigned E and the tabulated E°?

8. Discuss the concentration dependence of E.

9. Discuss the effect of complexation by $NH_3$ on E°.

# EXPERIMENT 19

# ELECTROCHEMISTRY: ELECTROLYTIC CELLS

## Introduction

In an *electrolytic cell*, an oxidation–reduction reaction that does not proceed spontaneously can be forced to occur by using an external device (a battery) to pull electrons from one electrode (the anode), causing oxidation to occur, and to pump them into the other electrode (the cathode), causing reduction to occur. The process occurring in the cell is called *electrolysis*. This experiment deals with the relationship between the amount of oxidation and reduction taking place during electrolysis and with the quantity of electricity passing through the cell.

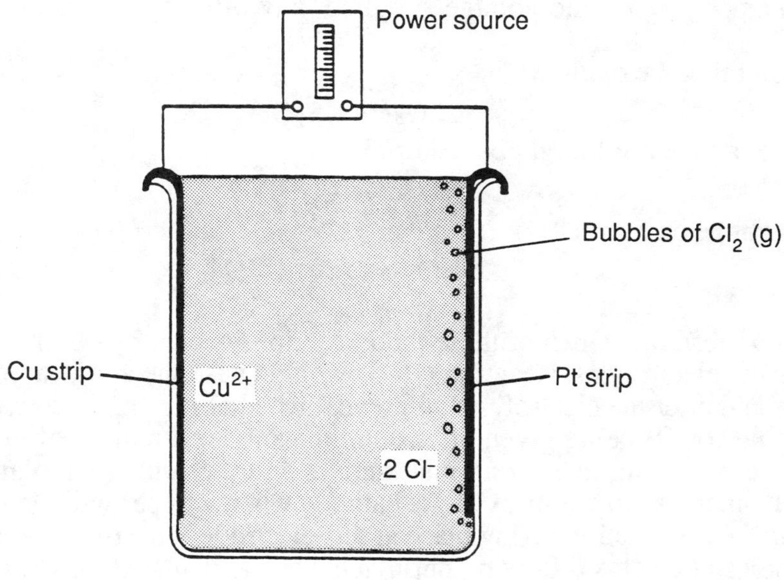


**Figure 19-1**

Electrolysis of $CuCl_2$.

Figure 19-1 is a diagram of a cell for the electrolysis of a concentrated $CuCl_2$ solution; a strip of copper is used as the cathode and a strip of platinum is used as the anode. Pt is a fairly inert metal and undergoes no reaction in this cell; it serves only as an electron-transfer medium. The reactions at each electrode are:

Anode: $$2Cl^- \rightarrow Cl_2(g) + 2e^- \text{ (oxidation)} \quad (19\text{-}1)$$

Cathode: $$Cu^{2+} + 2e^- \rightarrow Cu(s) \text{ (reduction)} \quad (19\text{-}2)$$

It would not be necessary to use a copper strip as the cathode. However, it is convenient to have the Cu plated out onto a surface of the same metal.

If the current (in amperes) is known and the time of electrolysis (in seconds) is recorded, then the quantity of electricity (in coulombs) passing through the cell can be calculated:

$$\text{amps} \times \text{sec} = \text{coul/sec} \times \text{sec} = \text{coul} \qquad (19\text{-}3)$$

Faraday's first law states that the quantity of a metal deposited at an electrode in an electrolytic cell is proportional to the number of coulombs of charge passing through the circuit. Thus, the greater the amount of charge that passes through the cell, the greater the weight of metal that is plated onto the cathode.

Faraday's second law states that the weights of different substances produced by the passage of a given amount of charge are proportional to the equivalent weights of those substances. The equivalent weight is the number of grams deposited by one mole of electrons (96,500 coul of charge). For example, the passage of 96,500 coul of charge would deposit different weights of silver, copper and lead.

Of course the laws describing electrolysis can also be applied to cells in which the reactions occurring do not involve plating out of a metal onto the cathode. The convenience of measuring by weighing the amount of reduced material (in this case, the metal) is the factor that leads us to use a cell involving plating of a metal in this experiment.

You will electrolyze a solution of $CuSO_4$ (copper sulfate), so that copper is deposited onto the cathode. In addition, oxidation at the anode must also take place. You will use copper for both electrodes. Depending on the voltage supplied, any one of the three following reactions might occur:

Sulfate ion might be oxidized:

$$2SO_4^{2-} \rightarrow S_2O_8^{2-} + 2e^- \qquad (19\text{-}4)$$

The copper anode itself might be oxidized:

$$Cu \rightarrow Cu^{2+} + 2e^- \qquad (19\text{-}5)$$

Water might be oxidized:

$$2H_2O \rightarrow O_2(g) + 4H^+ + 4e^- \qquad (19\text{-}6)$$

You must determine which of these three reactions occurs by observing the anode before, during and after the electrolysis. Reaction (19-5) would cause the anode to lose weight; weighing the anode before and after the electrolysis will reveal whether this has occurred. Reaction (19-6) would result in bubbles of $O_2$ being given off around the anode. If neither of these is observed, reaction (19-4) must be occurring (assuming that there is no other unknown material present that can be oxidized). The increase in the mass of the cathode where copper will plate out serves as a measure of the quantity of the reaction taking place at this electrode. This mass increase can be related to the number of moles of electrons flowing through the cell (calculated from the total electric charge and the charge on one mole of electrons). The equivalent weight of Cu (the weight deposited by one mole of electrons) can then be easily calculated.

Plating metals by electrolysis is best carried out under conditions of long plating time and low current. The currents used in this experiment are substantially higher than optimum in order to save time; consequently, some inaccuracies will be introduced. But your results will probably be within 10-15% of the theoretical results.

## Equipment Needed

alligator clips
ammeter
100-mL beaker
250-mL beaker
direct current power supply
sandpaper
variable resistance

## Chemicals Needed

$CH_3C(O)CH_3$; acetone (in wash bottle)
Cu strips
1M $CuSO_4$; copper sulfate

## Procedure

Clean two copper strips with sandpaper. Bend each strip about 2 cm from the end, so that it can be hooked over the side of a beaker. Handle the cleaned strips only by the edges to keep them as clean as possible. Weigh each strip as accurately as possible on the analytical balance. Be sure that you can distinguish between the strips. One will be the anode and one will be the cathode. You must know which is which, so that you can compare the weight of each strip after the electrolysis with its weight before the electrolysis.

Place the copper electrodes over the opposite sides of a 100-mL beaker; be sure that the strips remain parallel to one another. Assemble the electrical equipment as shown in Figure 19-2. Take care to connect the *cathode* to the *negative* terminal of the power source. Before proceeding further, ask your instructor to check your apparatus.

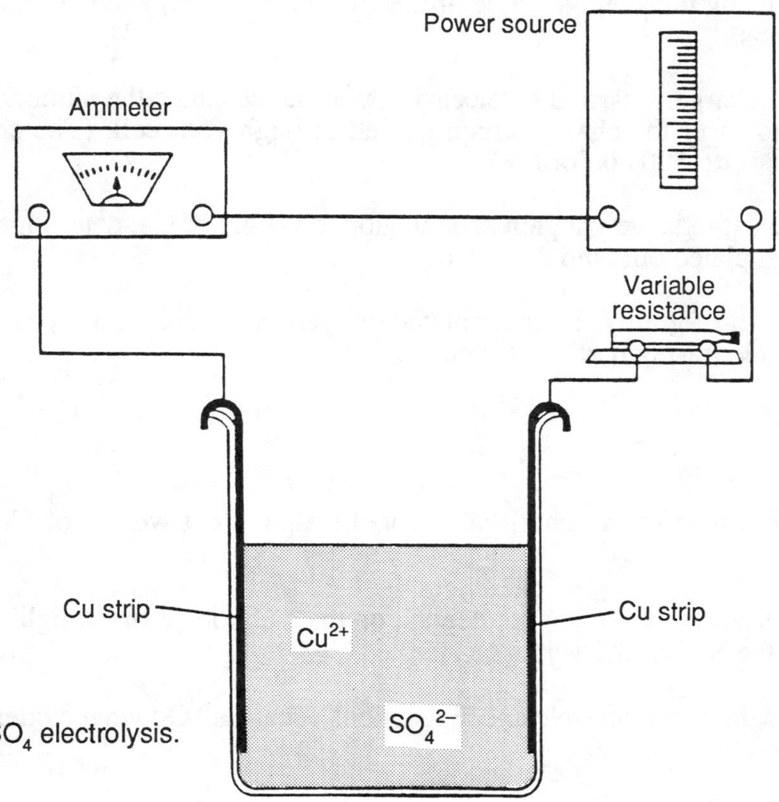


**Figure 19-2**

Apparatus for $CuSO_4$ electrolysis.

Pour 60 mL of $CuSO_4$ solution into the beaker. Adjust the variable resistance to its maximum value. Turn on the power supply switch; note the time. Immediately adjust the resistance so that a current of about 100 milliamp (read on the ammeter dial) flows in the cell.

Allow the current to flow for 25 minutes to carry out the electrolysis. Check and record in your notebook the current reading once a minute. If necessary, readjust the resistance to maintain the current at 100 milliamp.

At the end of 25 minutes, turn off the power supply. Then disconnect the clips from the electrodes. Lift each electrode by the edge and hang it over the lip of a dry 250-mL beaker. Wash each electrode carefully by dipping it into a beaker of distilled water or with water from your wash bottle. If you use a wash bottle, do not squirt the water directly onto the metal deposited during the electrolysis. This might cause the metal to fall off the surface. Instead, direct the water at the upper portion of each strip and allow it to run down the strip. Be sure to wash both sides of the strips. Repeat the washing procedure using acetone.

Allow the electrodes to dry for about 15 minutes, and then weigh them. Be careful to handle the electrodes only by the edges and above the area where the metal has been deposited.

Your instructor might ask you to carry out a second trial.

## Data and Results

1. To obtain the average value of the current, add all the current readings (in amps) and divide by the number of readings.

2. From Equation (19-3), calculate the amount of charge (in coulombs) that passed through the cell, using the average value of the current (in amps) and the time (in seconds) of your electrolysis.

3. Since the charge is carried by electrons, you can calculate the number of moles of electrons needed to carry the charge which passed through your cell. (The charge on one mole of electrons is 96,500 coulombs.)

4. From the cathode weight gain and the atomic weight of Cu, determine the number of moles of Cu that plated out onto the cathode.

5. The number of grams of copper plated out per mole of electrons passing through the cell is the equivalent weight of Cu. Calculate this.

## Questions

1. Compare your experimental value for the equivalent weight of Cu with the theoretical value.

2. If you lost some of the copper deposit on the cathode before weighing it, how would your value of the equivalent weight be affected?

3. Which of the three possible anode reactions occurred? On what evidence did you base your conclusion?

4. (a) If reaction (19-5) occurs at the anode, what should the relation be between the weight changes of the anode and cathode during the course of the electrolysis?

   (b) Can you think of any experimental problems which could lead to results which differ from the ones expected for part (a)? In what direction should the differences occur?

# EXPERIMENT 19

# REPORT SHEET

Name ______________________________ Date ______________

Section ______________________________

**Data**

| | Trial 1 | Trial 2 |
|---|---|---|
| Average Current | ______ | ______ |
| Electrolysis time (seconds) | ______ | ______ |
| Anode weight before electrolysis | ______ | ______ |
| Anode weight after electrolysis | ______ | ______ |
| Cathode weight before electrolysis | ______ | ______ |
| Cathode weight after electrolysis | ______ | ______ |
| Cathode weight gain | ______ | ______ |
| Moles Cu deposited | ______ | ______ |
| Coulombs of charge passed<br>Show calculation. | ______ | ______ |
| Moles of electrons passed<br>Show calculation. | ______ | ______ |
| Moles Cu deposited/mole of electrons<br>Show calculation. | ______ | ______ |

Average equivalent weight of Cu ______
Show calculation.

(On the back of the Report Sheet answer the Questions.)

# EXPERIMENT 20

# A COMPLETE RATE LAW

## Introduction

The rate at which any chemical reaction proceeds depends on several factors. One factor, over which an experimenter has no control, is the natures of the reacting species. Other factors, over which an experimenter does have some control, are the amount of contact between reactants, the temperature at which the reaction occurs, and whether or not a catalyst is present. In this experiment, you will investigate how reactant concentrations (that is, amount of contact between reactants in solution) and temperature influence reaction rate.

You will use the technique of measuring how long it takes for a particular amount of reaction to be completed; the longer the time required for completion, the slower the reaction and the lower the rate. Mathematically, the relationship is an inverse one:

$$\text{rate} \propto \frac{1}{\text{completion time}} \tag{20-1}$$

The reaction you will investigate is the oxidation of $I^-$ to $I_3^-$ (triiodide) by $H_2O_2$ (hydrogen peroxide) in an acidic solution:

$$H_2O_2 + 3I^- + 2H_3O^+ \rightarrow I_3^- + 4H_2O \tag{20-2}$$

The rate law for this reaction can be written

$$\text{rate} = k[H_2O_2]^x\,[I^-]^y\,[H_3O+]^z \tag{20-3}$$

Your task will be to determine the exponents $x$, $y$, and $z$, the order of the reaction with respect to each of the reactants; in other words, you will determine how the concentration of each of the reactants affects the reaction rate.

## Technique

You are to determine how the concentration of each of the three reactants affects reaction rate. A way to begin would be to vary $[H_2O_2]$ in two trials while keeping $[I^-]$ and $[H_3O^+]$ the same in both trials, and to measure and compare the rates for the two trials. For example, if $[H_2O_2]$ is twice as great in one trial as it is in another and the reaction rate is four times as great (when other concentrations remain the same for the two trials), we can conclude that $x = 2$, that is, that the rate depends on $[H_2O_2]^2$. A similar procedure could be used to determine $y$ and $z$, the orders with respect to $I^-$ and $H_3O^+$.

The rate of the reaction in Equation (20-2) can be easily measured by visual means. You determine how long it takes for the same $[I_3^-]$ to be produced using different concentrations of reactants. A small amount of $Na_2S_2O_3$ and some starch indicator are added to each reaction mixture. As the reaction product $I_3^-$ forms, it very rapidly reacts with $S_2O_3^{2-}$:

$$I_3^- + 2S_2O_3^{2-} \rightarrow S_4O_6^{2-} + 3I^- \tag{20-4}$$

Thiosulfate converts $I_3^-$ back to iodide and prevents any change in $[I^-]$ as long as $S_2O_3^{2-}$ is present.

When the small amount of $S_2O_3^{2-}$ is used up, $I_3^-$ will no longer react with anything and will begin to accumulate in solution. The $I_3^-$ will form a deep blue complex with the starch and the solution will suddenly change color. Hence, the length of time that it takes for the blue color to appear is a measure of how long it takes the reaction in Equation (20-2) to produce enough $I_3^-$ to consume a particular amount of $S_2O_3^{2-}$. This time interval is inversely related to the reaction rate.

## Review

Equipment and Techniques G
Experiment 1

## Equipment Needed

beakers
burets (3)
buret clamps
clock or watch with second hand
Erlenmeyer flasks
10-mL pipet
20-mL pipet
pipet bulb
ring stands
watch glasses
ice

## Chemicals Needed

Buffer A: $[H_3O^+] = 2.0 \times 10^{-5}M$ (contains 2 g starch/liter)
Buffer B: $[H_3O^+] = 1.0 \times 10^{-5}M$ (contains 2 g starch/liter)
0.020M $H_2O_2$; hydrogen peroxide
0.50M KI; potassium iodide
0.0050M $Na_2S_2O_3$; sodium thiosulfate

## Procedure

### A. Effect of Concentration

You will need to measure out volumes of the stock solutions of reactants in such a way that (1) the amount of $S_2O_3^{2-}$ is the same in every solution, and (2) the concentrations of the other reagents are varied systematically. Each of the four sets of volumes given in the table below has the same amount of $S_2O_3^{2-}$ and the same final total volume of solution. Each of the three reactants ($H_2O_2$, $H_3O^+$, and $I^-$) has the same concentration in three of the trials and a different concentration in a fourth trial.

| Solution | Volume (mL) | | | |
|---|---|---|---|---|
| $Na_2S_2O_3$ | 10 | 10 | 10 | 10 |
| $H_2O_2$ | 10 | 20 | 10 | 10 |
| Buffer A | 20 | 10 | 10 | - |
| Buffer B | - | - | - | 20 |
| KI | 10 | 10 | 20 | 10 |
| Trial | I | II | III | IV |

Clean three burets and set them up in a row. Label them for use with $H_2O_2$, buffer A, and KI.

Label four clean, dry beakers for $Na_2S_2O_3$, $H_2O_2$, buffer A, and KI. Obtain the amount of each solution that you will need for the entire experiment:

| | |
|---|---|
| $Na_2S_2O_3$: | about 100 mL |
| $H_2O_2$: | about 130 mL |
| Buffer A: | about 110 mL |
| KI: | about 130 mL |

Cover each beaker with a watch glass. Place the $Na_2S_2O_3$ beaker to the side of the burets. Place each of the other three beakers behind the appropriate buret. Rinse (3 x 3 mL) and fill each buret with the proper solution; reposition each beaker behind the appropriate buret. Now you are ready to begin the kinetics trials.

Use a 10-mL pipet and a pipet bulb to measure the $Na_2S_2O_3$ solution into a clean, dry Erlenmeyer flask. Use burets to measure the $H_2O_2$ solution and buffer solution into the same Erlenmeyer flask. Use a buret to measure the KI solution into a different clean, dry Erlenmeyer flask. To begin a trial, pour the KI solution into the other flask as you start timing. Pour the contents back and forth between the two flasks several times to mix the solution completely; as you do this, roll the flask from which you are pouring to wash solution off the entire wall of the flask. Place the flask on a piece of white paper. Stop timing at the appearance of a blue color. Record this completion time. (Hint! The first trial will take at least a minute.) Do two determinations of reaction completion time for each mixture of reactants.

As soon as you complete a determination, empty the flasks, rinse them thoroughly with tap water and distilled water, and turn them upside down to drain. If they are clean, they should drain dry enough to use after your next determination; if very much water remains in them, it will dilute the reaction mixture and slow the reaction.

Before you do trial IV, prepare the two solutions to mix for the low temperature trial. Let the two flasks stand in the ice bath while you do trial IV.

When you are ready for trial IV, get buffer B in a small, clean, dry beaker. Use a 20-mL pipet to measure it into the reaction mixture.

## B. Effect of Temperature

After you have completed your work with the first three trials, and before you do trial IV, choose the fastest one (or either of the fastest ones) for this part of the experiment. Measure $Na_2S_2O_3$, $H_2O_2$, and buffer A solutions into one flask. Measure KI solution into a different flask. Cool the solutions (in their separate flasks) to 0°C in an ice-and-water bath; while they chill, do trial IV at room temperature. When both solutions are at 0°C, mix the reactants, leave the mixture in the bath, and record the time required for appearance of the blue color. You are not required to repeat this determination of reaction completion at 0°C.

To dispose of the final reaction mixtures and any unused reactants, wash them down the drain with lots of water.

## Calculations and Results

### A. Effect of Concentration

1. Fill in the table giving the molar concentration of each reactant in each of the four solutions for the trials at room temperature. Remember that each of these calculations involves a dilution of the stock solution:

$$\frac{(M \text{ of stock solution})(V \text{ of stock solution})}{V \text{ of total solution}} \qquad (20\text{-}5)$$

The $[H_3O^+]$ is the same for all three trials with buffer A, because buffer solutions do not change $[H_3O^+]$ when they are diluted.

2. Record your reaction completion times in seconds and their average for each trial. Or, instead of recording the average, use the completion time for the determination that you think was done more accurately; cross out the rejected determination with one neat line.

3. Use your data from 1. and 2. to determine the order of the reaction with respect to each reactant, $H_2O_2$, $I^-$ and $H_3O^+$. Write a rate law showing these concentration dependencies.

4. Use the following instructions to calculate the value of the rate constant *k* for each trial (or your better determination for each of the four trials). Be sure to include units.

Express the reaction rate as the rate of formation of $I_3^-$. Begin with the concentration of $S_2O_3{}^{2-}$, which is the same for all four trials. According to Equation (20-4), the concentration of $I_3^-$ that forms in the reaction that is being timed is one-half as great as the concentration of $S_2O_3{}^{2-}$. Calculate the average rate, $[I_3^-]$/second, for each of the four trials (using your average time or better time for each trial).

Write the general rate law, Equation (20-3), and put in your values for the rate of formation of $I_3^-$; for $[H_2O_2]$, $[I^-]$, and $[H_3O^+]$; and for *x*, *y*, and *z*. Calculate *k*. Do this for each trial. Determine the average of your four values.

5. Write the complete rate law using your calculated value for *k* (including units) and your experimentally determined concentration dependencies.

### B. Effect of Temperature

Calculate the concentration of each reactant. Compare the rate of the reaction at 0°C with the rate of the same mixture of reactants at room temperature.

# EXPERIMENT 20

## REPORT SHEET

Name________________________________ Date ____________

Section______________________________

### A. Effect of Concentration

Room temperature ____________________

| | Trial I | Trial II | Trial III | Trial IV |
|---|---|---|---|---|
| $[S_2O_3^{2-}]$ | | | | |
| $[H_2O_2]$ | | | | |
| $[H_3O^+]$ | | | | |
| $[I^-]$ | | | | |
| time #1 | | | | |
| time #2 | | | | |
| ave. time | | | | |

Write the general rate law.

Explain how you determined the general rate law.

Calculation of rate constant, *k*

| | Trial I | Trial II | Trial III | Trial IV |
|---|---|---|---|---|
| $[S_2O_3^{2-}]$ | | | | |
| $[I_3^-]$ formed | | | | |
| time | | | | |
| $[I_3^-]$/sec | | | | |
| $k$ | | | | |

average value of *k*, including units ______________________

Show the calculation of $[I_3^-]$ formed.

For Trial I, show the calculation of rate of formation of $I_3^-$, $[I_3^-]$/sec.

For Trial I, show the calculation of *k*; include all units.

Write the complete rate law.

**B. Effect of Temperature**

Room temperature ______________________________

Low temperature ______________________________

$Na_2S_2O_3$ Volume ________________ Concentration ________________

$H_2O_2$ Volume ________________ Concentration ________________

Buffer A Volume ________________ $[H_3O^+] = 2.0 \times 10^{-5}M$

KI Volume ________________ Concentration ________________

Time for color change ________________

$[I_3^-]$ formed ________________

$[I_3^-]$/sec ________________

rate at room T/rate at low T ______________________________

Compare your results to the generalization that, as a rough average, the rates of reactions in solution at temperatures in the vicinity of room temperature approximately double to triple for each 10°C increase in temperature.

# EXPERIMENT 21

# CHEMICAL EQUILIBRIUM

## Introduction

Complete conversion of reactants into products is not characteristic of all chemical changes. It is more common for a reaction to reach a state in which both reactants and products are present and their concentrations no longer change. This situation is called *chemical equilibrium*. The equilibrium is a *dynamic* one; it is established when the rate at which products are being reconverted to reactants equals the rate at which products are forming, and so there is no net change in the amounts of reactants and products. Equations for such reactions are often written with two arrows to indicate that both the forward and reverse directions of reaction occur. For example

$$N_2(g) + O_2(g) \rightleftharpoons 2NO(g) \qquad (21\text{-}1)$$

An equilibrium constant *K* quantitatively describes the relative amounts of reactants and products present at equilibrium. This constant states the ratio of product concentrations to reactant concentrations with each concentration raised to the power that is the substance's coefficient from the balanced equation. The value of *K* is characteristic of a particular reaction at a particular temperature. For Equation (21-1)

$$K = \frac{[NO]^2}{[N_2][O_2]} = 0.05 \text{ at } 2100°C \qquad (21\text{-}2)$$

For Equation (21-1), $K < 1$, meaning that at equilibrium the amount of product is small compared to the amounts of reactants remaining unreacted. We say that the position of equilibrium lies to the left, meaning that when chemical equilibrium is established, the substances on the left side of the equation (reactants) predominate in the reaction mixture.

For many reactions, $K > 1$, meaning that product concentrations are greater than reactant concentrations when equilibrium is established. For these reactions we say that the position of equilibrium lies to the right.

A change in temperature changes the rates of forward and reverse directions of a reaction by different amounts. Consequently, a change in temperature changes the relative amounts of products and reactants in an equilibrium mixture and, therefore, the value of the equilibrium constant. For Equation (21-1), $K = 0.05$ at 2100°C. This temperature can be reached in the cylinders of an automobile engine; nitrogen and oxygen in the air that enters the cylinders can react extensively enough to form enough NO, nitrogen(II) oxide, to be a serious cause of air pollution. However, at normal room or outdoor temperatures, the equilibrium constant for this reaction is much smaller. At 25°C, $K = 10^{-31}$; the position of equilibrium is extremely far to the left.

This effect of temperature can be predicted using Le Chatelier's principle: If a stress is applied to a system in equilibrium, the system will change in a way that relieves the stress and establishes a new equilibrium state. If the "stress" is a temperature increase, the direction of reaction that requires heat (is endothermic) is favored and occurs more extensively. If the "stress" is a temperature decrease, the direction of reaction that evolves heat (is exothermic) is favored. We say that the position of equilibrium shifts with temperature, meaning that the amounts of reactants and products change with temperature. The forward direction of the reaction of Equation (21-1) is endothermic; consequently, it occurs more extensively as the temperature increases, and *K* increases as the temperature increases.

Other changes in conditions can also shift the position of equilibrium. If, to an equilibrium mixture, more of a reactant or product is added, reaction occurs more extensively in the direction using that substance. When equilibrium is re-established, the amounts of substances in the mixture will be different from those in the previous equilibrium mixture, but the concentration *ratio* described by $K$ will be the same. Only a change in temperature changes the value of $K$ for a particular reaction. Removing some of a substance from an equilibrium mixture also shifts the position of equilibrium; the direction of reaction producing more of that substance occurs more extensively.

For reactions involving gases, changes in the volume of a reaction container shift the position of equilibrium. A decreased volume favors the direction of reaction for which there is a decrease in the number of moles of gas. An increased volume favors the direction of reaction for which there is an increase in the number of moles of gas. But neither effect causes a change in the value of $K$.

In this experiment you will work with reversible reactions that occur in aqueous solution. You will investigate the effects that addition of a substance, removal of a substance, and temperature changes have on position of equilibrium.

## Equipment Needed

- balance
- beakers
- burner
- 10-mL graduated cylinder
- medicine droppers
- ring
- ring stand
- test-tube holder
- test-tube rack
- 20-mL test tubes
- wire gauze

## Chemicals Needed

- 2M $AgNO_3$
- 0.5M $Co(NO_3)_2$; cobalt(II) nitrate
- 0.5M $Cr(NO_3)_3$; chromium(III) nitrate
- 0.5M $Cu(NO_3)_2$; copper(II) nitrate
- 6M HCl; hydrochloric acid
- 12M HCl; concentrated hydrochloric acid
- 6M $HNO_3$; nitric acid
- ice
- $K_2CrO_4$; potassium chromate
- $K_2Cr_2O_7$; potassium dichromate
- KNCS; potassium thiocyanate
- 6M KNCS
- 6M NaOH; sodium hydroxide

**Caution! Concentrated hydrochloric acid (12M) and 6M NaOH are extremely corrosive.** Nitric acid reacts very quickly with proteins in skin leaving a yellow stain. Also, 6M HCl is very corrosive. If you spill any of these on your skin, wash it off immediately under lots of cold, running water. If you spill any of these on the outside of the bottle or on the table top, sponge it off and wash out the sponge.

## Technique and Procedure

### A. A Reversible Reaction

A solution that is prepared by dissolving a salt containing $CrO_4^{2-}$ *or* by dissolving a salt containing $Cr_2O_7^{2-}$ can actually have *both* of these ions present. If the solution is basic, $CrO_4^{2-}$ predominates; if the solution is acidic, $Cr_2O_7^{2-}$ predominates. Because the colors of these two ions are different, it is possible to tell which ion is present in a solution, or at least to tell which ion is present in a greater amount when both ions might be present.

Equations for the reversible conversion of one ion into the other are

$$Cr_2O_7^{2-} + OH^- \rightleftharpoons HCrO_4^- + CrO_4^{2-} \qquad (21\text{-}3)$$

$$HCrO_4^- + OH^- \rightleftharpoons CrO_4^{2-} + H_2O \qquad (21\text{-}4)$$

These reactions can be summarized in one equation that overlooks the presence of $HCrO_4^-$:

$$Cr_2O_7^{2-} + 2OH^- \rightleftharpoons 2CrO_4^{2-} + H_2O \qquad (21\text{-}5)$$

This simplification will be used in this experiment.

Equation (21-5) makes it clear that $Cr_2O_7^{2-}$ in solution can be converted to $CrO_4^{2-}$ by the addition of $OH^-$, that is, by adding a base. You will use NaOH and watch for a color change indicating that the forward reaction given in Equation (21-5) is occurring.

It is not as obvious from Equation (21-5) how $CrO_4^{2-}$ in solution can be converted to $Cr_2O_7^{2-}$. Keep in mind that a solution prepared from one of the ions will also contain at least a trace of the other ion because of the interconversion. If a solution is prepared from $CrO_4^{2-}$, the extent to which the reverse reaction of Equation (21-5) occurs will be increased if $OH^-$ is removed from solution; this is the same as saying that the position of equilibrium will be farther to the left. The easiest way to remove $OH^-$ is to add an acid (a source of hydronium ions, $H_3O^+$) to the solution, so that the following reaction will occur:

$$H_3O^+ + OH^- \rightarrow 2H_2O \qquad (21\text{-}6)$$

As $OH^-$ is removed, the extent to which the reverse reaction given in Equation (21-5) occurs will increase, and the accompanying color change will be your visible evidence of the formation of $Cr_2O_7^{2-}$.

Dissolve 0.3 g of $K_2Cr_2O_7$ in 20 mL of distilled water. Dissolve 0.4 g of $K_2CrO_4$ in 20 mL of distilled water. Record the color of each solid and of each solution. The two solutions contain different weights of solute and different concentrations of chromium-containing ions, but both solutions have the same concentration of chromium atoms.

1. Pour about one-fourth of the $K_2Cr_2O_7$ solution into a 20-mL test tube. Save the rest of the solution for color comparisons. To the solution in the test tube, add 6M NaOH dropwise; after each drop shake the tube to mix the contents thoroughly. Add NaOH until a definite color change occurs. Record the amount of base required. Account for the color change in terms of the preceding equations. Then add 6M $HNO_3$ dropwise to this basic solution; as before, mix the solution thoroughly after you add each drop. Add acid until a definite color change occurs; explain this color change in terms of the appropriate equations. Again, add base to the solution, and explain any color change. Remember that dilution will make the colors slightly lighter than they were in the original solutions.

2. Pour about one-fourth of the $K_2CrO_4$ solution into a 20-mL test tube. To this solution, add 6M $HNO_3$ dropwise until a definite color change occurs. Then add NaOH until a color change occurs. Then add nitric acid, as before. Account for your observations in terms of the appropriate equations.

When you complete this part of the experiment, dispose of any unused NaOH and $HNO_3$ by combining them and washing the solution down the drain. Dispose of the $K_2Cr_2O_7$ and $K_2CrO_4$ solutions as directed by your instructor; **do not wash these Cr-containing solutions down the drain.**

## B. Relative "Positions" of Equilibrium

Transition metal ions exist in aqueous solution as complex ions in which water molecules are bound by coordinate covalent bonds to the metal. For cobalt(II), chromium(III), and copper(II), these complex ions have the formulas $Co(H_2O)_6^{2+}$, $Cr(H_2O)_6^{3+}$, and $Cu(H_2O)_6^{2+}$. The water molucules can be replaced by other ions or molecules forming different complex ions. If two complex ions of a particular metal differ in color, it is easy to tell which ion is present in a solution.

You will attempt to convert the *aquo* complex ions listed above to *chloro* complex ions, using concentrated HCl as a source of $Cl^-$. For $Co^{2+}$, the reaction is

$$Co(H_2O)_6^{2+} + 4Cl^- \rightleftharpoons CoCl_4^{2-} + 6H_2O \qquad (21\text{-}7)$$

A general equation for various metal ions would be

$$M(H_2O)_x^{+n} + yCl^- \rightleftharpoons MCl_y^{(+n-y)} + xH_2O \qquad (21\text{-}8)$$

In working with the three metal ions chosen for this experiment, you must decide whether or not there is any visible evidence of formation of a complex ion within a short time period. You will then list the ions on the basis of how extensive the reaction of Equation (21-8) is for a particular amount of $Cl^-$, that is, on the basis of the size of the equilibrium constant for the reaction of Equation (21-8).

After comparing chloro complex ions of the three different metal ions, you will compare two different complex ions of one metal ion. You will investigate the reactions in Equations (21-7) and (21-9) to determine which reaction is more extensive for a particular amount of a complexing agent, $Cl^-$ or $NCS^-$; that is, which reaction has a larger equilibrium constant.

$$Co(H_2O)_6^{2+} + 4Cl^- \rightleftharpoons CoCl_4^{2-} + 6H_2O \qquad (21\text{-}7)$$

$$Co(H_2O)_6^{2+} + 4NCS^- \rightleftharpoons Co(NCS)_4^{2-} + 6H_2O \qquad (21\text{-}9)$$

1. Pour 5.0 mL of 0.5M $Co(NO_3)_2$ into each of two 20-mL test tubes. Pour 5.0 mL of 0.5M $Cr(NO_3)_3$ into each of a second set of two 20-mL test tubes. Pour 5.0 mL of 0.5M $Cu(NO_3)_2$ into each of a third set of 20-mL test tubes. Record the color of each solution. These solutions contain the aquo ions for which the formulas are given above. Set aside one test tube of each solution to use for color comparisons. To each of the three remaining test tubes, add 12M HCl. Use a medicine dropper to add the acid dropwise from a 10-mL graduated cylinder containing 10.0 mL of the acid. Record, to the nearest 0.5 mL, the volume of HCl that is required to produce a definite color change in each solution. After you think a color change has occurred, add another drop or two of acid to decide whether you are seeing a distinct color rather than a mixture of two colors. For example, if a solution changes color from red to orange, the orange might be the color of the product or it might be from a mixture of red-colored reactant and yellow-colored product. Do not add more than 10.0 mL of acid to each solution. If you are not sure whether the solution has changed color, add to the color comparison tube a volume of water equal to the volume of acid that you used; a slight color change will occur just from the dilution. Account for the differences among the three solutions on the basis of the position of equilibrium for the reaction of Equation (21-8). For which of the three metal ions does the position of equilibrium appear to be farthest to the right for a particular amount of $Cl^-$? For which of the three metal ions does the position of equilibrium appear to be farthest to the left for a particular amount of $Cl^-$?

2. Mix 3.0 mL of 0.5M $Co(NO_3)_2$ and 3.0 mL of 0.5M $Cu(NO_3)_2$ in a 20-mL test tube. Add 12M HCl dropwise until a definite color change occurs; record the volume of acid that you use. Does the chloro complex of cobalt or the chloro complex of copper form first? Does this agree with your previous conclusions about the positions of equilibrium for the reactions of these two metal ions with $Cl^-$? The color that you see in this mixture will not be *exactly* like the colors you observed for the solutions containing only one metal ion. However, the colors of $CoCl_4^{2-}$ and $CuCl_4^{2-}$ are sufficiently intense and different enough that you should be able to recognize either one despite the presence of another colored material.

3. pour 5.0 mL of 0.5M $Co(NO_3)_2$ into each of two 20-mL test tubes. To one test tube, add 6M KNCS dropwise until a definite color change occurs; continue to add KNCS solution until the color no longer is changing. To the second test tube, add 6M HCl until a definite color change occurs. Do not add more than 10 mL of either solution to the $Co(NO_3)_2$; record the volume of solution that is required in each case. The colors of the two complex ions $CoCl_4^{2-}$ and $Co(NCS)_4^{2-}$ are very nearly the same. Account for any difference in the amounts of complexing agent required in terms of the relative positions of equilibrium for the reactions of Equations (21-7) and (21-9). You should also be able to account for the difference in your results here and your results when you added 12M HCL

When you complete this part of the experiment, save any unused $Co(NO_3)_2$ and KNCS solutions for the next part of the experiment. All other solutions may be washed down the drain.

## C. "Shifting the Position" of Equilibrium

The colors of $Co(H_2O)_6^{2+}$ and $Co(NCS)_4^{2-}$ are quite distinct, and so you can easily tell which of these complex ions is present in the greater amount in a solution. If similar amounts of these two ions are present, the solution will have an intermediate color that can be described as reddish-purple. If anything is added to this solution that causes the forward reaction of Equation (21-9) to occur to a greater extent (that is, shifts the position of equilibrium to the right), the color becomes bluer. If a change in conditions causes the reverse reaction of Equation (21-9) to occur to a greater extent (that is, shifts the position of equilibrium to the left), the color becomes pinker. You will investigate how the position of equilibrium for this reaction is affected (1) by adding each of two materials that are involved in the reaction, (2) by removing one of the materials, and (3) by changing the temperature.

In a small beaker, add 7 mL of 6M KNCS to 3 mL of 0.5M $Co(NO_3)_2$. The solution should have the intense blue color of $Co(NCS)_4^{2-}$. Dilute this solution with 15-18 mL of distilled water until it has a purplish color that is intermediate between the colors of $Co(H_2O)_6^{2+}$ and $Co(NCS)_4^{2-}$. This intermediate color indicates that a moderate amount of each complex is present. Divide this solution as evenly as possible among seven 20-mL test tubes. Set tube 1 aside to use for color comparisons.

1. Effect of concentration of a reactant. To tube 2, add 1 mL of distilled water. Record any color change, and account for this change in terms of the forward or the reverse direction of the reaction of Equation (21-9), that is, in terms of a shift in the position of equilibrium when the material that is involved in the reaction is added to the reaction mixture.

To tube 3, add 5 mL of distilled water. Compare the colors of the solutions in tube 2 and 3; account for your observations.

To tube 4, add about 0.5 g of KNCS. Account for your observations in terms of the position of equilibrium for the reaction of Equation (21-9).

To tube 5, add 1 mL of 2M $AgNO_3$. The white precipitate that forms is insoluble AgNCS; the addition of $Ag^+$ causes the removal of $NCS^-$ from the solution. Mix thoroughly, allow the precipitate to settle, and compare the color of the clear solution in tube 5 to the color of the original solution in tube 1. Has the position of equilibrium shifted? In which direction? Why? Compare the color of the solution in tube 5 to the solution in tube 2, to which 1 mL of distilled water was added. Is the color change that you observe in tube 5 simply the change that you would expect after adding the water in the $AgNO_3$ solution, or has a greater change occurred?

2. Effect of temperature. Place tube 6 in a beaker of boiling water. After several minutes, compare the color of the solution in tube 6 to the original solution in tube 1. Explain what has happened.

Place tube 7 in a beaker of ice and water. After several minutes, compare the color of the solution in tube 7 to the color of the original solution in tube 1, which remained at room temperature, and to the color of the solution in tube 6, which was heated. Determine whether the two color changes in tubes 6 and 7 are reversible.

What can you conclude about the effect of temperature changes on the position of equilibrium for the reaction of Equation (21-9)? Is the forward or reverse reaction favored by an increased temperature? Is the forward or reverse reaction favored by a decreased temperature? Which direction of the reaction is endothermic? Which direction of the reaction is exothermic?

## Results

### A. A Reversible Reaction

On the Report Sheet, provide the requested information about the colors of the original solids and solutions and about the color changes that provide visible evidence of the reaction that is occurring. Account for each color change in terms of the forward or reverse direction of the reaction of Equation (21-5).

### B. Relative "Positions" of Equilibrium

1. Provide the requested information about colors that is the basis for comparing the extent of the reactions in Equation (21-8) for the three metal ions. In addition, calculate the concentration of $Cl^-$ that has been added to each solution; this is the total concentration of $Cl^-$ that would be present if none reacted with a metal ion. If no significant color change occurred, meaning that no significant amount of chloro complex formed, calculate the chloride ion concentration that is present after the addition of the entire 10 mL of HCl. As an example of such a calculation, assume that a color change occurs when 2.5 mL of 12M HCl is added to 5.0 mL of the original solution; the total volume is 7.5 mL. In the total solution

$$[Cl^-] = \frac{(12 \text{ moles } Cl^-/1 \text{ L HCl}) \times (0.0025 \text{ L HCl})}{0.0075 \text{ L of solution}} \qquad (21\text{-}10)$$

$$[Cl^-] = 4.0M$$

2. Because it is difficult to assign specific terms to the descriptions of colors, especially when they are mixtures, not all students will have the same answers here. However, your conclusions about positions of equilibrium should be the same, regardless of how you describe the colors of the solutions.

3. In addition to describing your observations and explaining them in terms of the relative positions of equilibrium for the reactions in Equations (21-7) and (21-9), calculate the concentrations of $Cl^-$ and $NCS^-$ that you put into the solutions.

### C. "Shifting the Position" of Equilibrium

Describe the appearance of the solution in each tube. Account for each observation in terms of the effect that each change has on the position of equilibrium of the reaction of Equation (21-9).

Write an equation for the reaction of Equation (21-9) putting the word "heat" on the proper side to indicate its effect on the reaction.

## Question

1. Is it possible to reach any definite conclusion about whether or not $Cr^{3+}$ reacts with $Cl^-$ in aqueous solution to form a chloro complex ion?

   (a) If the reaction is extensive but slow at room temperature, how could you show that this is the case?

   (b) If the reaction is extensive at some temperature other than room temperature, how could you show that this is the case?

   (c) If the reaction is extensive at room temperature but forms a chloro complex ion that is very nearly the same color as the aquo complex, how could you show that this is the case? (Hint: Refer to Experiment 27.)

# EXPERIMENT 21

# REPORT SHEET

Name________________________________________ Date ______________

Section ______________________________________

## A. A Reversible Reaction

$$Cr_2O_7^{2-} + 2OH^- \rightleftharpoons 2CrO_4^{2-} + H_2O \qquad (21\text{-}5)$$

1. Color of $Cr_2O_7^{2-}$ solid and in solution ________________
   Color after addition of __________mL of NaOH ________________
   Explanation of the effect of NaOH:

   Color after addition of $HNO_3$ to above solution ________________
   Explanation of the effect of $HNO_3$:

   Color after addition of NaOH to above solution ________________
   Explanation of the effect of NaOH:

2. Color of $CrO_4^{2-}$ solid and in solution ________________
   Color after addition of ___________mL of $HNO_3$ ________________
   Explanation of the effect of $HNO_3$:

   Color after addition of NaOH to above solution ________________
   Explanation of the effect of NaOH:

   Color after addition of $HNO_3$ to above solution ________________
   Explanation of the effect of $HNO_3$:

## B. Relative "Positions" of Equilibrium

$$M(H_2O)_6^{+n} + yCl^- \rightleftarrows MCl_y^{(+n-y)} + 6H_2O \qquad (21\text{-}8)$$

1.

| | $Co^{2+}$ | $Cr^{3+}$ | $Cu^{2+}$ |
|---|---|---|---|
| Color of aquo complex | ________ | ________ | ________ |
| Does chloro complex form? | ________ | ________ | ________ |
| If so, its color | ________ | ________ | ________ |
| mL of 12M HCl to form complex | ________ | ________ | ________ |
| $[Cl^-]$ to form complex | ________ | ________ | ________ |

Show calculation of $[Cl^-]$ added to $Co(H_2O)_6^{2+}$ when $CoCl_4^{2-}$ color appears.

Most extensive reaction with________________________

Least extensive reaction with________________________

Reaction with largest equilibrium constant______________

Reaction with smallest equilibrium constant_____________

2. Mixture containing $Co(H_2O)_6^{2+}$ and $Cu(H_2O)_6^{2+}$

Original color of solution__________________________

Color after addition of ______ mL of 12M HCl__________

Ion for which the chloro complex is present_____________

Conclusion about relative positions of equilibrium for chloro complexes of $Co^{2+}$ and $Cu^{2+}$:

Does this agree with your conclusions in Part 1?

3. $CoCl_4^{2-}$ and $Co(NCS)_4^{2-}$

Color of $Co(H_2O)_6^{2+}$______________________________

Color after addition of _______ mL of 6M HCl__________

Total [$Cl^-$] of the solution________________________

Color after addition of _______mL of 6M KNCS________

Total [$NCS^-$] of the solution_____________________

For the formation reaction of which complex ion is the position of equilibrium farther right for a particular amount of complexing agent?

For the formation reaction of which complex ion is the equilibrium constant larger?

Compare the results you obtained in this part of the experiment with the results you obtained in Part 1., using 12M HCl as a source of $Cl^-$. Why is the same color change not observed for the same volume of acid? What total [$Cl^-$] was necessary in Part 1. for the formation of $CoCl_4^{2-}$? Would it be possible to convert $Co(H_2O)_6^{2+}$ completely to $CoCl_4^{2-}$ using 6M HCl? Explain your answer.

## C. "Shifting the Position" of Equilibrium

$$Co(H_2O)_6^{2+} + 4NCS^- \rightleftharpoons Co(NCS)_4^{2-} + 6H_2O \qquad (21\text{-}9)$$

1. Effect of concentration of a reactant

Color in tube 1__________________________________

Color in tube 2__________________________________

Color in tube 3__________________________________

Comparison of tubes 1, 2, and 3; explanation of color changes:

Color in tube 4__________________________________

Explanation of color change:

Color in tube 5__________________________________

Comparison of tubes 1, 2, and 5; explanation of color changes:

2. Effect of temperature

Color in tube 6__________________________________

Is the change reversible?________________________

Color in tube 7__________________________________

Is the change reversible?________________________

Is the forward or reverse reaction endothermic?__________

Is the forward or reverse reaction exothermic?___________

Write an equation for the reaction putting "heat" on the proper side.

(Attach a separate sheet with the answers to the Question.)

# EXPERIMENT 22

# SPECTROPHOTOMETRIC DETERMINATION OF AN EQUILIBRIUM CONSTANT

## Introduction

The extent to which a chemical reaction occurs can be expressed in terms of the equilibrium constant for the reaction. This number is characteristic of the reaction at a particular temperature and states the ratio of product concentrations to reactant concentrations (each raised to a power that is the coefficient from the balanced equation for the reaction) in the reaction mixture when chemical equilibrium is attained, that is, when these concentrations no longer change. For example, for a general reaction

$$aA + bB \rightleftharpoons cC + dD \tag{22-1}$$

$$K = \frac{[C]^c\,[D]^d}{[A]^a\,[B]^b} \tag{22-2}$$

The greater the numerical value of $K$, the more extensive the reaction is and the greater the tendency for the reaction to occur.

You have done calculations involving $K$ and concentrations of reactants and products in a reaction mixture. Of course this type of calculation requires that you know the value for an equilibrium constant in the first place. The purpose of this experiment is to acquaint you with one method that can be used to determine the numerical value of an equilibrium constant. In problems in your chemistry textbook, you have calculated equilibrium constants from data stating the number of moles of each reactant and product in a reaction mixture. However, for many reactions it is difficult, or even impossible, to measure these concentrations directly. In this experiment, you will use an indirect method that makes it unnecessary to measure the concentration of each component of an equilibrium mixture.

The reaction you will investigate is the formation of a complex ion in aqueous solution from reactants in aqueous solution:

$$Fe^{3+}(aq) + NCS^{-}(aq) \rightleftharpoons FeNCS^{2+}(aq) \tag{22-3}$$

The concentration ratio you will determine is

$$K = \frac{[FeNCS^{2+}]}{[Fe^{3+}][NCS^{-}]} \tag{22-4}$$

You will need to know the total concentration of $Fe^{3+}$ and the total concentration of $NCS^{-}$ put into the solution. However, you will not need to know how much of either of these materials is in the form of the complex ion or how much remains uncomplexed. Nor will you ever need to know the actual concentration of $FeNCS^{2+}$ in the solution. Instead, the only piece of experimental data that you must determine is the *absorbance* $A$ of $FeNCS^{2+}$ in a solution prepared from known amounts of $Fe^{3+}$ and $NCS^-$.

*Absorbance* is a measure of the amount of light absorbed by a colored material. Its numerical value is directly related to the concentration of the absorbing material in solution $c$, to the depth or thickness of the sample through which light passes $l$, and to an intrinsic physical property of the absorbing material called its molar absorptivity $\varepsilon$:

$$A = \varepsilon c l \qquad (22\text{-}5)$$

In this experiment, you will not need to know the values of any of these terms except $A$; however, the symbols $\varepsilon$, $c$, and $l$ will be used in Treatment of Data.

## Technique

You will begin with an accurately known volume of a solution that contains an accurately known concentration of KNCS. To this solution, you will add an accurately measured volume of a solution that contains an accurately known concentration of $Fe(NO_3)_3$. You will measure the absorbance of the solution at a wavelength of visible light at which $FeNCS^{2+}$ absorbs strongly. Then you will add more $Fe(NO_3)_3$ and measure the absorbance of this new solution. The increased amount of $Fe^{3+}$ in the solution will cause the position of equilibrium for the reaction of Equation (22-3) to shift to the right; because more $FeNCS^{2+}$ is present in the second solution, the absorbance will increase. This process is repeated for increasing amounts of $Fe^{3+}$. Before you begin your work, your instructor will provide directions for the spectrophotometer that you will use to measure absorbance.

## Review

Equipment and Techniques G
Experiment 27
Appendix III

## Equipment Needed

beakers
cuvettes or test tubes (2)
1-mL pipet
10-mL pipet
spectrophotometer
100-mL volumetric flask

## Chemicals Needed

0.1M $Fe(NO_3)_3$ in 0.5M $HNO_3$; iron(III) nitrate in nitric acid
0.5 $HNO_3$; nitric acid
0.002M KNCS; potassium thiocyanate

## Procedure

Prepare a $2 \times 10^{-4}$M solution of KNCS in 0.5M $HNO_3$ in the following manner. Clean and rinse a 100-mL volumetric flask; then rinse it with three 5-mL portions of 0.5M $HNO_3$. Clean and rinse a 10-mL pipet giving it a final rinse with the $2 \times 10^{-3}$M KNCS that is provided. Pipet 10.00

mL of 2 x $10^{-3}$M KNCS into the volumetric flask, and add 0.5M $HNO_3$ to the 100-mL calibration mark on the neck of the flask. Invert the stoppered flask several times to ensure that the solution is homogeneous. Be sure to record the exact molarity of the KNCS solution that is provided.

In a small clean beaker obtain about 20 mL of the 0.1M solution of $Fe(NO_3)_3$ in 0.5M $HNO_3$ that is provided. If the beaker is not completely dry, rinse it several times with a few milliliters of the solution, so that the 20 mL of solution will not be diluted by any water remaining in the beaker. Clean and rinse a 1-mL pipet and then rinse it with the $Fe(NO_3)_3$ solution. Be sure to record the exact molarity of the $Fe(NO_3)_3$ solution.

Pour *all* the KNCS solution that you prepared into a clean, *dry*, 400-mL beaker. Assume that the volume of solution transferred is 0.100 L. Into this solution, pipet 1.00 mL of 0.1M $Fe(NO_3)_3$ solution. Gently swirl the beaker to ensure thorough mixing. Throughout the experiment, it is essential not to lose any solution, because each measurement and calculation in a series of eight is based on the assumption that all of the original solution has been retained.

Adjust the wavelength control of the spectrophotometer to 447 nm. If wavelengths are marked every 5 nm on your instrument, rather than every 1 nm, set the wavelength control at 445 to be sure of a reproducible setting; if wavelengths are marked only every 10 nm, set the control at 450 nm. Adjust the zero control. Rinse a clean sample cuvette or test tube several times with 0.5M $HNO_3$ and then fill it with 0.5M $HNO_3$. Place this tube in the sample holder and adjust the light control to "0" absorbance to ensure that the absorbance you measure is due only to reaction components and not due to solvent. It is assumed that you will share a spectrophotometer with other students. Before you use the instrument each time, check the wavelength setting and the zero setting; then use the tube of 0.5M $HNO_3$ to check the light control. The settings might drift or you or another student might accidentally bump the knobs and change the settings.

Into a clean and *dry* matching sample tube, pour some of the KNCS–$Fe(NO_3)_3$ solution; measure its absorbance. Pour this solution back into the beaker and retain the sample tube.

Pipet another 1.00 mL of $Fe(NO_3)_3$ into the solution; mix thoroughly by swirling the beaker gently. Pour some of this solution into the sample tube and then pour it back into the beaker; repeat this process three more times washing the solution over the entire inside wall of the tube to ensure that all of the solution has the same concentration of components. Fill the tube using the same volume of solution as you used for your first measurement; measure the absorbance. Pour the solution back into the beaker and retain the tube.

Repeat this procedure, adding 1.00 mL of $Fe(NO_3)_3$ at a time until a total of eight mL has been added. It is important not to lose any solution. If any solution ever drips down the side of the tube during pouring, rinse *only* the *outside* of the tube with distilled water and dry the tube so that no solid residue can remain to affect the absorbance; do not add this wash water to the solution.

Record the temperature in the laboratory.

When you complete your work, all the solutions may be washed down the drain.

## Treatment of the Data

You are trying to determine a value for the ratio

$$K = \frac{[FeNCS^{2+}]}{[Fe^{3+}][NCS^{-}]} \tag{22-4}$$

A way must be found to relate these three unknown concentrations to known quantities.

The concentration of $FeNCS^{2+}$ can be related to your one experimental measurement, absorbance $A$:

$$A = \varepsilon c l = \varepsilon l[FeNCS^{2+}] \tag{22-6}$$

Rearranging this equation gives

$$[FeNCS^{2+}] = A/\varepsilon l \tag{22-7}$$

You will know the total concentration of $NCS^-$, $T_n$, put into solution, but you will not know how it is divided between uncomplexed and complexed $NCS^-$.

$$T_n = [NCS^-]_{free} + [FeNCS^{2+}] \tag{22-8}$$

Rearranging this equation gives

$$[NCS^-]_{free} = T_n - [FeNCS^{2+}] \tag{22-9}$$

Substituting Equation (22-7) into Equation (22-9) gives

$$[NCS^-]_{free} = T_n - A/\varepsilon l \tag{22-10}$$

You will know the total concentration of $Fe^{3+}$, $T_f$, put into solution, but you will not know how it is divided between uncomplexed and complexed $Fe^{3+}$.

$$T_f = [Fe^{3+}]_{free} + [FeNCS^{2+}] \tag{22-11}$$

Rearranging this equation gives

$$[Fe^{3+}]_{free} = T_f - [FeNCS^{2+}] \tag{22-12}$$

Substituting Equation (22-7) into Equation (22-12) gives

$$[Fe^{3+}]_{free} = T_f - A/\varepsilon l \tag{22-13}$$

When Equations (22-7), (22-10), and (22-13) are substituted into Equation (22-4), the new expression for the concentration ratio becomes

$$K = \frac{A/\varepsilon l}{(T_f - A/\varepsilon l)(T_n - A/\varepsilon l)} \tag{22-14}$$

Multiplying by the denominator gives

$$KT_fT_n - KT_nA/\varepsilon l - KT_fA/\varepsilon l + K(A/\varepsilon l)^2 = A/\varepsilon l \tag{22-15}$$

These terms can be rearranged and combined into a more familiar form

$$K(A/\varepsilon l)^2 - (KT_n + KT_f + 1)\,(A/\varepsilon l) + KT_fT_n = 0 \tag{22-16}$$

which you should recognize as a quadratic equation of the general form

$$ax^2 + bx + c = 0 \tag{22-17}$$

In this case, the term $A/\varepsilon l$ is the $x$ in Equation (22-17). Quadratic equations with numerical coefficients are simple, although sometimes tedious, to solve. Here, the complexity of the coefficients makes the equation quite difficult to solve. An approximate solution is

$$A/\varepsilon l = \frac{T_fT_n}{T_f + T_n + 1/K} \tag{22-18}$$

A stepwise rearrangement of this equation puts it into the form that you will use:

$$\frac{T_f + T_n + 1/K}{\varepsilon l} = \frac{T_f T_n}{A} \qquad (22\text{-}19)$$

$$\frac{T_f + T_n}{\varepsilon l} + \frac{1/K}{\varepsilon l} = \frac{T_f T_n}{A} \qquad (22\text{-}20)$$

$$\frac{T_f + T_n}{\varepsilon l} + \frac{1}{\varepsilon l K} = \frac{T_f T_n}{A} \qquad (22\text{-}21)$$

The final form that we want for this equation is

$$\frac{T_f T_n}{A} = (1/\varepsilon l)(T_f + T_n) + (1/\varepsilon l K) \qquad (22\text{-}22)$$

This equation is now in the form of the general equation for a straight line

$$y = mx + b \qquad (22\text{-}23)$$

If $(T_f T_n/A)$ is plotted along the $y$ axis and $(T_f + T_n)$ is plotted along the $x$ axis, a straight line results. Its slope is $(1/\varepsilon l)$, and its $y$ intercept is $(1/\varepsilon l K)$. To obtain a value of $K$, the numerical value of the slope is divided by the numerical value of the intercept

$$\frac{\text{slope}}{\text{intercept}} = \frac{1/\varepsilon l}{1/\varepsilon l K} = K \qquad (22\text{-}24)$$

Thus, the value of $K$ can be determined by using two pieces of easily calculated data and one simple experimental measurement, without ever knowing the equilibrium concentrations of the components of the mixture.

## Calculations

For each of the eight mixtures, calculate the total concentration of $NCS^-$ on the basis of the dilution that has occurred

$$T_n = \frac{(\text{M of NCS}^- \text{ in stock solution}) \times (V \text{ of stock solution})}{V \text{ of total solution}} \qquad (22\text{-}25)$$

For each mixture, calculate the total concentration of $Fe^{3+}$ added to the solution

$$T_f = \frac{(\text{M of Fe}^{3+} \text{ in stock solution}) \times (V \text{ of stock solution})}{V \text{ of total solution}} \qquad (22\text{-}26)$$

In each calculation of concentration, be sure to retain the proper number of significant figures as given on the stock solutions of KNCS and $Fe(NO_3)_3$.

For each mixture, calculate the terms $T_f T_n/A$ and $(T_f + T_n)$ that are to be graphed. Be careful in placing the decimal points!

Plot $T_f T_n/A$ on the $y$ axis vs. $(T_f + T_n)$ on the $x$ axis. Use scales that enable you to plot data to three significant figures. Draw the best possible straight line through your plotted points. Determine the slope and the $y$ intercept of the straight line. You are to *read* the $y$ intercept *from the graph*; therefore, the $x$ axis must extend to zero. If your first graph does not have a scale on the $y$ axis that allows you to read the $y$ intercept, make a new graph with the $y$ axis extending to lower values. Calculate $K$ to three significant figures; $K$ = slope/intercept.

## Questions

1. Use your value of the equilibrium constant to calculate the concentration of each component in the eighth mixture. You will be solving a quadratic equation.

2. Use your answers to Question 1. to calculate the fraction of the total $NCS^-$ that is in the complex ion and the fraction of the total $Fe^{3+}$ that is in the complex ion.

# EXPERIMENT 22

## REPORT SHEET

Name________________________________ Date ____________

Section ______________________________

Temperature________________________________

Molarity of stock KNCS________________________

Molarity of stock $Fe(NO_3)_3$________________________

| | $T_f$ | $T_n$ | A | $T_fT_n$ | $T_fT_n/A$ | $T_f + T_n$ |
|---|---|---|---|---|---|---|
| 1. | | | | | | |
| 2. | | | | | | |
| 3. | | | | | | |
| 4. | | | | | | |
| 5. | | | | | | |
| 6. | | | | | | |
| 7. | | | | | | |
| 8. | | | | | | |

Show the calculation of $T_f$ for mixture 8.

Show the calculation of $T_n$ for mixture 8.

Attach your graph of $T_fT_n/A$ vs. $(T_f + T_n)$ to the Report Sheet.

Show the calculation of the slope of the straight line.

Give the value of the $y$ intercept of the straight line.

$K$ = ______________________________

Show the calculation of $K$.

Answer the Questions. Attach an extra sheet if you need more space.

# EXPERIMENT 23

# ANTACIDS

## Introduction

An antacid is a substance that prevents or neutralizes acidity. Most commonly, the term refers to a medication (usually nonprescription) used to neutralize acid in the stomach. Part of the digestion process takes place in the stomach where food is acted on by gastric juices that contain digestive enzymes and hydrochloric acid. Gastric juice is quite acidic, with pH about 1, but stomach contents consisting of a mixture of gastric juice and food are less acidic, with pH up to about 4. Hence, an "acid stomach" is normal, because the acidity is necessary for digestion to occur. However, under some circumstances such as emotional stress, a person might have an unusually high concentration of acid in the stomach. Highly acidic or spicy foods might also cause a feeling of abnormal acidity. The terms "excess stomach acid", "acid indigestion", "hyperacidity", "sour stomach", and "heartburn" are all used to describe these conditions.

Antacids contain compounds that are bases and will neutralize acid; these compounds either contain hydroxide ions or can react with water producing hydroxide ions. Usually these are compounds that have a low solubility in water; they dissolve more extensively in acid and neutralize the acid as they dissolve in it. Antacids can be purchased as tablets consisting of acid-neutralizing compounds, nonreactive "fillers", and flavoring agents pressed together, or as a water suspension of the ingredients. Common active ingredients are $Mg(OH)_2$ (magnesium hydroxide), $Al(OH)_3$ (aluminum hydroxide), $CaCO_3$ (calcium carbonate), $MgCO_3$ (magnesium carbonate), and $NH_2CH_2CO_2H$ (glycine). Just one or several of these ingredients might be present in a particular antacid.

## Technique

The antacids used in this experiment are tablets that can be purchased without a prescription. You will not know the brand name of the antacid unless you recognize identifying marks on the tablets and you will not know what acid-neutralizing compound(s) it contains. These antacids contain compounds that are only slightly soluble in water combined with fillers that are insoluble in both water and dilute acid solutions.

Because the stomach pH is normally about 3 to 4, you will determine how much "excess" acid a tablet can neutralize, that is, how much acid a tablet can neutralize while maintaining the pH around 3 to 4. To do this, you will treat a tablet with a quantity of HCl that is more than enough to dissolve the tablet and react completely with the acid-neutralizing compounds it contains. Then you will titrate the remaining HCl with NaOH. The acid-base indicator you will use is a mixture of methyl orange and xylene cyanole; it is purple at $pH < 3.2$ and green at $pH > 4.2$. By knowing the number of moles of HCl used, and the number of moles of NaOH required for the titration, you can calculate the moles of $OH^-$ provided by the antacid. Then you will convert this to a weight of hydroxide per tablet.

## Review

Equipment and Techniques G, H
Experiment 16

## Equipment Needed

analytical balance
burets (2)
clamps
Erlenmeyer flasks
microburner
ring
ring stand
wire gauze

## Chemicals Needed

antacids
standardized 1M HCl; hydrochloric acid
methyl orange–xylene cyanole indicator solution
standardized 0.5M NaOH; sodium hydroxide

## Procedure

To determine the hydroxide content of the antacid you will use three one-tablet samples. Weigh each tablet on the analytical balance by difference from the bottle containing the tablets and put it into an Erlenmeyer flask. This means that you weigh the bottle without the lid, pour a tablet into a flask, and weigh the bottle again; the difference between the two weights is the weight of the tablet. Before you weigh the tablets, put an identifying number on each flask; record this number in your notebook as you put a tablet into the flask. Weigh all three tablets when you begin the experiment instead of completing the analysis of one tablet before you weigh another. When you return to your desk, immediately put a few milliliters of distilled water into each flask to begin the disintegration and dissolution of the tablet.

From a buret measure about 15 mL of standardized 1M HCl into one flask. Be sure to take the initial buret reading, but do not take a final buret reading yet. Swirl the flask to speed the dissolving. It might help to crush the tablet with a glass rod (not a metal spatula); be sure to rinse all solution and solid particles off the rod and into the flask when you remove the rod. Ten to fifteen minutes might be required for complete dissolution of the active compound. However, undissolved solid might still be present; an inert filler might remain as a uniformly fine powder. It is essential to dissolve the antacid compound completely, because any hydroxide-producing compound that does not dissolve cannot be detected in the analysis. This might be the most time-consuming part of the experiment.

If the tablets you are analyzing contain carbonate ions, they will react with HCl producing $CO_2$

$$2H_3O^+(aq) + CO_3^{2-}(aq) \rightarrow CO_2(g) + 3H_2O \qquad (23\text{-}1)$$

As a tablet dissolves, it will effervesce. After the effervescence subsides, any $CO_2$ that remains in solution must be removed before you begin the titration, for it would react with NaOH.

$$CO_2(aq) + OH^-(aq) \rightarrow HCO_3^-(aq) \qquad (23\text{-}2)$$

To remove the $CO_2$, heat the mixture *almost* to boiling using a microburner; be careful not to let any solution spatter out of the flask. Continue to heat the mixture *gently* for about five minutes; then cool the flask to room temperature under cold tap water.

When you are sure that dissolution is as complete as possible, add four drops of methyl orange–xylene cyanole indicator to the mixture. This indicator is orange in a very acidic solution, and so the color that you see might be either orange or purple. So that you will know what color to expect at the end point in the titration, put about 20 mL of distilled water into a flask and add two drops of the indicator to it; the color should be the green that is characteristic of pH > 4.2.

From a buret rapidly add a stream of standardized 0.5M NaOH to the antacid mixture until the solution has the green color that is characteristic of pH > 4.2. This mixture will not exactly match the solution of water and indicator because this mixture might have undissolved material in it that makes the color appear lighter. Just be sure to add NaOH until the color is definitely green without any hint of purple or gray; intentionally go *past* the end point. Then rapidly add HCl dropwise until the solution has the purple color characteristic of ph < 3.2. Then carefully titrate the solution with NaOH just to the green end point; as you approach the end point, the color will fade from purple through gray to green. If you think you have passed the end point, add another drop or two of HCl and again titrate the solution with NaOH just until you reach the green end point. This kind of titration, called *back titration,* is often faster than the ones in other experiments in this manual; it can also be more accurate.

During the titration, material might precipitate from solution when you go past the end point. This material will redissolve when you add more acid and it will not be present when you titrate just to the end point.

After you complete the titration, read the NaOH buret and calculate the volume of NaOH that you used. If it is less than 10.00 mL, you will have only three figures in your calculations instead of the possible four figures. Figure out how much more HCl solution you need to require at least 10.00 mL of NaOH solution; keep in mind that the HCl solution is twice as concentrated as the NaOH solution. Add the necessary amount of HCl and titrate with NaOH. Record the final readings for both burets. Be sure to record the molarities of HCl and NaOH as they are given on the bottles.

Titrate the other two samples. After you complete the titrations the mixtures may be washed down the drain. Any unused HCl and NaOH may also be washed down the drain.

## Calculations

For each of the titrations with an antacid tablet

(a) Use the concentration and volume of HCl solution to calculate the number of moles of $H_3O^+$ used.

(b) Use the concentration and volume of NaOH solution to calculate the number of moles of $OH^-$ used.

(c) How many moles of $H_3O^+$ were neutralized by the NaOH solution?

(d) The rest of the $H_3O^+$ was neutralized by the antacid. Calculate the number of moles of $OH^-$ provided by the antacid.

(e) Convert moles of $OH^-$ from the antacid to weight of $OH^-$ from the antacid.

(f) Determine the percentage by weight of the antacid tablet that is $OH^-$.

(g) Report an average percentage of $OH^-$.

## Question

1. How will the calculated percentage of hydroxide be affected (high, low, no change) by each of the following? Explain each answer.

   (a) Three acid-neutralizing compounds, rather than just one, are present in the antacid.

   (b) Some of the acid-neutralizing compound remains undissolved.

   (c) The $CO_2$ produced by the reaction of a carbonate compound with acid is not fully removed.

   (d) You incorrectly record your initial buret reading for the HCl as 0.03 mL rather than 0.30 mL.

   (e) You incorrectly record your initial buret reading for the NaOH as 0.03 mL rather than 0.30 mL.

   (f) You go past the end point in the titration.

# EXPERIMENT 23

## REPORT SHEET

Name______________________________________ Date ______________

Section____________________________________

Unknown identification code_________________

| | Trial 1 | Trial 2 | Trial 3 |
|---|---|---|---|
| 1. Weight of tablet | ________ | ________ | ________ |
| 2. Volume of HCl used | ________ | ________ | ________ |
| 3. M of HCl | ________ | ________ | ________ |
| 4. Moles of HCl used | ________ | ________ | ________ |
| 5. Volume of NaOH used | ________ | ________ | ________ |
| 6. M of NaOH | ________ | ________ | ________ |
| 7. Moles of NaOH used | ________ | ________ | ________ |
| 8. Moles of HCl neutralized by NaOH | ________ | ________ | ________ |
| 9. Moles of HCl neutralized by antacid | ________ | ________ | ________ |
| 10. Moles $OH^-$ from tablet | ________ | ________ | ________ |
| 11. Weight of $OH^-$ in tablet | ________ | ________ | ________ |
| 12. % by weight of $OH^-$ in tablet | ________ | ________ | ________ |
| 13. Average weight % $OH^-$ | ________ | ________ | ________ |

On the back of the Report Sheet show the calculation of #4, 7, 9, 11, and 12.

(On the back of the Report Sheet, answer the Question.)

# EXPERIMENT 24
# ACIDIC AND BASIC SALTS

## Introduction

Water makes up about 60% of the weight of a human body; other animals contain a similar proportion of water, and some plants are almost 95% water. Water covers about 70% of the earth's surface; also, it is in the soil and is present as vapor in the atmosphere. Consequently, reactions that occur in aqueous solutions and reactions in which water participates are very important to us; many are essential to the life and growth of organisms. These reactions might affect the pH of water by using or producing hydronium or hydroxide ions.

It is apparent from the formulas of some compounds that they will produce $H_3O^+$ or $OH^-$ in aqueous solution. An Arrhenius acid contains hydrogen, which is usually written first in the formula to emphasize its ability to react with water producing $H_3O^+$. An ionic Arrhenius base contains hydroxide ions, which become separated from the positive ions when the base dissolves in water.

Salts are ionic compounds that contain a cation from a base and an anion from an acid. Even though most salts do not contain H or OH, many of them do react with water changing its pH. Such a reaction is called the *hydrolysis of a salt*, the reaction of an ion of a dissolved salt with water producing $H_3O^+$ or $OH^-$.

The hydrolysis of a negative ion produces the negative ion of water. This reaction occurs if the anion is the conjugate base of a weak acid. For example, HCN is a weak acid; $K_a = 4.9 \times 10^{-10}$:

$$HCN + H_2O \rightleftharpoons H_3O^+ + CN^- \tag{24-1}$$

The acid molecule HCN has little tendency to donate protons; its conjugate base, $CN^-$, is a good proton acceptor. Consequently, when a salt containing $CN^-$, such as NaCN, dissolves in water, a Bronsted-Lowry acid–base reaction occurs between $H_2O$ (acid) and $CN^-$ (base); that is, a proton is transferred from $H_2O$ to $CN^-$:

$$CN^- + H_2O \rightleftharpoons HCN + OH^- \tag{24-2}$$

The resulting solution is basic, because hydroxide ions in excess of those produced by the autoionization of water are present.

The hydrolysis of a positive ion produces the positive ion of water. This reaction occurs if the cation is the conjugate acid of a weak base. Molecular bases such as $NH_3$ are weak bases:

$$NH_3 + H_2O \rightleftharpoons NH_4^+ + OH^- \tag{24-3}$$

For this reaction, $K_b = 1.8 \times 10^{-5}$. The ammonium ion can donate $H^+$ to a chemical species that is capable of acting as a base. Consequently, when a salt containing $NH_4^+$ dissolves in water, $NH_4^+$ and $H_2O$ react in a Bronsted-Lowry fashion:

$$NH_4^+ + H_2O \rightleftharpoons NH_3 + H_3O^+ \tag{24-4}$$

Some hydrated metal ions also hydrolyze producing acidic solutions. The strong attraction between a small, perhaps highly charged, metal ion and the oxygen atom of a bound water molecule weakens an oxygen–hydrogen bond in the bound water molecule. As a result, the hydrate acts as a proton donor:

$$[M(H_2O)_x]^{-n} + H_2O \rightleftharpoons [M(H_2O)_{(x-1)}(OH)]^{+n-1} + H_3O^+ \tag{24-5}$$

The extent to which a hydrolysis reaction occurs is inversely related to the extent to which the parent acid or base reacts with water, that is, the weaker the acid or base, the stronger the conjugate base or acid. The constant describing the equilibrium concentration ratio is call the hydrolysis constant, $K_h$. For the hydrolysis of an anion $A^-$

$$K_h = \frac{[HA][OH^-]}{[A^-]} = \frac{K_w}{K_a} \tag{24-6}$$

where $K_w$ is the water ionization constant. For the hydrolysis of a cation $BH^+$

$$K_h = \frac{[H_3O^+][B]}{[BH^+]} = \frac{K_w}{K_b} \tag{24-7}$$

In the first part of this experiment you will measure the pH of aqueous solutions and account for the pH in terms of the reaction of water and solute.

The ability of ions from salts to behave as acids or bases is utilized in *buffer solutions*. A buffer solution can consist of a weak base and a salt of a weak base, such as a mixture of $NH_3$ and $NH_4Cl$. The $NH_3$ is a proton acceptor and can neutralize added acid by the reaction

$$NH_3 + H_3O^+ \rightleftharpoons NH_4^+ + H_2O \tag{24-8}$$

The $NH_4^+$ is a proton donor and can neutralize added base by the reaction

$$NH_4^+ + OH^- \rightleftharpoons NH_3 + H_2O \tag{24-9}$$

Or a buffer solution can consist of a weak acid and a salt of a weak acid, such as a mixture of acetic acid and sodium acetate; $HC_2H_3O_2$ neutralizes added base, and $C_2H_3O_2^-$ neutralizes added acid. In the second part of this experiment you will work with a buffer solution.

In the third part of this experiment, you will use the acidic nature of a cation to determine the molar weight of a salt in which it is present. You will dissolve a 1:1 electrolyte containing $NH_4^+$ in a carefully measured volume of standardized NaOH. The reaction of Equation (24-9) will occur; the $NH_3$ can be boiled out of the solution. The $OH^-$, which is not volatile and remains in solution, will be titrated with HCl. You will know the number of moles of $OH^-$ put into the solution from NaOH and the number of moles of $OH^-$ that remained to react with HCl after the reaction in Equation (24-9) occurred. The difference between these two quantities is the number of moles of $OH^-$ that reacted with $NH_4^+$ according to Equation (24-9); it equals the number of moles of $NH_4^+$ in the salt sample. Knowing the weight of the salt sample and the number of moles that it contains, you can calculate the molar weight of the salt.

## Review

Equipment and Techniques G, H
Experiment 16

## Equipment Needed

- beakers
- buret
- clamp
- microburner
- pH paper or sticks
- 20-mL pipet
- ring
- ring stand
- stirring rod
- test tubes
- watch glass
- wire gauze

## Chemicals Needed

### For Part A, Hydrolysis and pH

- 0.1M $Fe(NO_3)_3$; iron(III) nitrate
- 0.1M $HC_2H_3O_2$; acetic acid
- 0.1M HCl; hydrochloric acid
- 0.1M $NH_3$; ammonia
- 0.1M $NH_4C_2H_3O_2$; ammonium acetate
- 0.1M $NH_4Cl$; ammonium chloride
- 0.1M $NaC_2H_3O_2$; sodium acetate
- 0.1M NaCl; sodium chloride
- 0.1M $NaHCO_3$; sodium hydrogen carbonate
- 0.1M $NaH_2PO_4$; sodium dihydrogen phosphate
- 0.1M $Na_2HPO_4$; sodium hydrogen phosphate
- 0.1M NaOH; sodium hydroxide
- 1.0M solutions of unknown hydrolyzable salts

### For Part B, A Buffer Solution

- bromothymol blue solution
- 1.0M $HC_2H_3O_2$; acetic acid
- 1.0M HCl; hydrochloric acid
- $NaC_2H_3O_2$; sodium acetate
- 1.0M $NaC_2H_3O_2$
- 1.0M NaOH; sodium hydroxide

### For Part C, Molar Weight of an Ammonium Salt

- standardized 0.5M HCl; hydrochloric acid
- standardized 0.5M NaOH; sodium hydroxide
- thymol blue solution
- unknown salts

## Procedure

### A. Hydrolysis and pH

1. Known acids, bases, and salts. Use pH paper or sticks to measure the pH of 0.1M solutions of $Fe(NO_3)_3$, $HC_2H_3O_2$, HCl, $NH_3$, $NH_4C_2H_3O_2$, $NH_4Cl$, $NaC_2H_3O_2$, NaCl, $NaHCO_3$, $NaH_2PO_4$, $Na_2HPO_4$, and NaOH. Also measure the pH of distilled water; do not assume that it is exactly 7. If the pH of distilled water is not 7, then the pH of a solution can be different from 7 even if no acid–base reaction occurs between water and the solute.

Put a few drops of a solution into a 4-mL test tube. Place a piece of wide-range pH paper on a clean watch glass, dip a clean glass stirring rod into the solution, and transfer a drop of the solution onto the paper. (Do not dip the paper into the solution, for the indicators could be washed out of the paper leaving a nondescript, or no, color.) After determining the pH of the solution within one or two units, select the proper narrow-range pH paper and determine the pH of the solution within a few tenths of one pH unit. If necessary, use an educated imagination about color to make approximations between the colors on the comparison charts. If the color seems to fade, determine the pH on the basis of the first color that appears when you put the solution on the paper.

If you use plastic pH sticks instead of pH paper, the sticks can be dipped into the solutions without losing their colors. Be sure the entire colored portion of the stick is dipped into the solution.

2. An unknown salt. Measure the pH of a 1.0M solution of an unknown salt in which one of the ions hydrolyzes. The salt will contain only one hydrolyzable ion per formula unit. That is, the formula will be of the type $B_1X_x$ if $B^{x+}$ hydrolyzes. The formula will be of the type $X_xA_1$ if $A^{x-}$ hydrolyzes. The salt does not contain any hydrolyzable ion that you worked with in the first section of this part of the experiment.

When you complete this part of the experiment, all the solutions may be washed down the drain.

### B. A Buffer Solution

Thymol blue is an acid–base indicator that changes from red to yellow over the pH range 1.2-2.8 and from yellow to blue over the pH range 8.0-9.6. To see the colors characteristic of each form of thymol blue, put 5 mL of distilled water into each of three 20-mL test tubes and add 5 drops of thymol blue to each test tube. To one tube, add 1.0M HCl dropwise until the color of the solution is the light red that is characteristic of ph < 1.2. To the third tube, add 1.0M NaOH dropwise until the blue color that is characteristic of pH> 9.6 appears. Save these three tubes to make color comparisons later in the experiment.

1. In a small beaker, mix 10.0 mL of 1.0M $HC_2H_3O_2$ and 10.0 mL of 1.0M $NaC_2H_3O_2$. This gives a buffer solution having 0.50M of the acid and 0.50M of the salt. Divide this solution evenly among four 20-mL test tubes. To each 5.0-mL sample of the buffer solution add two drops of thymol blue. The color should be yellow, showing that the pH of the buffer solution is between 2.8 and 8.0. *Before you come to the laboratory*, calculate the pH of this solution.

2. To a 5.0-mL sample of the buffer solution add 1.0M NaOH dropwise until the color shows that the solution has become strongly basic; count the drops as you add them and record the number.

3. To a second 5.0-mL sample of the buffer solution add 1.0M HCl until the color shows that the solution has become strongly acidic; record the number of drops.

4. Combine 5.0 mL of 1.0M $HC_2H_3O_2$ and 5.0 mL of distilled water to give 0.50M $HC_2H_3O_2$. Divide this solution evenly between two 20-mL test tubes and add two drops of thymol blue to each test tube. *Before you come to the laboratory* calculate the pH of this solution. Note the indicator color in this solution and account for this color in terms of the pH of the solution. To show that the base-neutralizing capacity of the buffer is comparable to the base-neutralizing capacity of a solution containing only the same concentration of acid, add 1.0M NaOH dropwise to 5.0 mL of 0.50M $HC_2H_3O_2$ until the color of the solution is blue; record the number of drops.

5. To show that the acid-neutralizing capacity of the buffer results from $C_2H_3O_2^-$, an ion that can act as a proton acceptor, prepare 10.0 mL of 0.50M $NaC_2H_3O_2$ by diluting 1.0M $NaC_2H_3O_2$, divide this solution evenly between two 20-mL test tubes, and add two drops of thymol blue to each test tube. *Before you come to the laboratory*, calculate the pH of this solution. Note the color of the indicator in this solution and account for this color in terms of the pH. Then add 1.0M HCl dropwise to one test tube until the color of the solution changes to light red; record the number of drops.

6. To the third test tube containing 0.50M $HC_2H_3O_2$ and 0.50M $NaC_2H_3O_2$, add 1.0M HCl until the solution is strongly acid. Then add a few crystals of solid $NaC_2H_3O_2$. What happens to the color? Why?

When you complete this part of the experiment, all the solutions may be washed down the drain.

### C. Molar Weight of an Ammonium Salt

On the analytical balance, weigh out 0.1-0.3 g of your unknown ammonium salt into a small beaker or crucible. Quantitatively transfer this sample to an Erlenmeyer flask; use a small amount of distilled water from a wash bottle to complete the transfer. Pipet 20.00 mL of standardized 0.5M NaOH into the flask.

Heat the solution to boiling using a microburner; maintain a *gentle* boil for about 10 minutes to remove the ammonia that forms by the reaction of Equation (23-9). To determine whether ammonia is still present, hold a piece of moist red litmus paper over the neck of the flask. When the vapors no longer test basic, you may discontinue the heating. Cool the flask under cold tap water. If there is any solid material on the inside wall of the flask, use distilled water to wash it into the solution.

To the solution in the flask add four drops of bromothymol blue indicator solution. This acid–base indicator changes from yellow to blue over the pH range 6.0-7.6; hence, it will be blue in your original solution and will change to yellow when all the $OH^-$ has been neutralized. Place a piece of white paper under the flask. Titrate the solution with standardized 0.5M HCl to the yellow end point.

In this same manner, weigh out, treat with NaOH, heat, and titrate two more samples of your unknown ammonium salt. When you complete the titrations, the solutions and any unused NaOH and HCl solutions may all be washed down the drain.

## Results

### A. Hydrolysis and pH

1. Known acids, bases, and salts. List the solutions from most acidic to most basic; state the pH of each solution. State whether each compound is an Arrhenius acid, an Arrhenius base, or a salt. For each solution, write an equation for any reaction that occurs and affects pH. For a particular salt, more than one proton-transfer reaction might be possible; if this is the case, show all possible reactions and state which reaction is more extensive as indicated by the pH.

2. An unknown salt. Use the pH of the 1.0M solution to state whether it is the cation (Bronsted-Lowry acid) or the anion (Bronsted-Lowry base) of the salt that hydrolyzes. Use the pH to calculate the $[H_3O^+]$ or $[OH^-]$ that results from hydrolysis. Calculate $K_h$ for the hydrolyzable ion. State whether this hydrolysis constant is $K_a$ or $K_b$ for a Bronsted-Lowry conjugate acid and base pair.

### B. A Buffer Solution

State the pH of each solution and describe the color of the indicator in each solution: buffer solution, acid solution, and salt solution.

For each solution, state the quantity of acid or base that is required to cause the requested change in pH. Compare the acid- and base-neutralizing capacities of the solutions.

### C. Molar Weight of an Ammonium Salt

Calculate the number of moles of NaOH that you pipetted into the flask. Calculate the number of moles of HCl used in the titration, which equals the number of moles of NaOH that HCl neutralized. The difference between these two quantities is the number of moles of NaOH that reacted with $NH_4^+$ according to Equation (24-9); it equals the number of moles of $NH_4^+$ in the salt sample. Use the weight of the salt sample and the number of moles of $NH_4^+$ that it contains to determine the molar weight of the salt $NH_4X$.

If the results of your three trials agree within 1%, report the average of your results for the three trials. If they do not agree within 1%, report an average of the two trials with the closest agreement, or of the two trials that you think you did with the least error.

## Questions

1. Some compounds that are used as fertilizers are $NH_3$, $NH_4NO_3$, and $(NH_4)_3PO_4$. What effect would each of these compounds have on the pH of ground water (that is, water that is in the soil and in which the compound would dissolve)?

2. Some compounds that are used for the specific purpose of altering soil pH are $CaCO_3$, $FeSO_4$, and $Al_2(SO_4)_3$. What effect would each of these compounds have on the pH of ground water?

3. Equimolar amounts of NaOH and $HC_2H_3O_2$ are combined in aqueous solution. Will the resulting solution be neutral? Explain.

4. Describe a way to prepare an acetic acid/sodium acetate buffer solution that has an acid-neutralizing capacity twice as great as its base-neutralizing capacity.

5. Describe a way to prepare an acetic acid/sodium acetate buffer solution that has a base-neutralizing capacity three times as great as its acid-neutralizing capacity.

# EXPERIMENT 24

# REPORT SHEET

Name____________________________________ Date _____________

Section____________________________________

## A. Hydrolysis and pH

1. Known acids, bases, and salts

   Attach a separate sheet of lined paper with your results and equations.

2. Unknown salt

   pH = ____________________________________________

   Does cation or anion hydrolyze?___________________________

   $[H_3O^+]$ or $[OH^-]$ resulting from hydrolysis =__________________

   Show the complete calculation.

   $K_h$ = ____________________________

   Show a general chemical equation for the hydrolysis of the ion and the complete calculation of $K_h$.

**B. A Buffer Solution**

1. Buffer = 0.50M $HC_2H_3O_2$/0.50M $NaC_2H_3O_2$

   pH =__________; color of thymol blue________________

2. Buffer +______ drops of 1.0M NaOH changes pH to > 9.6.

   Write a net equation for the neutralization of NaOH by a component of the buffer.

3. Buffer +______ drops of 1.0M HCl changes pH to < 1.2.

   Write a net equation for the neutralization of HCl by a component of the buffer.

Compare your results in 2. and 3. Are the base-neutralizing capacity and the acid-neutralizing capacity of the buffer similar? Explain why or why not.

4. 0.50M $HC_2H_3O_2$

   pH =__________; color of thymol blue________________

   0.50M $HC_2H_3O_2$ +_______ drops of 1.0M NaOH changes pH to > 9.6.

   Write a net equation for the neutralization of NaOH.

Compare your results in 2. and 4. Are the base-neutralizing capacities of buffer and acid comparable? Explain why or why not.

5. 0.50M $NaC_2H_3O_2$

pH = ______; color of thymol blue ____________________

0.50M $NaC_2H_3O_2$ + ______ drops of 1.0M HCl changes pH to < 1.2.

Write a net equation for the neutralization of HCl.

Compare your results in 3. and 5. Are the acid-neutralizing capacities of buffer and salt comparable? Explain why or why not.

6. Buffer + 1.0M HCl + $NaC_2H_3O_2(s)$

State the sequence of colors and color changes.

Account for these changes in terms of pH and the reactions that occur.

(On the back of this Report Sheet answer the Questions.)

# EXPERIMENT 24C

# REPORT SHEET

## C. Molar Weight on an Ammonium Salt

Unknown identification code ____________________

| | | | |
|---|---|---|---|
| Weight of Salt | ________ | ________ | ________ |
| Molarity of NaOH | ________ | ________ | ________ |
| Volume of NaOH used | ________ | ________ | ________ |
| Moles of NaOH | ________ | ________ | ________ |
| Molarity of HCl | ________ | ________ | ________ |
| Volume of HCl used | ________ | ________ | ________ |
| Moles of HCl | ________ | ________ | ________ |
| Moles of NaOH neutralized by HCl | ________ | ________ | ________ |
| Moles of NaOH neutralized by $NH_4^+$ | ________ | ________ | ________ |
| Moles of $NH_4^+$ | ________ | ________ | ________ |
| Moles of $NH_4X$ | ________ | ________ | ________ |
| Molar weight of $NH_4X$ | ________ | ________ | ________ |

Average molar weight ____________________________

# EXPERIMENT 25

# SOLUBILITY OF A SLIGHTLY SOLUBLE ELECTROLYTE

## Introduction

Electrolytes are compounds that are present in solution as ions. They are more likely to be soluble in water than in most other liquids that are commonly used as solvents. For example, NaCl (sodium chloride) is fairly water soluble; about 6.1 moles of NaCl will dissolve in a liter of water at 25°C. But some electrolytes are only slightly water soluble; only about $10^{-4}$ mole of $CaCO_3$ (calcium carbonate, the main component of limestone) will dissolve in a liter of water at 25°C.

In this experiment you will determine the solubility of the slightly soluble electrolyte $Cu(IO_3)_2$, copper(II) iodate, for which the dissolving equilibrium is

$$Cu(IO_3)_2(s) \rightleftharpoons Cu^{2+}(aq) + 2IO_3^-(aq) \qquad (25\text{-}1)$$

This equilibrium can be described by a solubility product constant, $K_{sp}$, which is calculated from molar concentrations of ions in a saturated solution of $Cu(IO_3)_2$:

$$K_{sp} = [Cu^{2+}][IO_3^-]^2 \qquad (25\text{-}2)$$

## Technique

A saturated solution of $Cu(IO_3)_2$ in distilled water is prepared. Carefully measured 20-mL portions of this solution are taken, and the amounts of $Cu^{2+}$ and $IO_3^-$ contained in 20 mL are determined by a titration. Using this information, the solubility of the salt in moles/liter and the solubility product can be calculated.

An indirect procedure is used to determine the amounts of $Cu^{2+}$ and $IO_3^-$ in the saturated solution. Solid KI is added to the solution causing the following reactions to occur:

$$Cu^{2+} + 2I^- \rightarrow CuI(s) + \tfrac{1}{2}I_2 \qquad (25\text{-}3)$$

$$IO_3^- + 5I^- + 6H_3O^+ \rightarrow 3I_2 + 9H_2O \qquad (25\text{-}4)$$

For each mole of $Cu(IO_3)_2$ that dissolves and reacts with $I^-$, $6\tfrac{1}{2}$ moles of $I_2$ are produced: 1/2 mole from the reaction of one mole of $Cu^{2+}$ and six moles from the reaction of two moles of $IO_3^-$. By determining the amount of $I_2$ produced, the quantity of $Cu(IO_3)_2$ originally present in the saturated solution is calculated.

The amount of $I_2$ produced in the reactions of Equations (25-3) and (25-4) can be determined by titration with a standardized $Na_2S_2O_3$ solution using a starch indicator. Iodine forms a deep-blue complex with starch; the color will disappear when all the $I_2$ has been reconverted to colorless $I^-$ by its reaction with $S_2O_3^{2-}$

$$2S_2O_3^{2-} + I_2 \rightarrow S_4O_6^{2-} + 2I^- \qquad (25\text{-}5)$$

Equation (25-5) states that for each mole of $I_2$ in the solution, two moles of $S_2O_3^{2-}$ will be needed for the titration. Since each mole of $Cu(IO_3)_2$ produces 6 ½ moles of $I_2$, each mole of $Cu(IO_3)_2$ originally present in the solution will require 13 moles of $S_2O_3^{2-}$. Knowing the molarity and the volume of the $Na_2S_2O_3$ solution used in the titration permits the calculation of the number of moles of $S_2O_3^{2-}$ used and of the number of moles of $Cu(IO_3)_2$ originally present in the solution.

But first you will determine the molarity of the $Na_2S_2O_3$ solution; this is called *standardizing* the solution. For a $Na_2S_2O_3$ solution, this must be done shortly before the solution is used, for $S_2O_3^{2-}$ is not stable in solution for extended periods of time. The solution that will be supplied is ~0.08M, but its molarity must be determined to a greater number of significant figures. Weighed samples of $KIO_3$ are dissolved in distilled water and then treated with $I^-$ to produce $I_2$ via the reaction of Equation (25-4). The $I_2$ is then titrated with $Na_2S_2O_3$ solution. Since the number of moles of $IO_3^-$ is known from weighing, the number of moles of $S_2O_3^{2-}$ required for the titration can be calculated from Equations (25-4) and (25-5). Because this number of moles must be contained in whatever volume of $Na_2S_2O_3$ solution was used in the titration, the molarity can then be calculated.

## Review

Equipment and Techniques E, G, H
Experiment 1

## Equipment Needed

- balances
- beaker
- buret
- clamp
- Erlenmeyer flasks
- filter paper
- funnel
- microburner
- 10-mL pipet
- 20-mL pipet
- ring
- ring stand
- stoppered bottle or flask
- wire gauze

## Chemicals Needed

- $Cu(IO_3)_2$; copper(II) iodate
- $HC_2H_3O_2$; glacial acetic acid
- KI; potassium iodide
- $KIO_3$; potassium iodate
- 0.08M $Na_2S_2O_3$; sodium thiosulfate
- 0.2% starch solution

## Procedure

### A. Preparation of Saturated $Cu(IO_3)_2$ Solution

Omit this part of the experiment if a saturated solution is supplied. On the laboratory balance weigh 0.3-0.4 g of $Cu(IO_3)_2$ into a small beaker; record the weight even though it is not required for your calculations. Wash the $Cu(IO_3)_2$ into a 250-mL Erlenmeyer flask using 100 mL of distilled water. Slowly warm this mixture to 50-60°C using a microburner; swirl the flask repeatedly during this warming. (Put a thermometer into the flask only briefly to meaure temperature; do not leave it in the flask during the entire warming and do not use it as a stirring rod.)

Stopper the flask and let the mixture cool to room temperature in your desk until the next laboratory period.

## B. Standardization of $Na_2S_2O_3$ Solution

Obtain about 400 mL of ~0.08M $Na_2S_2O_3$ in a flask or bottle that is provided. The flask or bottle must be clean, but it need not be dry; if it is wet, swirl the solution over the entire inside wall to wash all the water into the solution homogeneously. This is all of the $Na_2S_2O_3$ solution that you will use for the entire experiment and is more than you should need. After the standardization, save the remaining $Na_2S_2O_3$ solution; stopper the flask or bottle. Clean a buret and rinse it with three 3-mL portions of the $Na_2S_2O_3$ solution; then fill the buret with this solution.

Weigh 0.10-0.12 g of $KIO_3$ into a small beaker or crucible on the analytical balance. Transfer this sample to an Erlenmeyer flask and dissolve it in 25-30 mL of distilled water. To ensure that this transfer is quantitative (that all the weighed solid is transferred), use a stream of distilled water from a wash bottle to rinse all the $KIO_3$ out of the weighing container into the flask; count this water as part of the total 25-30 mL that you are to use.

Weigh about 1.5 g of KI into a beaker, add it to the $KIO_3$ solution, and swirl the flask until all of it dissolves. Add 2 mL of glacial acetic acid to this mixture and swirl the flask to mix it in homogeneously. The mixture should be brown in color because $I_2$ is formed in the reaction of $IO_3^-$ and $I^-$. (**Caution! Glacial acetic acid is corrosive.** If you get it on your skin, wash it off immediately under lots of cold, running water; if you spill any on the outside of the bottle or on the table top, sponge it off and wash out the sponge.)

Immediately begin to titrate this mixture with the $Na_2S_2O_3$ solution. The brown color will gradually become lighter as $S_2O_3^{2-}$ converts $I_2$ to colorless $I^-$. When the solution is pale yellow, add 2 mL of starch solution to the mixture. (The cylinder that you used to measure acid does not have to be rinsed out before you measure starch, but it does have to be rinsed out before you measure acid again after measuring starch.) The remaining $I_2$ will form a deep-blue complex with the starch; if the color is brownish instead of blue, you can continue with the titration, but for the next titration you should let the solution become a lighter yellow before you add the starch. Titrate until the blue color just disappears showing that all of the $I_2$ has reacted.

Weigh out and titrate two more portions of $KIO_3$. If you do not have two consecutive titrations that agree within 1% for the molarity of the $Na_2S_2O_3$ solution, repeat the titration with new portions of $KIO_3$ until you do have this agreement.

To discard a solution after you finish a titration, wash it down the drain with lots of water.

## C. Determination of $Cu(IO_3)_2$ Solubility

Record the temperature in the laboratory.

If you prepared your own saturated solution of $Cu(IO_3)_2$, begin by removing any undissolved solid from the solution by filtration. Be sure the funnel and all glassware that you use in this filtration are clean and dry; do not wet the filter paper with water before you begin the filtration. Because you want to know the concentration of the solute in a *saturated* solution, it is essential that you do not dilute the solution. If the filtrate is cloudy, repeat the filtration. Filter all of the solution before you begin the titrations.

Pipet 20.00 mL of the clear filtrate, the saturated $Cu(IO_3)_2$ solution, into an Erlenmeyer flask (which must be clean, but not necessarily dry). Add about 15 mL of distilled water, 1.5 g of KI, and 2 mL of glacial acetic acid. Immediately titrate this solution with your standardized $Na_2S_2O_3$ solution following the same procedure you used in the standardization. Repeat the titration with two more 20.00-mL portions of the saturated $Na_2S_2O_3$ solution. If you do not have two consecutive titrations that agree within 0.1 mL, carry out another titration.

To discard a solution after you finish a titration, wash it down the drain with lots of water. Any unused $Na_2S_2O_3$ solution and $Cu(IO_3)_2$ solution may also be washed down the drain.

## Calculations and Results

### B. Standardization of $Na_2S_2O_3$ Solution

Calculate the number of moles of $KIO_3$ in each sample that you weighed. Equation (25-5) shows that two moles of $S_2O_3^{2-}$ are required to titrate each mole of $I_2$, and Equation (25-4) shows that three moles of $I_2$ are produced by each mole of $IO_3^-$ in your sample of $KIO_3$. Thus, the number of moles of $S_2O_3^{2-}$ used in a titration is *six times* the number of moles of $KIO_3$. This is the number of moles of $S_2O_3^{2-}$ contained in the volume of $Na_2S_2O_3$ solution used in the titration. The molarity of the $Na_2S_2O_3$ solution can be calculated from the definition

$$M = \frac{\text{moles } S_2O_3^{2-}}{\text{liters of solution}} = \frac{6(\text{moles } KIO_3)}{\text{liters of } Na_2S_2O_3 \text{ solution}} \qquad (25\text{-}6)$$

Average your trials that agree within 1% and use this value as the molarity of the $Na_2S_2O_3$ solution in the remaining calculations.

### C. Determination of $Cu(IO_3)_2$ Concentration

From the volume of $Na_2S_2O_3$ solution required to titrate the 20-mL portions of saturated $Cu(IO_3)_2$ and from the molarity of the $Na_2S_2O_3$ solution, calculate the number of moles of $S_2O_3^{2-}$ used:

$$\text{moles } S_2O_3^{2-} = M \text{ (liters } S_2O_3^{2-} \text{ used)} \qquad (25\text{-}7)$$

From the information given in the Introduction

$$\text{moles } Cu(IO_3)_2 = \frac{\text{moles } S_2O_3^{2-}}{13} \qquad (25\text{-}8)$$

This is the number of moles of $Cu(IO_3)_2$ contained in 20.00 mL of solution; use this to calculate the number of moles of $Cu(IO_3)_2$ in 1.000 L of solution, that is, the molarity of $Cu(IO_3)_2$ in a saturated solution. This is the molar solubility of $Cu(IO_3)_2$.

Write the concentration expression for $K_{sp}$ of $Cu(IO_3)_2$, first in terms of molar concentrations of $Cu^{2+}$ and $IO_3^-$, and then in terms of molar solubility, *s*.

For each of the titrations you carried out with the saturated $Cu(IO_3)_2$ solution, calculate $K_{sp}$ using the value of *s* that you determined. For your reported value of $K_{sp}$ take the average of the values you obtained for titrations that agree within 0.1 mL.

## Questions

1. State and explain how each of these errors would affect (high, low, no change) your calculated values of solubility and $K_{sp}$.
   (a) The saturated solution of $Cu(IO_3)_2$ is not filtered to remove undissolved $Cu(IO_3)_2$.
   (b) The saturated solution of $Cu(IO_3)_2$ is filtered into a wet flask.
   (c) You use only 10.00 mL of the saturated solution instead of 20.00 mL.

2. At the end point of a titration in the standardization of the $Na_2S_2O_3$ solution, you have a colorless (and clear) solution. At the end point in the determination of $Cu(IO_3)_2$ solubility, you have a colorless (but cloudy) heterogeneous mixture. Explain this difference.

## EXPERIMENT 25

## REPORT SHEET

Name________________________________________ Date ______________

Section_______________________________________

**B. Standardization of $Na_2S_2O_3$ solution**

| | Trial 1 | Trial 2 | Trial 3 |
|---|---|---|---|
| Weight of $KIO_3$ | ________ | ________ | ________ |
| Moles of $KIO_3$ | ________ | ________ | ________ |
| Moles of $S_2O_3^{2-}$ needed | ________ | ________ | ________ |
| Volume of $Na_2S_2O_3$ used | ________ | ________ | ________ |
| Molarity of $Na_2S_2O_3$ | ________ | ________ | ________ |

Average molarity ______________________________

For Trial 1, show the calculation of molarity of the $Na_2S_2O_3$ solution.

## C. Determination of $Cu(IO_3)_2$ Solubility

Room temperature ____________________

| | Trial 1 | Trial 2 | Trial 3 |
|---|---|---|---|
| Molarity of $Na_2S_2O_3$ | | | |
| Volume of $Na_2S_2O_3$ used | | | |
| Moles of $S_2O_3^{2-}$ used | | | |
| Moles of $Cu(IO_3)_2$ present | | | |
| Volume of $Cu(IO_3)_2$ | | | |
| Molarity of $Cu(IO_3)_2$ | | | |
| Molar solubility of $Cu(IO_3)_2$ | | | |
| $K_{sp}$ | | | |

Average $K_{sp}$ ____________________

For Trial 1, show the calculation of moles of $Cu(IO_3)_2$ present.

For Trial 1, show the calculation of molar solubility of $Cu(IO_3)_2$.

For Trial 1, show the calculation of $K_{sp}$.

(Attach a separate sheet with the answers to the Questions.)

# EXPERIMENT 26

# METAL COMPLEXES WITH MONODENTATE AND CHELATING LIGANDS

## Introduction

Even before studying transition metal complexes in your textbook, you encountered several of these in your laboratory work. For example, Experiment 7 is the synthesis of $[Co(NH_3)_6]Cl_3$. This compound contains $Co^{3+}$ bound to six ammonia molecules which are arranged octahedrally around it as shown in Figure 7-1. These surrounding molecules are called *ligands*. Each ammonia molecule has only one atom, N, that can bond to a metal cation; such a ligand is *monodentate*. Some ligands have more than one atom that is capable of bonding to a metal cation. One example of such a ligand is ethylenediamine, $NH_2CH_2CH_2NH_2$. Both of the N atoms can bond to the same metal ion simultaneously; the ligand is *bidentate*. The resulting complex contains rings of atoms and is called a *chelate*. An ion such as $Co^{3+}$ can be surrounded octahedrally by six coordinated N atoms by forming a complex with six $NH_3$ molecules or by forming a complex with three ethylenediamine molecules as shown in Figure 26-1.

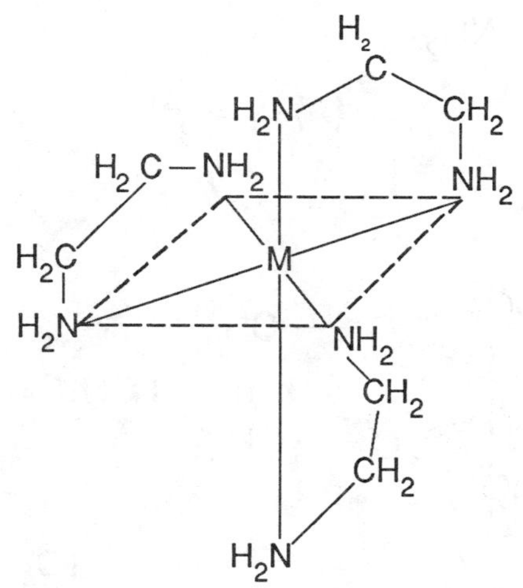


**Figure 26-1**

An ethylenediamine complex.

Chelating ligands form complexes that are quite stable. The equilibrium constants for their formation are extremely large in comparison to the equilibrium constants for formation of similar complexes containing ligands with only one donor atom. For example, the reaction shown in Equation (26-1) has the equilibrium constant $K \cong 10^{34}$,

$$Co(H_2O)_6^{3+} + 6NH_3 \rightarrow Co(NH_3)_6^{3+} + 6H_2O \qquad (26\text{-}1)$$

whereas the reaction shown in Equation (26-2) has $K \cong 10^{49}$

$$Co(H_2O)_6^{3+} + 3NH_2CH_2CH_2NH_2 \rightarrow Co(NH_2CH_2CH_2NH_2)_3^{3+} + 6H_2O \qquad (26\text{-}2)$$

This difference is especially surprising when one considers that $Co^{3+}$ is in essentially the same environment in both complexes, that is, it is surrounded octahedrally by six coordinated N atoms. This greater stability of complexes containing chelating ligands, compared to similar complexes containing nonchelating ligands, is called the *chelate effect*.

The fundamental reason for the added stability of chelate complexes is a statistical (entropy) one. After the first N of ethylenediamine bonds to the metal cation, the second N atom is held quite near. The metal cation is far more likely to find this second N and coordinate to it than to encounter and bond to a different ligand molecule. This statistical effect shows up in the entropy of complex formation. Another way of viewing the particularly favorable entropy change when chelates form is to note that seven particles are produced from four in the reaction of Equation (26-2), whereas reaction with a monodentate ligand produces seven particles from seven, Equation (26-1). The chelate effect occurs with many ligands besides ethylenediamine; the only requirement is that the ligand has more than one atom that can coordinate to one metal cation.

The purpose of this experiment is to study the consequences of chelate formation in comparison to formation of complexes with monodentate ligands. The chelating ligands you will use are shown in Figure 26-2.

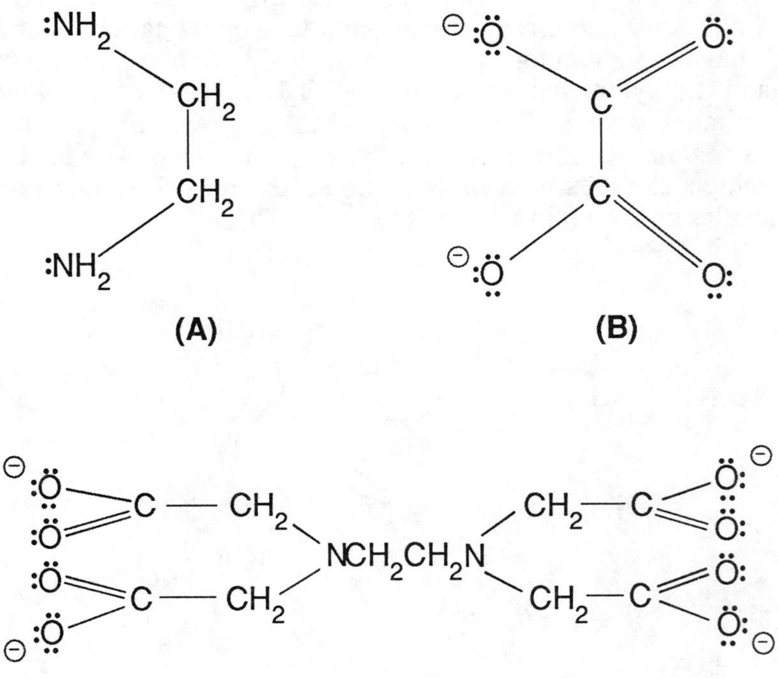

(A) (B)

(C)

**Figure 26-2**

(A) Ethylenediamine, (B) Oxalate, (C) Ethylenediaminetetraacetate (EDTA)

You will compare the behavior of these ligands with the behavior of the monodentate ligands water, ammonia, and chloride. Note that the ligands ethylenediamine and oxalate are bidentate; each has two atoms capable of coordinating simultaneously to one metal ion. Ethylenediaminetetraacetate (EDTA) is *sexadentate*; it has six atoms that can coordinate, two N atoms and four O atoms. Figure 26-3 shows that ethylenediaminetetraacetate can coil around a cation occupying all six octahedral coordination positions.

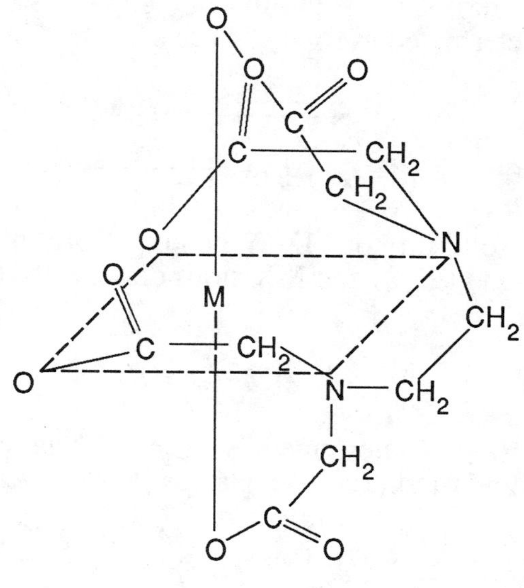


**Figure 26-3**

A six-coordinate EDTA complex.

One of the main practical uses of chelate formation is the dissolution of nearly insoluble salts. This aspect of chelate behavior is emphasized in this experiment.

Consider the equilibrium for complex formation between a metal cation $M^{2+}$ and ligands L:

$$M^{2+}(aq) + nL \rightleftharpoons ML_n^{2+} \tag{26-3}$$

As with any chemical equilibrium, this can be described by an equilibrium constant; for this type of reaction the euqilibrium constant is called a *formation constant.*

$$K_f = \frac{[ML_n^{2+}]}{[M^{2+}]\,[L]^n} \tag{26-4}$$

By adding sufficient L, the position of equilibrium can be shifted quite far to the right and essentially all the $M^{2+}$ can be converted to $ML_n^{2+}$. If $ML_n^{2+}$ has a characteristic color, it is easy to determine when this reaction occurs. Because chelating ligands have especially large values of $K_f$, a lower concentration of L will be required to push the equilibrium to the right if L is chelating than if L is monodentate (assuming that the same kind of atom is bonded to the metal ion in each case so that the ligands are as much alike as possible).

Now consider the process of dissolving a slightly soluble salt MX in a solution of some ligand L. The salt will dissolve according to Equation (26-5):

$$MX(s) \rightleftharpoons M^{2+}(aq) + X^{2-}(aq) \tag{26-5}$$

It will stop dissolving when $[M^{2+}]$ and $[X^{2-}]$ are high enough that their product is equal to $K_{sp}$ for the salt:

$$K_{sp} = [M^{2+}][X^{2-}] \tag{26-6}$$

If the solution also contains L, $M^{2+}$ will be continuously converted to $ML_n^{2+}$ as MX dissolves, and the solubility equilibrium, Equation (26-5), will be shifted to the right by this removal of $M^{2+}$, Equation (26-3). The more effective L is at complexing $M^{2+}$ (that is, the larger the formation constant for $ML_n^{2+}$), the smaller the amount of L that is required to push the equilibrium far to the right and to dissolve MX completely.

In more mathematical terms, the solubility of a slightly soluble salt in the presence of a complexing ligand L can be determined as follows. The equilibria in solution are

$$MX(s) \rightleftharpoons M^{2+}(aq) + X^{2-}(aq) \tag{26-5}$$

$$M^{2+}(aq) + nL \rightleftharpoons ML_n^{2+} \tag{26-3}$$

Let $s$ stand for the molar solubility of MX. When equilibrium is established for both of the reactions in Equations (26-5) and (26-3) and MX stops dissolving (that is, when the solution is saturated), $[X^{2-}] = s$ and

$$[M^{2+}]s = K_{sp} \tag{26-7}$$

The concentration of $M^{2+}(aq)$ is determined by the equilibria of both Equations (26-5) and (26-3). Rearranging Equations (26-4) and (26-7) to express $[M^{2+}]$ gives

$$[M^{2+}] = \frac{[ML_n^{2+}]}{K_f\,[L]^n} = \frac{K_{sp}}{s} \tag{26-8}$$

Rearranging the second and third terms to solve for $s$ gives

$$s = \frac{K_f K_{sp}\,[L]^n}{[ML_n^{2+}]} \tag{26-9}$$

Equation (27-9) shows that the solubility of MX in the presence of a complexing ligand L depends on two factors: (1) how effectively L complexes $M^{2+}$, which depends on $K_f$ and [L]; (2) how soluble MX is in water, as shown by $K_{sp}$. In this experiment, you will observe how solubility depends on each of these factors.

In the first part of the experiment, you will investigate the complexing ability of monodentate and bidentate ligands with $Ni^{2+}$ and $Cu^{2+}$. Complexing ability will be reflected by the ease with which solutions of the ligands dissolve precipitates of $NiC_2O_4$ (nickel oxalate) and $CuC_2O_4$ (copper oxalate). If you begin with several test tubes containing equal amounts of precipitate, the least amount of the ligand that is the best complexing agent (that is, the ligand with the largest $K_f$) will be required to dissolve the precipitate. In contrast, if $K_f$ is small, a large amount of the ligand solution will be required to dissolve the precipitate. By measuring the volume of each ligand solution required to dissolve the precipitate, you can order the ligands according to the size of $K_f$ for complexation with each metal.

In the second part of the experiment, you will compare the stabilities of two $Ni^{2+}$ complexes, one containing a monodentate ligand and the other containing a chelating ligand. You make this comparison by attempting the interconversion of the two complexes.

In the third part of the experiment, you will examine the effect of solubility differences on complex formation by dissolving $Cu(IO_3)_2$, for which $K_{sp} = 10^{-7}$, and $Cu(OH)_2$, for which $K_{sp} = 10^{-19}$, in solutions of monodentate and chelating ligands.

## Equipment Needed

| | |
|---|---|
| beakers | 20-mL test tubes |
| medicine droppers | test-tube rack |

## Chemicals Needed

1.0M $CuSO_4$; copper(II) sulfate
0.1M EDTA; ethylenediaminetetraacetic acid
1.0M $K_2C_2O_4$; potassium oxalate
0.25M $KIO_3$; potassium iodate
6.0M $NH_3$; ammonia
3.0M $NH_2CH_2CH_2NH_2$; ethylenediamine
6.0M NaCl; sodium chloride
0.25M NaOH; sodium hydroxide
1.0M $NiSO_4$; nickel(II) sulfate

## Procedure

Throughout the experiment you will combine drops of solutions. Try to release each solution from the medicine dropper straight into the bottom of the test tube; if a drop of solution falls onto the wall of the tube, tip the tube to wash the drop into the rest of the mixture.

Some solubility equilibria are established slowly. When you are trying to dissolve a precipitate, mix the contents of the test tube thoroughly after each drop of solution and wait at least a second or two before adding another drop. If the ligand solution causes any noticeable effect, such as a slight color change or a decrease in the amount of precipitate, allow several seconds between drops and shake the tube vigorously to mix the contents. If a precipitate does not dissolve in 50 drops of a ligand solution, begin adding the solution 1.0 mL (20 drops from a medicine dropper) at a time. If about 10 mL (about one-half the volume of the test tube) does not dissolve all of the precipitate, you may stop adding solution.

When making color comparisons, hold the test tubes against a piece of white paper. You might find it helpful to tip the tubes nearly horizontal so that you can compare the colors of equal depths of solutions.

When you are working with several solutions in one part of the experiment, be sure to use a different medicine dropper for each solution. Rinse each dropper thoroughly before it is used again.

### A. Varying the Ligand

Into each of six test tubes put five drops of 1.0M $CuSO_4$; then add five drops of 1.0M $K_2C_2O_4$ to each tube. A precipitate of $CuC_2O_4$ should form within seconds. To the first tube, add 6.0M $NH_3$ dropwise (as described above) until the precipitate dissolves completely. Attempt to dissolve the precipitates in the next four tubes in 3.0M $NH_2CH_2CH_2NH_2$, 1.0M $K_2C_2O_4$, 0.1M EDTA, and 6.0M NaCl. For each ligand solution, count and record the number of drops required to dissolve the precpitate. If the precipitate dissolves, record the color of the solution that results.

A dissolution or a color change might result simply from the water in the solution, rather than from a reaction with a ligand. To determine whether this is the case, add distilled water to the sixth test tube until the volume of material in the tube equals the maximum volume in the five preceding tubes.

If any ligand does not completely dissolve the precipitate, allow the tube to stand undisturbed until the precipitate settles. Then compare the color of the solution above the precipitate to the color of the solution in the tube to which you added water; any color difference would mean that at least a slight reaction has occurred between the precipitate and the ligand. A colored solid in the bottom of a tube can give a colored reflection in the meniscus; therefore, when you make this comparison, hold your hand or a piece of paper against the tubes at the top of the solution and look at the middle portions of the solutions against a white background.

Prepare six tubes containing $NiC_2O_4$ by combining five drops of 1.0M $NiSO_4$ and five drops of $K_2C_2O_4$ in each tube; allow the tubes to stand until precipitates form. Investigate the ability of each of the five ligand solutions to dissolve these precipitates; follow the procedure given for $CuC_2O_4$.

### B. Shifting the Position of Equilibrium: Monodentate vs. Chelating Ligands

Prepare a set of test tubes for color comparison as follows. Put two drops of 1.0M $NiSO_4$ into each of three test tubes. To the first tube, add 20 drops of distilled water. To the second tube, add 20 drops of 6.0M $NH_3$. To the third tube, add 20 drops of 3.0M $NH_2CH_2CH_2NH_2$. The tubes will contain $[Ni(H_2O)_6]^{2+}$, $[Ni(NH_3)_6]^{2+}$, and $[Ni(NH_2CH_2CH_2NH_2)_3]^{2+}$, respectively; they show the colors that are characteristic of these complexes.

Prepare a test tube containing $[Ni(NH_3)_6]^{2+}$ by putting two drops of $NiSO_4$ into a test tube followed by *just enough* $NH_3$ solution to produce the color that is characteristic of the complex. Then add 200% excess of $NH_3$ to ensure that essentially all the $Ni^{2+}$ is present as $[Ni(NH_3)_6]^{2+}$ rather than as $[Ni(H_2O)_6]^{2+}$; record the total volume of $NH_3$ used. To this solution, add $NH_2CH_2CH_2NH_2$ solution dropwise until $[Ni(NH_3)_6]^{2+}$ is converted to $[Ni(NH_2CH_2CH_2NH_2)_3]^{2+}$, as shown by the color. Record the number of drops required.

Prepare a solution containing $[Ni(NH_2CH_2CH_2NH_2)_3]^{2+}$ by combining two drops of 1.0M $NiSO_4$ and enough $NH_2CH_2CH_2NH_2$ solution to produce the color of this complex; then add a 200% excess. Add 6.0M $NH_3$ dropwise to try to convert this complex to $[Ni(NH_3)_6]^{2+}$.

### C. Varying the Precipitate for One Metal Ion

Prepare a precipitate of $Cu(IO_3)_2$ by combining in a test tube three drops of 1.0M $CuSO_4$ and 24 drops of 0.25M $KIO_3$. Prepare $Cu(OH)_2$ by combining three drops of 1.0M $CuSO_4$ and 24 drops of 0.25 NaOH. Dissolve each precipitate in 6.0M $NH_3$; record the number of drops of $NH_3$ solution that each precipitate requires.

Prepare second precipitates of $Cu(IO_3)_2$ and $Cu(OH)_2$ as you did previously. Dissolve each precipitate in 3.0M $NH_2CH_2CH_2NH_2$ and record the volume of solution that is required.

## Results and Discussion

1. The ligand solutions that you used in this experiment were not all of the same concentration. The ligand concentrations were different for reasons that include solubility difficulties ($K_2C_2O_4$ and EDTA) and the necessity to avoid side reactions that would occur with less concentrated solutions of $NH_3$ and $NH_2CH_2CH_2NH_2$. Therefore, to obtain a comparison of the complexing abilities of the ligands you used, you must make some concentration and statistical corrections in the volumes you measured. It is convenient to convert all the volumes to the comparison basis of 6.0M $NH_3$. Such a solution contains 6.0 moles of N atoms that can coordinate in each liter of solution.

Each molecule of $NH_2CH_2CH_2NH_2$ has two atoms that can coordinate. Therefore, a 3.0M $NH_2CH_2CH_2NH_2$ solution contains 6.0 moles of N atoms that can coordinate in each liter of solution. Hence, one drop of this solution is equivalent to one drop of 6.0M $NH_3$.

The 1.0M $K_2C_2O_4$ contains bidentate ligands having two O atoms that can coordinate. Calculate the concentration of coordinating atoms in this solution. How may drops of this solution contain the same number of coordinating atoms as are present in one drop of 6.0M $NH_3$? Use this relationship as a conversion factor to "correct" the volumes; that is, multiply the volume that you actually used by this ratio.

Each EDTA ion has six atoms that can coordinate, four O atoms and two N atoms. (It is likely that only five of these six atoms will coordinate with $Ni^{2+}$ and that only four of these six atoms will coordinate with $Cu^{2+}$, but you can ignore this refinement for the purposes of this experiment.) What is the concentration of coordinating atoms in 0.1M EDTA? How many drops of this solution contain the same number of coordinating atoms as are contained in one drop of 6.0M $NH_3$? When you use this relationship as a conversion factor, be sure to put all the proper units into the problem.

2. Assume that all complexes of a particular metal ion have the same number of ligand atoms coordinated from the added ligands. You can then rank the ligands that coordinate with $Cu^{2+}$ and $Ni^{2+}$ according to their formation constants for metal complexes by noting which requires the smallest corrected volume to dissolve precipitates. Do this separately for the ligands reacting with $CuC_2O_4$ and with $NiC_2O_4$. Write each series with the best coordinating ligand first. (Although the assumption that all complexes of a particular metal ion have the same number of ligand atoms coordinated from the added ligands is not strictly true, departures from this assumption will not significantly affect your results and you do not have to take this into account.)

3. Explain how your observations for the $NH_3$ and $NH_2CH_2CH_2NH_2$ reactions with $CuC_2O_4$ and $NiC_2O_4$ illustrate the chelate effect. Why is this a better comparison than the reactions of $NH_3$ and $C_2O_4^{2-}$ with these salts?

4. Account for the results you obtained in Part B. Can the $NH_3$-complex/$NH_2CH_2CH_2NH_2$-complex equilibrium be shifted both from left to right and from right to left using the concentrations of materials that you had? Explain why this is so.

5. Account for the difference in volume of $NH_3$ solution required to dissolve $Cu(IO_3)_2$ and $Cu(OH)_2$. Does the difference in volume of $NH_2CH_2CH_2NH_2$ solution show this same effect?

6. Account for the difference in volume of $NH_3$ and $NH_2CH_2CH_2NH_2$ solutions required to dissolve the two copper salts.

## EXPERIMENT 26

## REPORT SHEET

Name_______________________________ Date _______________

Section_______________________________

**A. Varying the ligands:**

| $CuC_2O_4$ treated with | Volume | Corrected volume | Color |
|---|---|---|---|
| 6.0M $NH_3$ | | | |
| 3.0M $NH_2CH_2CH_2NH_2$ | | | |
| 1.0M $K_2C_2O_4$ | | | |
| 0.1M EDTA | | | |
| 6.0M NaCl | | | |
| Distilled water | | | |

Rank the ligands on the basis of complexing ability as directed in Results and Discussion, point 2.

| $NiC_2O_4$ treated with | Volume | Corrected volume | Color |
|---|---|---|---|
| 6.0M $NH_3$ | | | |
| 3.0M $NH_2CH_2CH_2NH_2$ | | | |
| 1.0M $K_2C_2O_4$ | | | |
| 0.1M EDTA | | | |
| 6.0M NaCl | | | |
| Distilled water | | | |

Rank the ligands on the basis of complexing ability as directed in Results and Discussion, point 2.

In the space provided at the top of the next page, show the calculation of the correction of the $K_2C_2O_4$ volume and the EDTA volume.

Show the calculation of the correction of the $K_2C_2O_4$ volume and the EDTA volume.

**B. Shifting the Position of Equilibrium: Monodentate vs. Chelating Ligands**

| Complex | Color | Volume of ligand solution |
|---|---|---|
| $[Ni(H_2O)_6]^{2+}$ | | |
| $[Ni(NH_3)_6]^{2+}$ | | |
| $[Ni(NH_2CH_2CH_2NH_2)_3]^{2+}$ | | |

Volume of $NH_2CH_2CH_2NH_2$ solution to convert $NH_3$ complex to $NH_2CH_2CH_2NH_2$ complex

____________

Volume of $NH_3$ solution to convert $NH_2CH_2CH_2NH_2$ complex to $NH_3$ complex

____________

**C. Varying the Precipitate for One Metal Ion**

| Salt dissolved | $Cu(IO_3)_2$ | $Cu(OH)_2$ |
|---|---|---|
| Volume of 6.0M $NH_3$ | | |
| Volume of 3.0M $NH_2CH_2CH_2NH_2$ | | |

(Attach a separate sheet with the conclusions and discussions requested in Results and Discussion, points 3-6)

# EXPERIMENT 27

# SYNTHESIS AND IDENTIFICATION OF COORDINATION COMPOUNDS USING SPECTROPHOTOMETRY

## Introduction

Several features of synthetic chemistry are discussed in the Introduction to Experiment 7. You should review this material in conjunction with this experiment.

Here you will gain experience in synthetic techniques by preparing a synthetic intermediate, isolating it, and using it as starting material to synthesize another complex. This intermediate is related to the cobalt complex of Experiment 7; the difference is that two of the neutral $NH_3$ molecules coordinated to $Co^{3+}$ have been replaced by the dinegative carbonate ion $CO_3^{=}$. See Figure 27-1. This decreases the charge on the complex ion to +1, and so only one nitrate is needed to balance the charge. The formula for this compound is often written as $[Co(NH_3)_4(CO_3)](NO_3)$ to show that the molecules and ions within the square brackets are attached to $Co^{3+}$. The name of this compound is tetraamminecarbonatocobalt(III) nitrate.

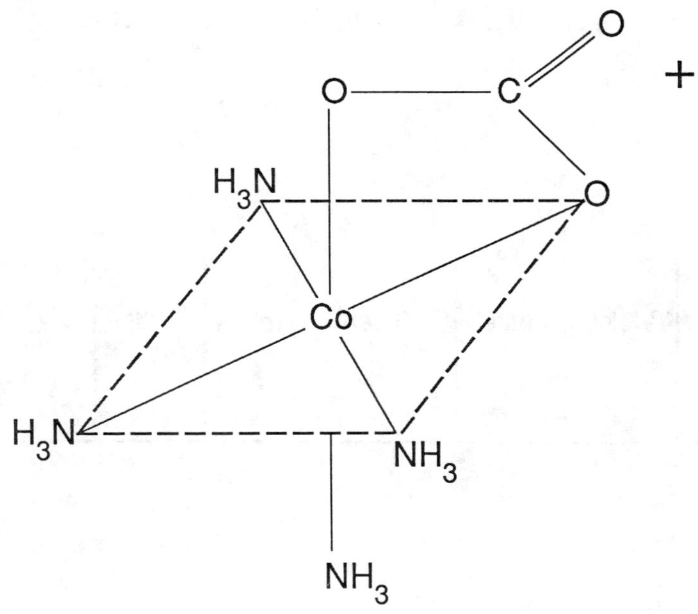


**Figure 27-1**

The method of preparation is given in Equation (27-1):

$$2Co(NO_3)_2{\bullet}6H_2O + 2(NH_4)_2CO_3 + 6NH_3 + H_2O_2 \rightarrow$$

$$2[Co(NH_3)_4(CO_3)](NO_3) + 2NH_4NO_3 + 14H_2O \qquad (27\text{-}1)$$

The purpose of the hydrogen peroxide is to oxidize the $Co^{2+}$ in the starting material to $Co^{3+}$.

Excess $H_2O_2$ is disposed of by heating:

$$2H_2O_2 \xrightarrow{50\text{-}60^\circ} O_2\,(g) + 2H_2O \qquad (27\text{-}2)$$

After you have prepared $[Co(NH_3)_4(CO_3)](NO_3)$, your instructor will give you an unknown substance that will react with it replacing the carbonate ion by some groups $X^-$ or by $X^-$ and $H_2O$ producing one of the complex ions, $[Co(NH_3)_4X_2]^+$, or $[Co(NH_3)_4(H_2O)X]^{2+}$. The cations will pick up some of the $X^-$ ions in solution forming solid $[Co(NH_3)_4X_2]X$ or $[Co(NH_3)_4(H_2O)X]X_2$.

The reaction in which an $X^-$ group or $H_2O$ molecule replaces carbonate is carried out in acidic solution; $H^+$ destroys the $CO_3^=$ ion causing it to bubble off as $CO_2$ gas:

$$CO_3^=(aq) + 2H^+\,(aq) \rightleftharpoons H_2CO_3\,(aq) \rightarrow H_2O + CO_2(g) \qquad (27\text{-}3)$$

Once you have isolated a solid containing the $[Co(NH_3)_4X_2]^+$ or $[Co(NH_3)_4(H_2O)X]^{2+}$ cation, you will identify it. Since these cations are brightly colored, they can be identified by their characteristic colors. You might remember that in Experiment 2 the property color was not used as a possible means of identification because there was no precise numerical way of measuring it. You will now learn about a way of measuring quantitatively the color of a compound so that you can identify the complex you prepared.

The technique is called *spectrophotometry* and makes use of the fact that substances are colored because they absorb visible light of certain colors and transmit the remaining colors. As you probably remember from high school science courses, ordinary white light is composed of light of all colors visible to the eye. The color of light depends on its wavelength, $\lambda$. Figure 27-2 shows approximate wavelengths for various colors. Ordinary visible white light is a mixture of light having all wavelengths between 4000 Å and 7000 Å. Light having wavelengths shorter than 4000 Å is called *ultraviolet* (UV) *light* and and cannot be seen by the human eye. Similarly, *infrared* (IR) *light* with wavelengths longer than 7000 Å is invisible to human eyes.

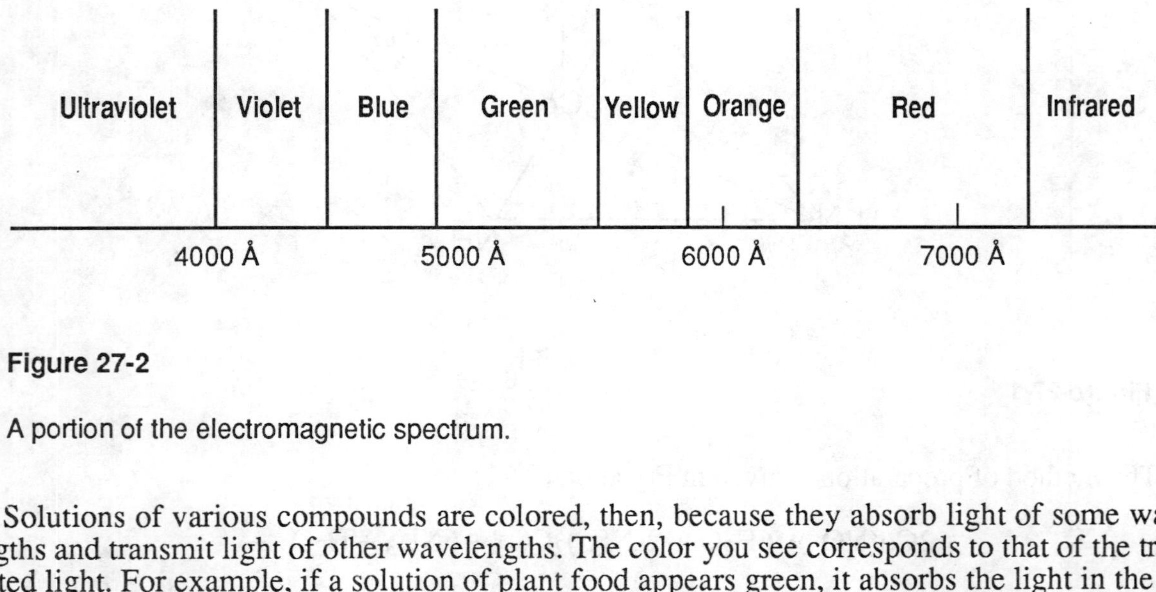


**Figure 27-2**

A portion of the electromagnetic spectrum.

Solutions of various compounds are colored, then, because they absorb light of some wavelengths and transmit light of other wavelengths. The color you see corresponds to that of the transmitted light. For example, if a solution of plant food appears green, it absorbs the light in the red, orange, yellow, blue, and violet wavelength regions and transmits green light. Light in the visible and ultraviolet regions of the spectrum causes electrons to jump to higher energy levels when it is absorbed.

Figure 27-3 diagrams a quantitative way of measuring light absorption by a solution. Some of the sample is placed in a small vessel called a cell or cuvette. Light of a single wavelength λ having intensity $I_0$ is passed into the cell. The solution might absorb some light of this wavelength and transmit the rest with reduced intensity $I$. The quantity $I/I_0$ is called the transmittance $T$. It is the fraction of the original intensity transmitted and is obviously related to how strongly the solution absorbs the light. If most of the light is absorbed, $I$ will be quite small and $T = I/I_0$ will also be quite small. Conversely, if most light is transmitted, $I$ will be large and $T = I/I_0$ will also be large. The quantity $I/I_0$ x 100% is called the percent transmittance, $\%T$:

$$\%T = \frac{I}{I_0} \times 100\% \qquad (27\text{-}4)$$

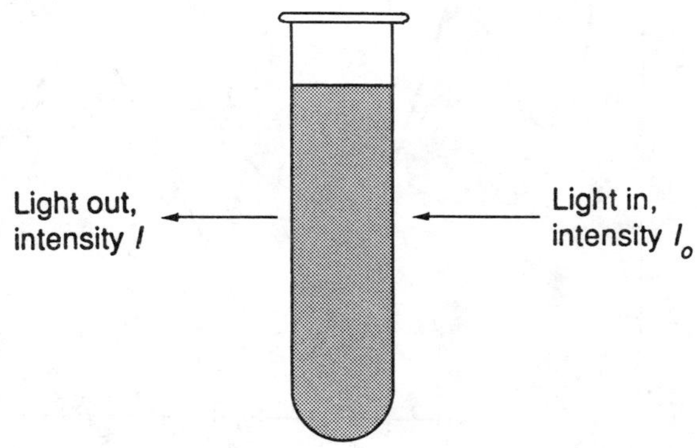


**Figure 27-3**

Light passing through a cell.

A related quantity is $\log\frac{I_0}{I}$ which is called the absorbance $A$:

$$A = \log\frac{I_0}{I} = -\log\frac{I}{I_0} = -\log T = 2.00 - \log\%\,T \qquad (27\text{-}5)$$

From the definition of $A$ you can see that, at any wavelength where the sample absorbs strongly $I_0/I$ A will be large. When little light is absorbed, $A$ will be small. A plot of the absorbance of a solution at various wavelengths is called the *absorption spectrum* and shows what colors are strongly absorbed and what ones are transmitted. The spectrum of $[Co(NH_3)_4(CO_3)]\ (NO_3)$ is depicted in Figure 27-4. The strong absorption of light in the orange, yellow, green, blue and violet regions gives the complex the red-violet color of the transmitted light. The color of the solution is complementary to the colors absorbed.

The spectrum of a compound is a characteristic property and may be used to identify the compound. Rather than reproducing the entire spectrum, it is customary to report the wavelengths of maximum absorbance. These are indicated on the spectrum of $[Co(NH_3)_4(CO_3)](NO_3)$ in Figure 27-4. You can identify your compound by measuring its absorption spectrum and comparing it to the spectra of the possible products which are given at the end of this experiment. This sort of color comparison is quite like doing flame tests which you probably did in high school. However, the more precise technique used here allows an easy distinction between different shades of red, for example, because they will have slightly different wavelengths of maximum absorption.

Locating the exact position of the maximum is sometimes rather difficult since the absorption curves are rather flat. A good technique for doing this is shown in Figure 27-4. Lines are drawn tangent to both sides of the curve and extrapolated to the point where they intersect. This point is taken as the maximum.

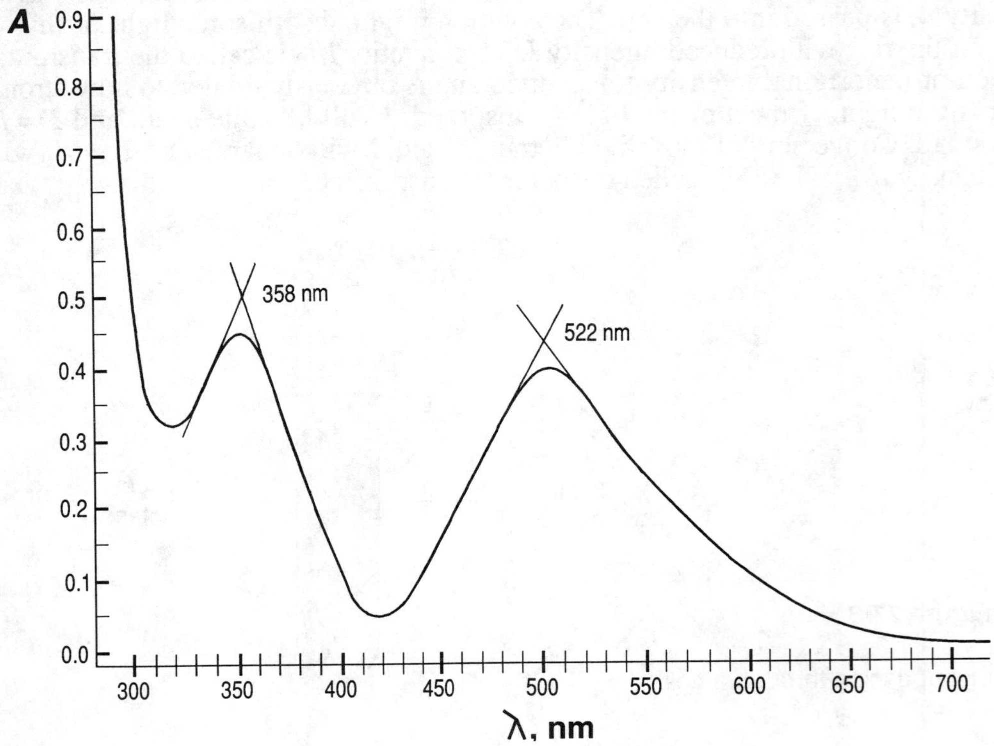


**Figure 27-4**

Spectrum of $[Co(NH_3)_4(CO_3)]NO_3$

The measured value of the absorbance *A* is related to several experimental factors by the Beer-Lambert Law:

$$A = \varepsilon c l \tag{27-6}$$

where $\varepsilon$ is the molar absorptivity
$c$ is the concentration in moles $L^{-1}$
$l$ is the thickness of the cell

This law simply states that the absorbance at any wavelength λ depends on how many absorbing molecules a light beam encounters in passing through the cell. This will obviously be greater as the cell thickness ($l$) and concentration ($c$) increase. Also, it depends on intrinsic ability of the molecules to absorb at the wavelength through $\varepsilon$.

## Review

Equipment and Techniques D, E, F, G

## Equipment Needed

beakers - 50 mL
Büchner funnel
burner
Erlenmeyer flasks
- 125 mL, 250 mL
filter flask
filter paper
graduated cylinder - 100 mL
medicine dropper
mortar & pestle
pneumatic trough
ring stand
spectrophotometer
(Bausch and Lomb Spectronic 21)
thermometer
universal clamp
vacuum tubing
volumetric flask - 250 mL

## Chemicals Needed

$Co(NO_3)_2 \cdot 6H_2O$ (20g); cobalt(II) nitrate hexahydrate
$(NH_4)_2CO_3$ (30 g); ammonium carbonate
conc. $NH_3$ (aq); aqueous ammonia
30% $H_2O_2$; hydrogen peroxide
$CH_3OH$; methanol
ice

Unknown Solution
(provided by instructor)

## Procedure

**A. Synthesis of Carbonatotetraamminecobalt(III)nitrate** $[Co(NH_3)_4(CO_3)]NO_3 \cdot 1/2\ H_2O$

Weigh to the nearest 0.1 g a clean empty 50-mL beaker and record the weight. Now add about 20 g $Co(NO_3)_2 \cdot 5H_2O$ and record the weight of the beaker and salt to the nearest 0.1 g. The weight of $Co(NO_3)_2 \cdot 6H_2O$ is the one on which your yield calculation will be based, and it is the only one which must be known precisely.

To a 250-mL Erlenmeyer flask add about 30 g of ammonium carbonate $(NH_4)_2CO_3$ and dissolve this in about 30 mL of distilled water. It is not easy to dissolve ammonium carbonate. Grind the salt in your mortar to break up any lumps. Swirl the mixture, warming with your hand. Do not heat the mixture with your burner. Heat decomposes carbonates: $(NH_4)_2CO_3 \xrightarrow{\Delta} 2NH_3\ (aq) + CO_2(g) + H_2O$. Add about 50 mL conc. $NH_3$ (aq) to the solution. Dissolve the $Co(NO_3)_2 \cdot 6H_2O$ in 10 mL of water and pour this solution into the Erlenmeyer flask being sure to rinse the beaker in order to get all the cobalt salt out. Mix the contents of the flask thoroughly by continued gentle swirling.

Let the flask stand while you obtain an ice bath. (This ice bath can be shared by two students.) Set the flask in the ice bath to cool. Carefully pour 25 mL 30% hydrogen peroxide ($H_2O_2$) from the reagent shelf into a small beaker. **Caution: Keep $H_2O_2$ off skin and eyes; it is a powerful oxidizing agent and can cause painful burns.** Using a medicine dropper, add the hydrogen peroxide to the solution in the Erlenmeyer flask a little at a time with stirring. Add the peroxide slowly enough that the heat from the oxidation of $Co^{2+}$ to $Co^{3+}$ does not raise the solution temperature above 10°C.

Now you can destroy the excess $H_2O_2$ by heating. Set up a ring stand with ring and wire gauze along with your burner. Place the flask on the wire gauze and heat it gently swirling it every now and then for about 20 minutes. *Do not let the temperature vary outside the range 50°-60°C.*

Cool the flask again in the ice bath until no more crystals of the red product seem to be forming. If you have trouble crystallizing the product, you may add some methanol to the solution. Then collect the crude product by filtration on a Büchner funnel and wash it with a little ice water followed by cold methanol. The compound should be collected during the laboratory period in which it is synthesized. If the solution is allowed to stand for a week, $[Co(NH_3)_5(CO_3)](NO_3)$ sometimes precipitates.

Dry the compound in air by placing it in your evaporating dish in your desk and leaving it for several days. Weigh the dry product.

## B. Synthesis of $[Co(NH_3)_4X_2]X$ or $[Co(NH_3)_4(H_2O)X]X_2$

Your instructor will provide you with enough of a compound which is a source of the acid HX to react with 3.0 g of $[Co(NH_3)_4(CO_3)](NO_3)$.

Place 3.0 g of your $[Co(NH_3)_4(CO_3)](NO_3)$ in your 250-mL Erlenmeyer flask. Add about 15 mL distilled water. Then (under a hood) add the unknown solution provided by your instructor a little at a time. **(Caution! This solution contains concentrated acid!)** Set up a ring with wire gauze and your burner. Heat the solution and maintain the temperature in the 50°-65°C range for one-half hour either with a burner or on a steam bath. Cool the solution in an ice bath and collect the precipitate by filtration on a Büchner funnel. Wash the precipitate first with cold water and then with methanol. Air dry the solid and then weigh it.

## C. Product Indentification

You will identify the product you have prepared by measuring its absorption spectrum. Weigh out as precisely as possible (to three or four significant figures) about 0.5 g of your dry unknown compound. Transfer it quantitatively to a clean 250-mL volumetric flask. After dissolving the compound completely in a few milliliters of distilled water, dilute to the mark on the neck. Then invert the flask several times to obtain complete mixing.

Place some of the solution in the cuvette of the spectrophotometer and read the percent transmittance at 250 Å intervals over the visible range. Directions for using the spectrophotometer are given below.

(If the 0.5-gram quantity of your product will not dissolve completely in 250 mL, discard the solution and rinse the volumetric flask with distilled water. Then weigh out as precisely as you can about 0.2 g of your dry unknown and proceed as before.)

To make readings on the Spectronic 21, you will need to do the following:

1. Turn the instrument on and let it warm up for about 20 minutes.
2. Select a pair of cuvettes. Fill one with distilled water; rinse the other with several portions of your solution and fill it.
3. Select the wavelength. The scale is in nanometers (nm); nm = $10Å = 10^{-9}m$.
4. Insert the reference cuvette. Adjust the meter needle to read 100% with the Light Control Knob (right front knob).
5. Remove the reference solution and insert the sample solution. Read the % T to the nearest 0.1% on the meter. (Although the meter is calibrated in percent transmittance and absorbance, it is better to read the % T and convert to A. This is because the A scale is non-linear and hard to read accurately.)
6. Steps 2-6 must be repeated for each wavelength at which a reading is desired.

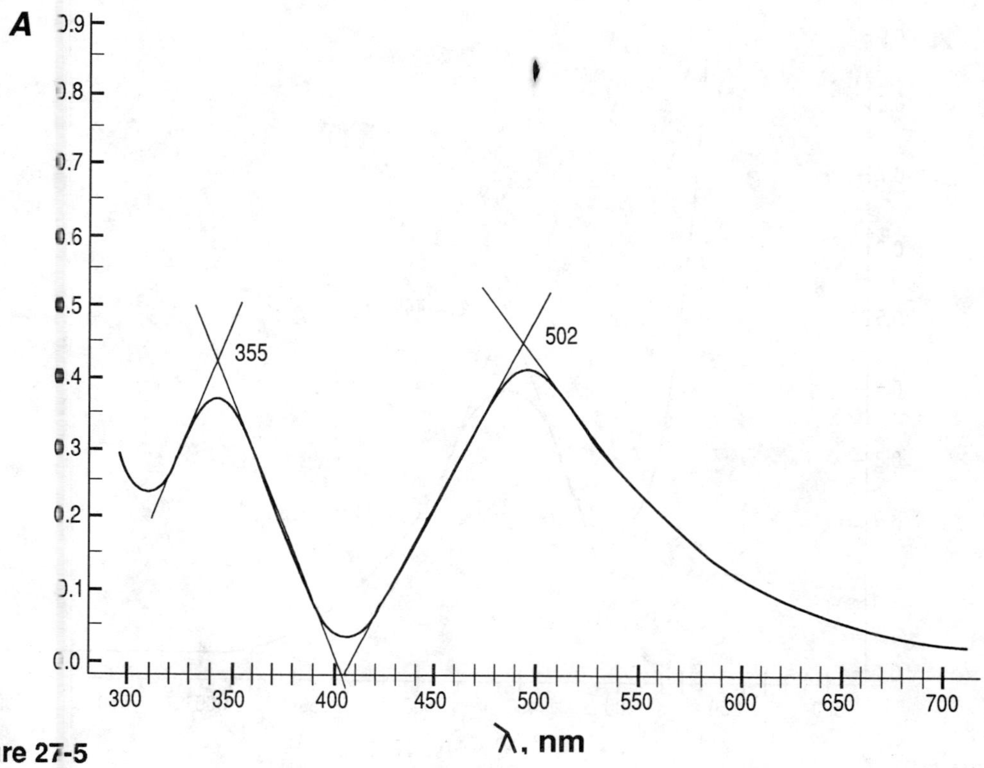


**Figure 27-5**

Spectrum of [*trans*-$Co(NH_3)_4Cl_2$]Cl

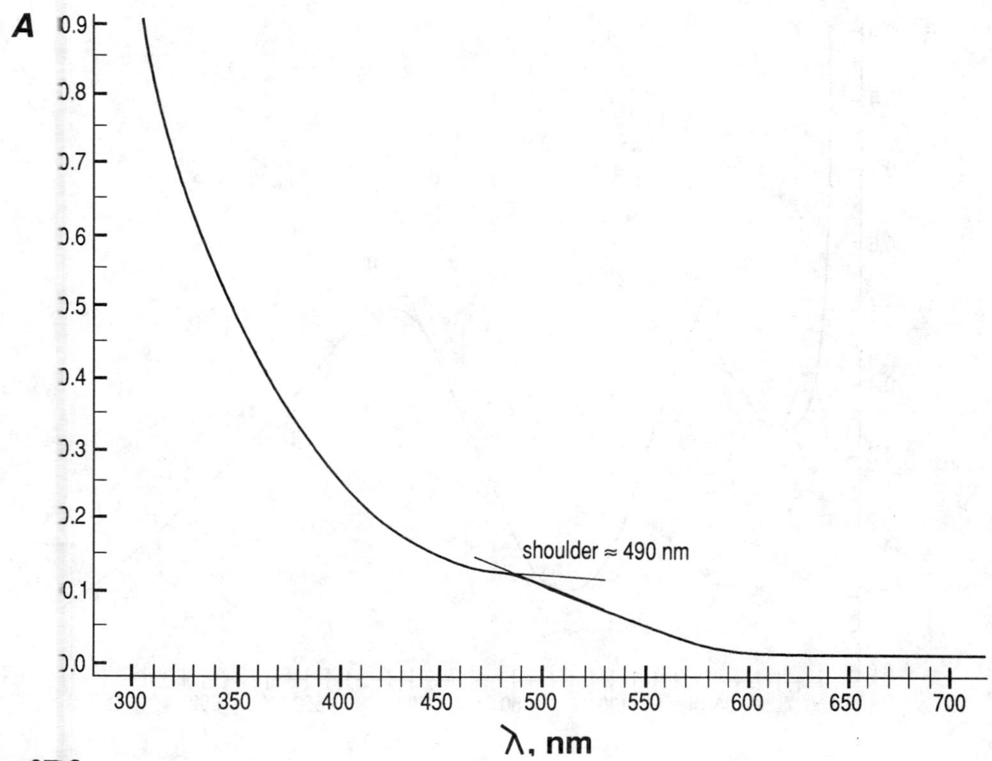


**Figure 27-6**

Spectrum of [$Co(NH_3)_4Br_2$]Br

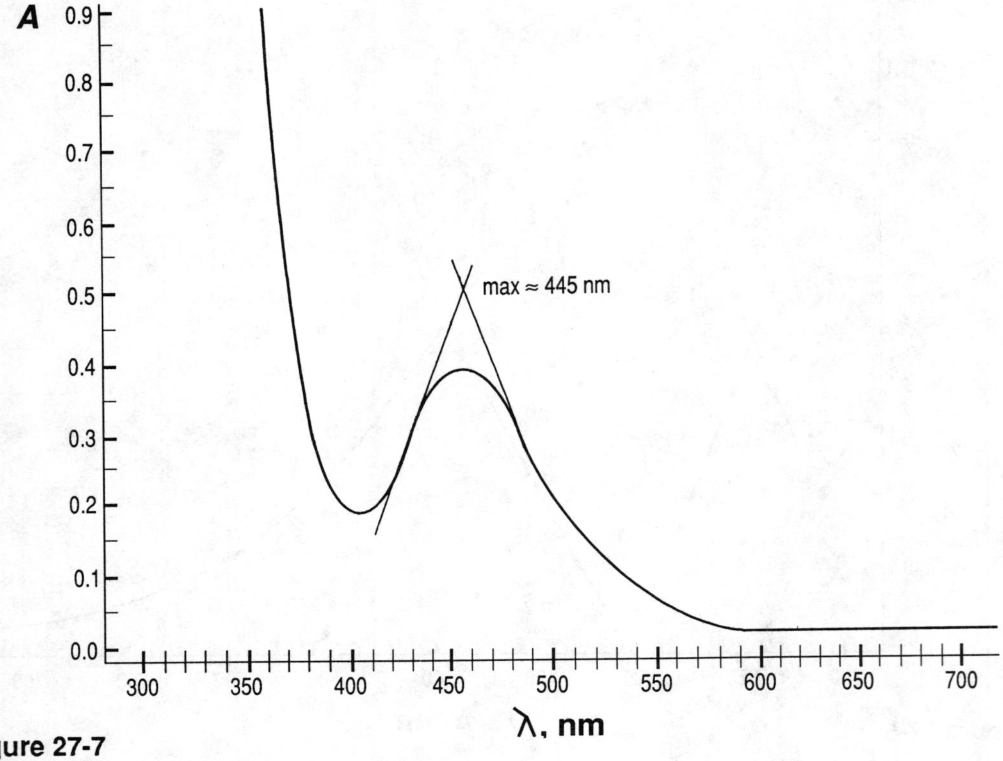


**Figure 27-7**

Spectrum of $[Co(NH_3)_4(NO_2)_2]NO_2$

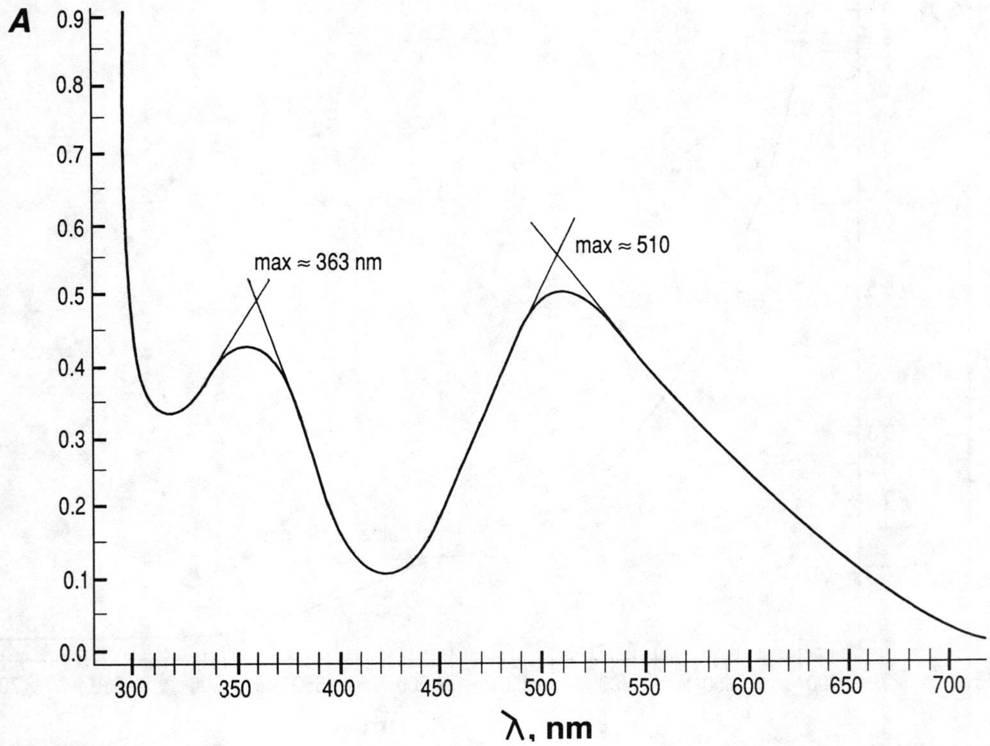


**Figure 27-8**

Spectrum of $[Co(NH_3)_4(H_2O)Cl]Cl_2$

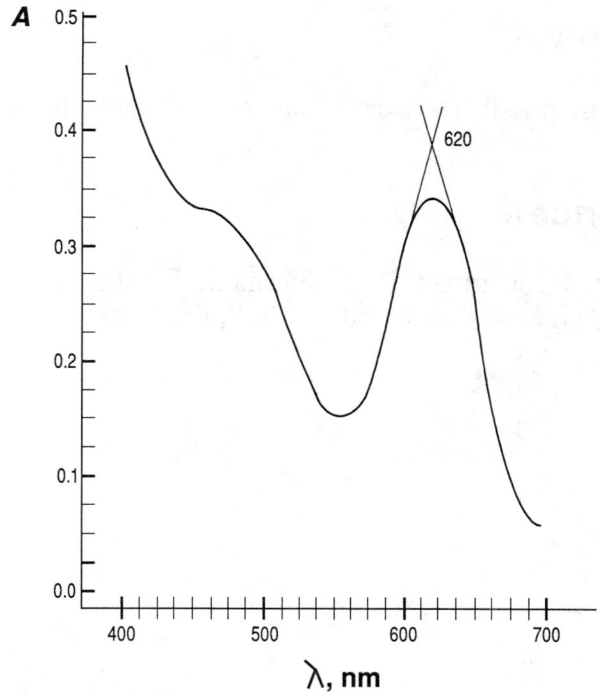


**Figure 27-9**

Spectrum of [*cis*-$Co(NH_3)_4Cl_2$]Cl

## Results

1. Using the weight of $Co(NO_3)_2 \cdot 6H_2O$ you started with, the weight of your product and equation (27-1). calculate the percent yield of $[Co(NH_3)_4(CO_3)](NO_3)$.

2. Make a plot on 8-1/2 x 11" graph paper of Absorbance vs. Wavelength. Draw a smooth curve through your experimental points to obtain the spectrum of your unknown compound in the visible region.

3. Locate the position(s) of maximum absorbance.

4. By comparing your spectrum with those for known cobalt complexes, identify the final product you prepared. Consult Figures 27-5 through 27-9.

5. (a) Knowing the identity of your product, compute its formula weight using a table of atomic weights.

   (b) From the formula weight and the amount of product weighed out, calculate the concentration, $c$, of the solution you prepared.

   (c) Read the absorbance values at points of maximum absorbance from your plot. Use this information together with the solution concentration to compute $\varepsilon$, the molar absorptivity, at each maximum from Equation (27-6).

6. Knowing the identity of the final product you synthesized, write an equation for its production from $[Co(NH_3)_4(CO_3)](NO_3)$. Assume that any necessary anions are provided by the acid HX. Recall that the solution used in the preparation contains acid. Compute the percent yield of this compound.

7. Turn in your products in test tubes labeled with your name, section number, and TA's name.

## Question

1. What are possible sources of random or systematic error in the values of $\varepsilon$ you calculated?

## References

1. Basolo, F., Murman, R. K. Inorganic Syntheses 1948, 4, 171.
2. Olson, G. L. J. Chem. Educ. 1969, 46, 508.

# EXPERIMENT 27

## REPORT SHEET

Name ______________________________ Date ____________________

Section ______________________________

**Data**

| | |
|---|---|
| Weight of Beaker + $Co(NO_3)_2{\bullet}6H_2O$ | ________________ |
| Weight of Beaker | ________________ |
| Weight of $Co(NO_3)_2{\bullet}6H_2O$ Taken | ________________ |
| Weight of Beaker + Dry $[Co(NH_3)_4(CO_3)]NO_3$ | ________________ |
| Weight of Beaker | ________________ |
| Weight of $[Co(NH_3)_4(CO_3)]NO_3$ Produced | ________________ |
| Weight of Beaker + Final Product | ________________ |
| Weight of Beaker | ________________ |
| Weight of Final Product | ________________ |

**Absorbance Data**

| Wavelength ($\lambda$) | $\%T$ | $A = 2.00 - \log \% T$ |
|---|---|---|

## Results

1. Percent yield of $[Co(NH_3)_4(CO_3)]NO_3$ ________________
   Sample Calculation

2. Attach your plot of Absorbance vs. Wavelength.

3. Wavelength(s) of maximum absorbance

   ____________ ____________ ____________

4. Indentity of product ________________________________

5. (a) Formula Weight of Product ________________

   (b) Solution Concentration, *c* ________________
   Sample Calculation

   (c) values

| $\lambda$ | $\varepsilon$ |
|---|---|
| ________ | ________ |
| ________ | ________ |
| ________ | ________ |

Sample Calculation

6. Equation for product synthesis

Percent Yield of Product ____________________
Sample Calculation

In the space below answer the Question.

# INTRODUCTION II

# INORGANIC QUALITATIVE ANALYSIS

## A. Chemical Systems for Study

Qualitative analysis is the study of how to identify the components of a mixture. The particular kinds of mixtures that will concern you are aqueous solutions of inorganic salts. It would be impossible in this section to study all the known inorganic ions, so we shall confine our attention to 15 cations and 7 anions.

The cations you will study here are metallic. This is not surprising if you recall that the most characteristic feature of metals is their *electropositivity*–the ease with which they give up electrons.

It is also worth noting that both cations and anions probably exist in aqueous solution as more complex species than the formulas used in this section indicate. For example, the $Co^{2+}$ ion is known to be closely surrounded by exactly six molecules of water, arranged in a regular fashion with the negative end of the water dipole pointing toward the positively charged cobalt cation. Other cations are also surrounded by several water molecules, but the exact number is not always known. Anions are also surrounded by water molecules, but these molecules are arranged so that the positive end of the water dipole points toward the negatively charged anion. Again, the precise number of water molecules bound in this way is not always known. For this reason, it is more convenient to refer to the ions as $Co^{2+}$ or $Br^-$, for example, even when they are in solution. Sometimes anions and cations are labeled $Co^{2+}(aq)$, as an example, if emphasis is to be placed on the surrounding water molecules.

## B. Identifying Ions

The cations and anions you will be asked to identify can be recognized by the characteristic reactions that they undergo. For instance, a solution containing silver ions will react with chromate ions to form a dark maroon compound, silver chromate. This compound is not very soluble in water, and so it precipitates from solution. The reaction between silver and chromate is represented by the equation

$$2Ag^+(aq) + CrO_4^{2-}(aq) \rightarrow \underline{Ag_2CrO_4} \tag{1}$$

The underlining indicates that the compound is a *solid* precipitate. The silver solution could not be composed merely of dissolved silver cations; enough anions must also be present to balance the positive charge. In fact, solutions containing $Ag^+$ ions are usually made by dissolving silver nitrate

$$AgNO_3(s) \rightarrow Ag^+(aq) + NO_3^-(aq) \tag{2}$$

Because the nitrate ion remains in solution and does not participate in reaction (1), it is of no interest. The nitrate ion is therefore omitted from the equation showing the reaction of silver ion with chromate ion. For the same reason, the cation in the chromate solution is omitted from the reaction, and we speak of the solution as if it contained only chromate.

To summarize, one reaction characteristic of the silver ion is the formation of a dark maroon compound with chromate ion. Thus, if you have a solution that forms such a precipitate when you add chromate, you can conclude that the solution contains $Ag^+$. This reaction can serve as a test for the presence of the ion.

Similarly, the barium ion forms a bright yellow precipitate of barium chromate when chromate ion is added, by the reaction

$$Ba^{2+}(aq) + CrO_4^{2-}(aq) \rightarrow \underline{BaCrO_4} \qquad (3)$$

This reaction can serve as a test for barium. The procedure explained here shows you the reasoning used in qualitative analysis: A solution that is known to contain some ion is found to give a characteristic reaction, such as the precipitation of $BaCrO_4$. If some other solution of an unknown composition behaves similarly, then this behavior is regarded as evidence that the $Ba^{2+}$ ion is present. Obviously, if no reaction is observed with $CrO_4^{2-}$, $Ba^{2+}$ must be absent.

But suppose that a solution contains both $Ba^{2+}$ and $Ag^+$; if we try to confirm the presence of either of these ions by adding chromate, both maroon silver chromate and yellow barium chromate will precipitate at the same time

$$2Ag^+(aq) + CrO_4^{2-}(aq) \rightarrow \underline{Ag_2CrO_4} \qquad \textit{(maroon)} \qquad (1)$$

$$Ba^{2+}(aq) + CrO_4^{2-}(aq) \rightarrow \underline{BaCrO_4} \qquad \textit{(yellow)} \qquad (3)$$

The resulting precipitate will be brownish in color, and the test will be indeterminate. Here, silver and barium are said to "interfere" with each other. To confirm the presence of either ion using the chromate test, the two ions would have to be separated, and the chromate test would have to be performed on each. It is easy to see, then, that the identification of ions in solution requires (1) the separation of interfering ions, and (2) the confirmation of each ion using a separate test reaction for each (usually the formation of a precipitate).

## C. Group Separation

To make it easier to detect the 15 cations, they are separated into more manageable groups that have some solubility similarities; each ion in a group is then separated from the others. As an example, all the ions in Group I ($Ag^+$, $Hg_2^{2+}$, and $Pb^{2+}$) form insoluble chlorides. A solution of all 15 cations can be treated with hydrochloric acid; this treatment immediately precipitates the insoluble chlorides of Group I cations, leaving the ions of all the other groups in solution. The white chloride precipitate can then be tested for $Ag^+$, $Hg_2^{2+}$, and $Pb^{2+}$ by separating each ion from the others and then confirming the presence of each ion individually.

Group II ions all form sulfides that are insoluble in acidic solutions. Thus, treating an acidic solution containing ions in Groups II-III with ammonium sulfide produces a black precipitate of Group II sulfides while the ions in Group III remain in solution.

## D. Technique

Your laboratory work in qualitative analysis will be carried out on a somewhat smaller scale than in most of the other experiments in this manual. In addition to the obvious savings in quantities of reagents, working on a smaller scale will allow you to complete the qualitative experiments in less time, because you will not have to wait for large quantities of precipitate to form, transfer large volumes of liquid, and so forth. This small-scale work is called *semimicro qualitiative analysis*. The special laboratory techniques and equipment used in semimicro work will be described in this section.

**1. Equipment.** Your set of desk equipment includes a micro burner, several semimicro test tubes in a rack, a metal test-tube support, and a set of dropping bottles for reagent solutions. Volume measurements are conveniently made by counting numbers of drops instead of by mL when working on a semimicro scale. Unless the dropping bottles have been filled in the stockroom, your instructor will tell you which reagent solutions to place in these dropping bottles. Be sure that the bottles have been washed and thoroughly rinsed twice with distilled water. Label each bottle.

In addition to the droppers provided in these bottles, you should have a few droppers for withdrawing liquids from small test tubes. The most convenient kind of dropper is a capillary pipet, a piece of glass tubing drawn to a very long and narrow constriction on one end and equipped with a rubber bulb for suction (see Figure ii-1). These will be supplied by your instructor.

You should make four or five stirring rods, each not longer that 4 in. (the technique for breaking and fire-polishing glass is described on pages 8-9). You will be using these stirring rods continually for several weeks, so be sure to fire-polish the ends to avoid injury.

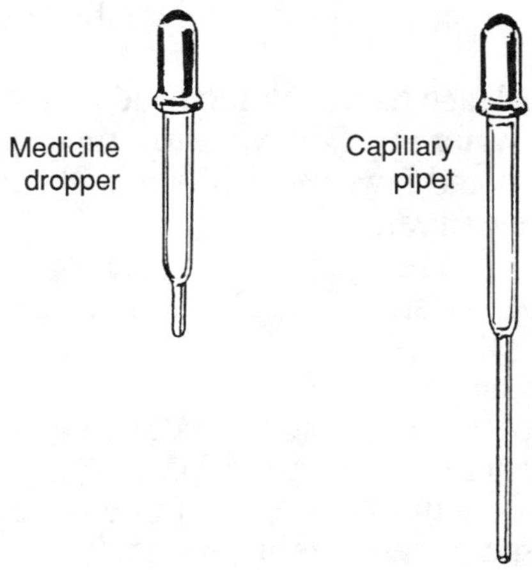


Figure ii-1

**2. Avoiding Contamination.** The anions and the cations that you will work with in qualitative analysis will be supplied in concentrations of only $\leq$ 10 mg/mL. If you do not clean your glassware carefully and completely, you can introduce quantities of contaminants that are almost this large in themselves into your analysis. Added reagents could then react with contaminants to produce spurious results.

Tap water is a particularly common source of contaminant ions. After washing any glassware to be used in qualitative analysis, rinse the equipment at least twice with distilled water. Capillary pipets should be rinsed twice with distilled water after each use; if you develop the habit of automatically washing your glassware after each use, you will be certain to pick up a clean pipet whenever you need one. Keeping a beaker of distilled water on your desk will provide a convenient way to rinse your pipets.

It is almost unnecessary to say that any water you add to solutions must be distilled water.

A clean towel on your working area will provide a place to lay stirring rods and pipets so that they will not pick up any contaminants from the desk surface.

As usual, remember never to introduce your pipets into community vessels. Most of these vessels will be equipped with their own droppers for removing solutions. The droppers you use to add reageants to test tubes should be held above the tube and not inserted into it.

**3. Adding Reagents.** When working on a semimicro scale, liquid reagents are added from droppers or capillary pipets, and volumes are measured by counting the number of drops. A good rule of thumb is that 1 drop = 0.05 mL; this means, of course, that 1 mL = 20 drops.

Occasionally, you will need to measure small volumes accurately. For this purpose, you will need to calibrate two capillary pipets. To do this, fill a 10-ml graduated cylinder to the 5.0 mL mark with distilled water, and count the number of drops of distilled water required from the pipet to bring the level up 2 ml to the 7.0-ml mark on the cylinder. Obviously, half this number of drops will equal 1 mL. Label the pipet with this calibration information.

**4. Mixing Solutions.** Several techniques are available for mixing solutions in semimicro test tubes. The most obvious one is to use a stirring rod. This is really not an efficient stirring technique, however. It is more satisfactory (if the tube is not too full) to hold the test tube in the thumb and forefinger of one hand and flick the bottom of the tube with the forefinger of the other hand. Alternatively, the contents may be partially sucked up in a capillary pipet (but not up to the bulb) and then squirted back into the tube. This procedure should be repeated two or three times for each mixing.

**5. Centrifuging.** Many tests in qualitative analysis produce precipitates that must be separated from the remaining solution (called the *supernatant liquid*). This is commonly done by placing the test tube in the head of a centrifuge and spinning it at high speed. The centrifugal force packs the precipitate at the bottom of the tube many times faster than it would take if you allowed the precipitate to settle naturally.

When you centrifuge, be sure to balance the revolving head by placing a similar test tube that is filled to the same level with distilled water opposite your test tube in the centrifuge. If you do not do this, the centrifuge will be damaged and fewer centrifuges will be available for your laboratory work if some must be removed to be repaired.

Once the centrifuge is turned on, it should spin at top speed for 30 seconds or so. Then turn off the switch and let the centrifuge come to a stop by itself. Do *not* attempt to slow it down or stop it with your hand.

**6. Testing for Completeness of Precipitation.** After you remove a tube from the centrifuge, you should always test to be sure the precipitation is complete. To do this, add an additional drop of the precipitating reagent and note whether more precipitate forms. If it does, add a few more drops of the reagent until precipitation seems to stop. Then centrifuge and test again.

**7. Separating the Supernatant Liquid from a Precipitate.** A capillary pipet can be used to draw off the solution above a precipitate. Before you insert the pipet into the test tube, be sure to squeeze the air out of the bulb.

Some precipitates *flocculate* (remain partly dispersed in solution) even after centrifuging. If this happens, you can insert a tiny plug of cotton into the capillary pipet, and then filter the supernatant liquid through the cotton plug.

**8. Washing Precipitates.** After a supernatant liquid has been withdrawn as completely as possible with a pipet, some of the liquid will still cling to the precipitate. This liquid will contain ions that can interfere with further tests you will make on the precipitate. Thus, this solution must be removed by washing.

The wash liquid is usually distilled water; sometimes another solution may be specified. The wash liquid is added to the precipitate, and the packed precipitate is broken up by stirring it until it is completely dispersed in the liquid. The tube is then centrifuged, and the liquid is drawn off. This wash liquid is usually discarded, but if it contains an appreciable quantity of ions, you may be directed to add it to the first solution you removed from the precipitate.

**9. Heating Solutions.** Test tubes containing solutions are usually heated in a water bath, as shown in Figure ii-2.

**10. Evaporating Solutions.** When the volume of a solution is to be reduced appreciably or when a solution is to be taken to dryness, it is placed in a porcelain casserole or crucible and heated under a *hood*, so that any toxic fumes can be expelled from the laboratory. This heating must be conducted in such a way that the solution does not spatter. This can be accomplished by using a 1/2 in. flame from the micro burner; such a flame is referred to as a "microflame". The gas level can be adjusted properly by using a screw clamp on the burner hose.

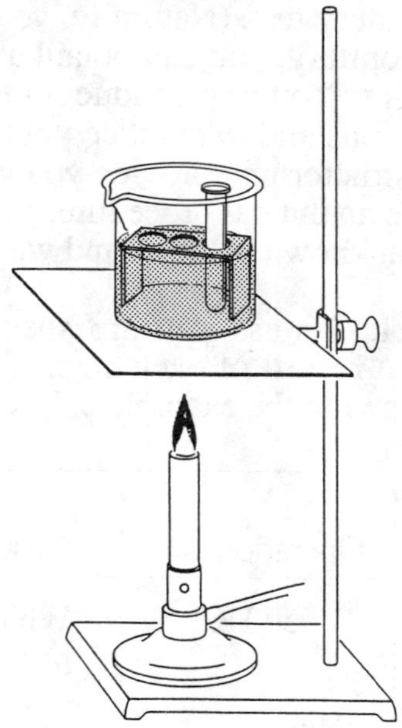

**Figure ii-2**

Heating test tube in a water bath.

During heating, the crucible (held with tongs) or casserole should be continuously swirled in the microflame. As dryness approaches, the bottom of the vessel will dry first. Keep bringing fresh solution to the dry bottom by swirling and tilting the vessel. Just before complete dryness is reached, remove the casserole or crucible from the flame. The heat of the vessel will complete the evaporation process.

If the residue is baked in the flame after it is dry, volatile substances may be lost or the residue may be converted to an insoluble form.

**11. Dissolving Precipitates.** If a precipitate is to be redissolved, do it as soon as possible. If it must wait, cover the precipitate with a few drops of the precipitating agent or the liquid into which it will be dissolved.

## E. Your Laboratory Work in Qualitative Analysis

After so many general instructions, it would probably be helpful to comment more specifically on exactly what you will do in the laboratory.

**1. First Qualitative Analysis Period.** Before you can work in the laboratory, you will need to:

(a) Clean your semimicro test tubes (and dropping bottles, if they are empty), rinsing each twice with distilled water.

(b) Obtain five or six capillary pipets and squeeze bulbs.

(c) Make four or five stirring rods, each 4 in. or so in length; fire-polish the ends.

(d) Calibrate two capillary pipets.

(e) Label and fill your dropping bottles with the solutions provided by your instructor. If the stockroom provides these reagents, be sure to obtain a complete set.

**2. Subsequent Laboratory Periods.** You will analyze each group of ions separately. Before coming to class, carefully study the Introduction to each qualitative experiment. You will be expected to know the analytical procedure for each group and to be able to write an equation for each separation and confirmatory test. (Naturally, you do not need to memorize quantities of reagents to be used.) In the laboratory, you can obtain a so-called "known" solution containing ~10 mg/mL each of the group ions. You will conduct the group analysis procedure using this known solution to observe the reactions and to practice your skills. After you complete your analysis of the known solution, your instructor will provide you with an unknown that may contain any of the group ions. By following the analytical procedure, you will decide and report which of the possible ions are present in your unknown solution and which are absent.

**3. Keeping a Notebook.** Unless you are specifically directed to do otherwise by your instructor, you are to keep a record of each step in the analysis of both your known and your unknown in a notebook, following the example below:

| Solution or precipitate | Operation | Observation | Conclusion |
|---|---|---|---|
| Group I known | Added HCl | White precipitate | $Ag^+$, $Pb^{2+}$, $Hg_2^{2+}$ possibly present |
| Precipitate 1.1 | Added hot water | Some precipitate dissolved | $Pb^{2+}$ probably present |

A complete record will help your instructor to surmise where you may have made any errors and will assure you that you have reached warranted conclusions.

If your notes are to be of any value, they must be recorded during the experiment–*not* at the end of the period or later in the day.

You may find it helpful to consult more complete treatments of qualitative analysis. [For example, see E.J. King, *Ionic Reactions and Separations: Experiments in Qualitative Analysis* (New York: Harcourt Brace Jovanovich, Inc., 1973); and C.H. Sorum and J.J. Lagowski, *Introduction to Semimicro Qualitative Analysis,* Fifth Edition (Englewood Cliffs, N.J.: Prentice-Hall, 1977).]

## Solubility Rules

1. All sodium, potassium, and ammonium salts are water soluble.
2. All metal nitrates, acetates, and chlorates are water soluble. Silver acetate is slightly soluble.
3. Hydroxides of the alkali metals are water soluble. Hydroxides of Ca and Ba are slightly soluble. All other hydroxides are insoluble.
4. Chlorides, bromides and iodides of all the metals except Pb, Ag, and Hg(I) are soluble. $PbCl_2$, $PbBr_2$, and $PbI_2$ are soluble in hot water. $HgI_2$ is insoluble.
5. Sulfates of all metals are soluble except for $CaSO_4$, $BaSO_4$, $PbSO_4$, and $(Hg_2)_2SO_4$. $Ag_2SO_4$ is slightly soluble.
6. Carbonates, phosphates, arsenates, sulfites, and chromates of all metals are insoluble except for those of Na, K, and ammonium.
7. All sulfides except those of Na, K, Mg, Ca, Ba and ammonium are insoluble. MgS, CaS, and BaS are slightly soluble.

# EXPERIMENT 28

# QUALITATIVE ANALYSIS OF CATION GROUP I

## Introduction

Separation of the Group I cations $Ag^+$, $Pb^{2+}$, and $Hg_2^{2+}$ from the cations in later groups depends on the fact that chlorides of Group I cations are insoluble in acid solution. This is a rather unusual occurrence as you can see from the general solubility rules given on page 262. All other cations form soluble chlorides which remain in solution when HCl is added.

## Separation of Group I Cations

The precipitate consisting of white, insoluble AgCl, $PbCl_2$, and $Hg_2Cl_2$ forms when 6M HCl is added to the solution containing the cations. Only a slight excess of HCl is added to obtain the precipitate.

$$Ag^+ + Cl^- \rightleftharpoons \underline{AgCl} \qquad (28\text{-}1)$$

In equations of experiments 28 to 31 solid precipitates are indicated by underlining the formula of the compound. Nonunderlined compounds or ions are soluble and remain dissolved in solution.

Although Equation (28-1) and similar equations which can be written for $Pb^{2+}$ and $Hg_2^{2+}$ suggest that the equilibrium can be shifted as far to the right as desired by adding sufficient $Cl^-$ ions, the solubilities of the chlorides begin to increase after a large enough $[Cl^-]$ is reached, because of complex ion formation:

$$\underline{AgCl} + Cl^- \rightleftharpoons AgCl_2^-$$

$$\underline{PbCl_2} + Cl^- \rightleftharpoons PbCl_3^- \qquad (28\text{-}2)$$

$$\underline{Hg_2Cl_2} + 2Cl^- \rightleftharpoons HgCl_4^{2-} + \underline{Hg(l)}$$

Even a slight excess of $Cl^-$ will cause the essentially complete removal of these cations as their chlorides. In the solutions used here, $[Ag^+] = 0.02M$ before the addition of HCl. At the conclusion of the group precipitation, sufficient $Cl^-$ is still present so that $[Cl^-] = 0.35M$. The maximum amount of $Ag^+$ that can remain in solution is given by:

$$[Ag^+]\ (0.35) = K_{sp} = 1.8 \times 10^{-10} \qquad (28\text{-}3)$$

which gives $[Ag^+] = 5 \times 10^{-10}M$. Thus $[Ag^+]$ has been decreased from 0.02M to $5 \times 10^{-10}M$; this is certainly complete enough removal of $Ag^+$ for practical purposes.

Although $Hg_2^{2+}$ is also completely precipitated under these conditions, $PbCl_2$ is more soluble, and a sufficiently high $[Pb^{2+}]$ remains in solution to give a precipitate of PbS with Group II cations.

## Separation and Confirmation of the Cations

The $PbCl_2$ can be separated from the AgCl and $Hg_2Cl_2$ precipitates by taking advantage of a difference in solubility characteristics. All three chlorides increase in solubility as the temperature of the water is raised; however, since $PbCl_2$ is much more soluble than the other chlorides even at room temperature, hot water will dissolve all the $PbCl_2$, thus separating it from AgCl and $Hg_2Cl_2$:

$$\underline{PbCl_2} \xrightarrow[\Delta]{H_2O} Pb^{2+} + 2Cl^- \qquad (28\text{-}4)$$

After the precipitates are heated and stirred and the $PbCl_2$ has dissolved, the AgCl and $Hg_2Cl_2$ must be quickly centrifuged to the bottom of the test tube. The supernatant liquid containing $Pb^{2+}$ is poured off or drawn off with a capillary pipet before the solution cools sufficiently to reprecipitate $PbCl_2$.

The presence of $Pb^{2+}$ is confirmed by adding $K_2CrO_4$. A yellow precipitate of lead chromate, $PbCrO_4$ ($K_{sp} = 2 \times 10^{-16}$), forms if $Pb^{2+}$ is present:

$$Pb^{2+} + CrO_4^{2-} \rightleftharpoons \underline{PbCrO_4} \qquad (28\text{-}5)$$

The precipitate of AgCl and $Hg_2Cl_2$ is treated with conc. (15M) ammonia, and the silver and mercurous ions confirmed by the different behavior of their chlorides with $NH_3$.

Silver chloride dissolves in $NH_3(aq)$ to form a complex ion:

$$\underline{AgCl} + 2NH_3(aq) \rightarrow Ag(NH_3)_2^+ + Cl^- \qquad (28\text{-}6)$$

Mercurous chloride undergoes a *disproportionation reaction* with $NH_3(aq)$. In a disproportionation reaction simultaneous oxidation and reduction of atoms of a single kind occurs. In this reaction, mercurous ions $Hg_2^{2+}$ are oxidized to mercuric ions $Hg^{2+}$ and reduced to liquid metallic mercury $Hg^0$ (l):

$$\underline{Hg_2Cl_2} + \underline{2NH_3} \rightarrow \underline{Hg(NH_2)Cl} + \underline{Hg(l)} + NH_4Cl \qquad (28\text{-}7)$$

The $Hg^{2+}$ forms a white precipitate of $Hg(NH_2)Cl$ (mercuric amidochloride), while the metallic mercury in finely divided form is black. The mixture of the two precipitates appears gray. The appearance of this gray color is taken as evidence for the presence of $Hg_2^{2+}$.

The supernatant liquid from the above reaction contains $Ag(NH_3)_2^+$ ions. Their presence is confirmed by acidifying the solution with $HNO_3$. Protons combine with basic $NH_3$ to form $NH_4^+$, destroying the silver ammine complex in the process:

$$Ag(NH_3)_2^+ + 2H_3O^+ \rightarrow Ag^+ + 2NH_4^+ + 2H_2O \qquad (28\text{-}8)$$

The $Ag^+$ ions then react with the $Cl^-$ ions in solution to form a white precipitate of AgCl:

$$Ag^+ + Cl^- \rightarrow \underline{AgCl} \qquad (28\text{-}9)$$

## Chemicals Needed

$Ag^+$ (2mg/mL)
6M $HC_2H_3O_2$; acetic acid
6M HCl; dil. hydrochloric acid
12 M HCl; conc. hydrochloric acid
$Hg_2^{2+}$ (10 mg/mL)
6M $HNO_3$; dil. nitric acid
16M $HNO_3$; conc. nitric acid
0.5M $K_2CrO_4$; potassium chromate
litmus paper
6M $NH_3(aq)$; dil. aqueous ammonia
15M $NH_3(aq)$; conc. aqueous ammonia
$Pb^{2+}$ (15 mg/mL)

## Procedure

### Analysis of Known and Unknown Samples

1. Use 5 drops of a known solution containing about 15 mg/mL $Pb^{2+}$, 10 mg/mL $Hg_2^{2+}$, and 2 mg/mL $Ag^+$.

2. The unknown solution may contain cations of Group I and perhaps of later groups also. If the solution is basic, make it acid to litmus with a few drops of 6M HCl.

3. If the unknown is a solid, weigh about 50 mg of the homogeneous mixture into a small casserole or crucible and dissolve it in a few drops of 6M $HNO_3$. Transfer the solution to a 4-mL test tube. Add 1 mL of 12M HCl, mix thoroughly, and heat in the water bath for several minutes. Remove the tube from the water bath, add 1 mL distilled water, and cool the tube under cold running water. Centrifuge. The white solid is the mixture of Group I chlorides, Precipitate 1.1 in Outline 1. If the solid contained only Group I cations, you can discard the supernatant liquid after centrifuging. If the solid was a General Unknown containing ions of all groups, the supernatant liquid will contain cations of Groups II and III and should be saved for analysis of these groups.

## Outline 1: Cation Group I Analysis

To your sample of known or unknown solution in a centrifuge tube, add 1 drop of 6M HCl and let the precipitate settle. Add one more drop of HCl and see whether any more precipitate forms. Continue this test for completeness of precipitation until no more precipitate forms on HCl addition. Stir and allow time for the $PbCl_2$ to precipitate before centrifuging. Then decant the supernatant liquid into another test tube or draw it off with a capillary pipet. Be sure to remove as much of the liquid as you can. After removal, test it again for completeness of precipitation. Then wash the precipitate with 3-4 drops of a solution made by adding 1 drop of 6M HCl to 8 mL of distilled water.

**Precipitate 1.1. AgCl, $Hg_2Cl_2$, $PbCl_2$.** Add about 10 drops of distilled water and warm for 2-3 minutes in a boiling water bath with occasional stirring. Centrifuge and rapidly separate the supernatant liquid from the precipitate. Transfer the hot solution to a test tube, and immediately test it for $Pb^{2+}$.

**Solution 1.1. Ions of later groups.** Place the solution in a stoppered test tube until you are ready to analyze it.

**Precipitate 1.2. AgCl, $Hg_2Cl_2$, ($PbCl_2$).** If you found $Pb^{2+}$ in Solution 1.2, extract the precipitate with hot water until the extract gives only faint cloudiness when treated with 0.5M $K_2CrO_4$. Add 3-4 drops of 15M $NH_3$ to the precipitate. Stir with a stirring rod, centrifuge, and transfer the supernatant liquid to a test tube.

**Solution 1.2. $Pb^{2+}$.** Add 1 drop 6M $HC_2H_3O_2$ [1] followed by 3-5 drops of 0.5M $K_2CrO_4$ and cool to room temperature. A yellow precipitate of $PbCrO_4$ confirms the presence of $Pb^{2+}$.

**Precipitate 1.3. $Hg(NH_2)Cl$, Hg.** The gray precipitate (a mixture of black Hg and white $Hg(NH_2)Cl$) confirms the presence of $Hg_2^{2+}$.

**Solution 1.3. $Ag(NH_3)_2^+$, $Cl^-$.** Make the solution acid to litmus with 6M $HNO_3$. A white precipitate of AgCl confirms the presence of $Ag^+$.

## Note to Outline 1

1. If the intial precipitate of a General Unknown were not sufficiently washed, $CuCrO_4$ and $(BiO)_2CrO_4$ might precipitate at this point. Addition of acetic acid protonates chromate to give $HCrO_4^-$, thus lowering the $[CrO_4^{2-}]$ to the point where only the very insoluble $PbCrO_4$ precipitates.

## Results

Submit a 3x5 card for your unknown solution, giving your name, desk number, ions present, and ions absent. Have your notebook available for your instructor to examine on request.

### References


1. King, Edward J. "Ionic Reactions and Separations: Experiments in Qualitative Analysis", Harcourt Brace Jovanovich, New York, 1973. pp. 129-142.

2. Sorum, C.H. and Lagowski, J.J. "Introduction to Semimicro Qualitative Analysis", 5th ed., Prentice Hall, Englewood Cliffs, 1977. pp. 143-156.

# EXPERIMENT 29

# QUALITATIVE ANALYSIS OF CATION GROUP II

## Introduction

Group II has several cations: $Hg^{2+}$, $Pb^{2+}$, $Bi^{3+}$, $Cu^{2+}$, Sb(III), Sb(V), $Sn^{2+}$, and Sn(IV). You will undoubtedly find it more difficult to separate and confirm the Group II cations than you did for the Group I cations. If you are to obtain good results, you must be thoroughly familiar with the chemistry described in this section before coming to the laboratory. Only then can you be sure that you understand exactly what is happening when you perform each step. It is also important that your work be done very carefully. With so many ions present, the possibility for error is considerably increased.

You will notice that some of the cations are identified by their oxidation states (with Roman numerals) instead of by an ionic charge. This is because the free cations do not exist in solution. Rather, some more complex species (and sometimes a mixture of more complex species) actually exists in solution. For example, there is no $Sb^{3+}$ ion. Instead, Sb(III) is found as $SbO^+$ in acid solution and as $SbO_2^-$ in basic solution.

Some of the cations might be present in more than one oxidation state, depending on what compounds were used to prepare solutions or solid unknowns. For example, $Sn^{2+}$ and Sn(IV) might be present in the same solution. In the course of the analysis, both of the oxidation states will be converted to the same compound which can then be reported as the oxidation state detected. All the $Sn^{2+}$ present in solutions is first oxidized to Sn(IV) and subsequently reduced to $Sn^{2+}$ which is then detected by its reaction with $Hg^{2+}$. This obviously makes it impossible to test separately for the presence of $Sn^{2+}$ and Sn(IV). Hence, a positive test for Sn is reported as indicating the presence of $Sn^{2+}$. (Note that the two oxidation states of mercury can be separately identified: $Hg_2^{2+}$ appears in Group I while $Hg^{2+}$ appears in Group II.)

## Separation of Group II Cations from Later Groups

The Group II cations all have the common property that they form sulfides that are insoluble in acid solution. Table 29-1 gives the solubility products and colors of several sulfides including those of Group II.

Table 29-1
$K_{sp}$'s and Colors of Some Sulfides

| Sulfide | $K_{sp}$ | Color | Sulfide | $K_{sp}$ | Color |
|---|---|---|---|---|---|
| $Bi_2S_3$ | $8 \times 10^{-101}$ | brown-black | CoS | $8 \times 10^{-23}$ | black |
| | | | ZnS | $3.2 \times 10^{-23}$ | white |
| $Sb_2S_3$ | $1 \times 10^{-91}$ | orange | NiS | $8 \times 10^{-19}$ | black |
| HgS | $4 \times 10^{-54}$ | red | FeS | $6 \times 10^{-19}$ | black |
| | $2 \times 10^{-53}$ | black | | | |
| CuS | $6 \times 10^{-37}$ | black | | | |
| PbS | $3 \times 10^{-28}$ | black | | | |
| SnS | $1 \times 10^{-28}$ | brown | | | |

Other cations that form insoluble sulfides include several from Group III. The important question to be considered here is how the Group II cations can be cleanly separated by precipitation as sulfides from the Group III cations that also form slightly soluble sulfides. The answer lies in a consideration of equilibria between solid sulfide precipitates and the sulfur-containing species present in aqueous solution.

$S^{=}$ is too strong a base to exist in any appreciable concentration in aqueous solutions. It is immediately protonated by water to form $HS^-$ (just as $O^{=}$ does not exist in solution, being immediately protonated to form $OH^-$). The values of the acid dissociation constants for $H_2S$ show that this is so:

$$H_2S(aq) + H_2O \rightleftharpoons H_3O^+ + HS^- \tag{29-1}$$

$$K_1 = \frac{[H_3O^+][HS^-]}{[H_2S]} = 1.0 \times 10^{-7} \tag{29-2}$$

$$HS^- + H_2O \rightleftharpoons H_3O^+ + S^{=} \tag{29-3}$$

$$K_2 = \frac{[H_3O^+][S^{=}]}{[HS^-]} = 10^{-17\pm2} \tag{29-4}$$

Even in very basic solutions where maximum deprotonation of $HS^-$ would be expected to occur, $[S^{=}]$ is negligible in comparison to the concentrations of protonated species. For example, in 0.1M NaOH (pH 13) rearranging Equation (29-4) shows that

$$\frac{[S^{=}]}{[HS^-]} = \frac{10^{-17\pm2}}{[H_3O^+]} = \frac{10^{-17\pm2}}{10^{-13}} = 10^{-4\pm2} \tag{29-5}$$

Thus, at a maximum, only 1% of the sulfide ion remains unprotonated in a pH 13 solution. The rest (99%) is protonated to give $HS^-$. We seldom work with such basic solutions, so the concentration of $S^{=}$ will be negligible in all our solutions.

You can see from the relatively large value of $K_1$ that, if the solution is very acidic, by far the greatest fraction of $HS^-$ will also be protonated to $H_2S$. For example, at pH 0.5,

$$\frac{[HS^-]}{[H_2S(aq)]} = \frac{K_1}{[H_3O^+]} = \frac{1.0 \times 10^{-7}}{10^{-0.5}} = 3 \times 10^{-7} \tag{29-6}$$

This means that when a metal sulfide dissolves, we can think of the following equilibria as occurring in acid solution:

$$\underline{MS} + H_2O \rightleftharpoons M^{2+}(aq) + HS^- + OH^- \qquad K_{sp} \tag{29-7}$$

$$HS^- + H_3O^+ \rightleftharpoons H_2S(aq) + H_2O \qquad K = \frac{1}{K_{a1}} \tag{29-8}$$

$$H_3O^+ + OH^- \rightleftharpoons 2H_2O \qquad K = \frac{1}{K_w} \tag{29-9}$$

The net result is the sum of Equations (29-7), (29-8) and (29-9):

$$\underline{MS} + 2H_3O^+ \rightarrow M^{2+}(aq) + H_2S(aq) + 2H_2O \tag{29-10}$$

The equilibrium constant for Equation (29-10) is:

$$K_{spa} = \frac{[M^{2+}(aq)]\,[H_2S(aq)]}{[H_3O^+]^2} = \frac{K_{sp}}{K_{a1}K_w} \qquad (29\text{-}11)$$

where the subscript "spa" stand for solubility product in acid solution.

We can now see what this has to do with separation of metal ions by precipitating their sulfides. The precipitation equilibrium is just the solubility equilibrium considered from the other direction. We will now show that by proper adjustment of the pH in a solution saturated with $H_2S$, we can precipitate the sulfides of Group II while leaving those of Group III in solution. The equilibrium established when a metal sulfide is precipitated out of solution is:

$$M^{2+}(aq) + H_2S + 2H_2O \rightleftharpoons \underline{MS} + 2H_3O^+ \qquad (29\text{-}12)$$

and we have for the equilibrium constant of Equation (29-12):

$$K_{eq} = \frac{[H_3O^+]^2}{[M^{2+}(aq)][H_2S(aq)]} = \frac{1}{K_{spa}} = \frac{K_{a1}K_w}{K_{sp}} \qquad (29\text{-}13)$$

Suppose we adjust the pH to 0.5 and saturate the solution with $H_2S$. It is an experimental fact that the $[H_2S]_{sat.} = 0.1M$. Consider the possibility of precipitating CuS under these conditions. Plugging relevant values into Equation (29-13):

$$\frac{[H_3O^+]^2}{[Cu^{2+}(aq)][H_2S(aq)]_{sat.}} = \frac{(10^{-0.5})^2}{[Cu^{2+}]\,(0.1)} = \frac{K_{a1}K_w}{K_{sp}} = \frac{(1.0 \times 10^{-7})(1.0 \times 10^{-14})}{K_{sp}} \qquad (29\text{-}14)$$

$$[Cu^{2+}(aq)] = \frac{K_{sp}(10^{-0.5})^2}{(0.1)(1.0 \times 10^{-7})(1.0 \times 10^{-14})} = \qquad (29\text{-}15)$$

$$\frac{K_{sp}}{1.0 \times 10^{-21}} = \frac{6.0 \times 10^{-37}}{1.0 \times 10^{-21}} = 6 \times 10^{-16}M$$

Hence, the calculation shows that at pH 0.5 in a saturated $H_2S$ solution, the maximum possible $[Cu^{2+}(aq)]$ is around $10^{-15}M$, the remainder of any $Cu^{2+}$ originally present being precipitated as CuS. If we start with a solution 0.01M in $Cu^{2+}$, you can see that essentially all the Cu is precipitated out as black CuS.

Suppose that the solution also intially contained 0.01M $Co^{2+}$. A similar calculation shows that in this solution, the maximum $[Co^{2+}]$ that can be present if the solution is saturated with $H_2S$ at pH 0.5 is:

$$[Co^{2+}(aq)] = \frac{K_{sp}(10^{-0.5})^2}{(0.1)(1.0 \times 10^{-7})(1.0 \times 10^{-14})} = \qquad (29\text{-}16)$$

$$\frac{K_{sp}}{1.0 \times 10^{-21}} = \frac{8.0 \times 10^{-23}}{1.0 \times 10^{-21}} = 0.08M$$

Hence, the maximum possible $[Co^{2+}]$ is greater than we started with and no CoS will precipitate. This effectively separates the $Cu^{2+}$ and $Co^{2+}$ ions.

A look at Equations (29-15) and (29-16) shows that, in a pH 0.5 solution saturated with $H_2S$ where $[M^{2+}] = 0.01$, any ion whose $K_{sp}$ for MS is $\leq 10^{-25}$ will be soluble to the extent of $\leq 10^{-4}$M, and so will precipitate. Any ion whose $K_{sp}$ for MS is $\geq 10^{-23}$ will be soluble to the extent of $\geq 10^{-2}$M and will not precipitate as the sulfide. (Similar calculations can be made for sulfides of different stoichiometry such as $Sb_2S_3$.) It is this behavior that allows the separation of Group II ions (sulfide $K_{sp}$'s $\leq 10^{-28}$) from Group III ions (sulfide $K_{sp}$'s $\geq 10^{-23}$).

Once the pH is adjusted to 0.5, the solution is kept saturated by adding $(NH_4)_2S$. The sulfide ions of this compound are protonated in solution and so act as a source of $H_2S$.

## Subdivision of Group II Cations

There are so many ions to test for in Group II that this group is separated into two subgroups after initial precipitation. The separation into subgroups is possible because of the amphoteric nature of some of the metal sulfides. As you may recall, an amphoteric compound can function both as a base and as an acid. When treated with KOH, $Sb_2S_3$ and $SnS_2$ behave as acids and dissolve with formation of complex ions:

$$\underline{Sb_2S_3} + 6OH^- \rightarrow SbS_3^{3-} + SbO_3^{3-} + 3H_2O \qquad (29\text{-}17)$$

$$\underline{SnS_2} + OH^- \rightarrow SnS_2OH^- \qquad (29\text{-}18)$$

HgS is weakly acidic, and some HgS may also dissolve. The remaining precipitate consists of the very weakly acidic sulfides HgS, PbS, CuS, and $Bi_2S_3$. These two subgroups are referred to as the antimony and the copper subgroups of Group II, respectively.

## Separation and Confirmation of Copper Subgroup Ions

**Mercury.** All the copper subgroup sulfides except HgS can be dissolved by heating them in $HNO_3$. Nitric acid oxidizes sulfide ion to elemental sulfur:

$$\underline{MS} + 2NO_3^- + 4H_3O^+ \rightleftharpoons M^{2+} + \underline{S} + 2NO_2(g) + 6H_2O \qquad (29\text{-}19)$$

The oxidation of sulfide shifts the solubility equilibria to the right as $H_2S$ is continually removed and the ion product drops below the value of $K_{spa}$. For example,

$$\underline{CuS} + 2H_3O^+ \rightleftharpoons Cu^{2+} + H_2S + 2H_2O \xrightarrow{NO_3^-} \underline{S} \qquad (29\text{-}20)$$

HgS does not dissolve in $HNO_3$ because its $K_{sp}$ is so small ($10^{-54}$) that sufficient $H_2S$ cannot be removed by oxidation to shift the equilibrium to the right.

After separating the solution containing $Pb^{2+}$, $Bi^{3+}$, and $Cu^{2+}$, the black precipitate of HgS can be dissolved in aqua regia. This reagent not only lowers the $[H_2S]$ by oxidizing it to S, but also lowers the $[Hg^{2+}]$ by converting it to the complex ion $HgCl_4^{2-}$, thus shifting the solubility equilibrium to the right:

$$\underline{HgS} + 2H_3O^+ \rightleftharpoons Hg^{2+} + H_2S + 2H_2O \qquad (29\text{-}21)$$

$$\begin{array}{ll} Hg^{2+} \;\; \Big\Updownarrow Cl^- & H_2S \;\; \Big\downarrow NO_3^- \\ HgCl_4^{2-} & \underline{S} \end{array}$$

$$\underline{HgS} + 2NO_3^- + 4Cl^- + 4H_3O^+ \rightarrow HgCl_4^{2-} + \underline{S} + 2NO_2(g) + 6H_2O \qquad (29\text{-}22)$$

The solution is evaporated to remove excess HCl and $HNO_3$, and the residue is dissolved. Addition of $SnCl_2$ leads to the reactions:

$$2HgCl_4^{2-} + Sn^{2+} \rightleftharpoons \underline{Hg_2Cl_2} + SnCl_6^{2-} \qquad (29\text{-}23)$$

$$\underline{Hg_2Cl_2} + Sn^{2+} + 4Cl^- \rightleftharpoons \underline{2Hg(l)} + SnCl_6^{2-} \qquad (29\text{-}24)$$

The gray precipitate of $Hg_2Cl_2$ and Hg confirms the presence of mercury.

**Lead.** The lead ion is separated from the solution of $Pb^{2+}$, $Bi^{3+}$, and $Cu^{2+}$ by precipitating it as white $PbSO_4$. Because the sulfate is somewhat soluble in nitric acid, the $HNO_3$ present from dissolving the sulfides must be destroyed. Because $HNO_3$ is volatile, it can be removed by evaporation after adding $H_2SO_4$:

$$NO_3^- + H_2SO_4 \xrightarrow{\Delta} HNO_3(g) + HSO_4^- \qquad (29\text{-}25)$$

When all the $HNO_3$ has evaporated, dense white $SO_3$ will appear and you can stop heating:

$$H_2SO_4 \xrightarrow{\Delta} SO_3(g) + H_2O \qquad (29\text{-}26)$$

The lead sulfate residue is then separated from the solution containing $Bi^{3+}$ and $Cu^{2+}$. Because the residue might be contaminated with small quantities of bismuth and copper sulfates, it is dissolved in hot ammonium acetate. The acetate ion lowers $[Pb^{2+}]$ by converting it to a complex ion, so that the ion product drops below $K_{sp}$ for $PbSO_4$ and the solubility equilibrium is shifted to the right:

$$\begin{array}{l} \underline{PbSO_4} \rightleftharpoons Pb^{2+} + SO_4^{2-} \\ \qquad\qquad\;\; \Big\Updownarrow \; C_2H_3O_2^- \\ \qquad\quad\; Pb(C_2H_3O_2)_3^- \end{array} \qquad (29\text{-}27)$$

Impurities are left behind as solids. Potassium chromate is added to the solution to give a precipitate of yellow lead chromate which confirms the presence of lead:

$$Pb(C_2H_3O_2)_3^- + CrO_4^{2-} \rightleftharpoons \underline{PbCrO_4} + 3C_2H_3O_2^- \qquad (29\text{-}28)$$

**Bismuth.** Precipitation of $Bi(OH)_3$ separates $Bi^{3+}$ from $Cu^{2+}$. Copper also forms an insoluble hydroxide, but if the base used is $NH_3(aq)$, $Cu^{2+}$ will be converted to an ammine complex and remain in solution.

$$Cu^{2+} + 4NH_3 \rightleftharpoons Cu(NH_3)_4^{2+} \qquad K = 2.1 \times 10^{13} \qquad (29\text{-}29)$$

As usual, formation of a complex ion reduces $[Cu^{2+}]$ to a value sufficiently low that:

$$[Cu^{2+}][OH^-]^2 < K_{sp} = 4.8 \times 10^{-20} \qquad (29\text{-}30)$$

The white $Bi(OH)_3$ precipitate can be reduced to black metallic Bi by basic stannite solution:

$$\underline{2Bi(OH)_3} + 3Sn(OH)_3^- + 3OH^- \rightarrow \underline{2Bi} + 3Sn(OH)_6^{2-} \qquad (29\text{-}31)$$

**Copper.** The copper tetraammine complex $Cu(NH_3)_4^{2+}$ is deep blue. So, the presence of $Cu^{2+}$ can usually be detected from the color of the solution after precipitation of $Bi(OH)_3$.

If this test seems inconclusive, a portion of the solution can be used to obtain a precipitate of maroon copper ferrocyanide. The ammine complex is first destroyed by protonating the $NH_3$:

$$Cu(NH_3)_4^{2+} + 4H_3O^+ \rightarrow Cu^{2+} + 4NH_4^+ + 4H_2O \qquad (29\text{-}32)$$

$$2\ Cu^{2+} + Fe(CN)_6^{4-} \rightarrow \underline{Cu_2Fe(CN)_6} \qquad (29\text{-}33)$$

## Separation and Confirmation of the Antimony Subgroup

The antimony group cations are first reprecipitated as $Sb_2S_3$ and $SnS_2$ by acidifying the solution containing complex ions. This simply reverses the reactions that occurred when the sulfides were dissolved in KOH.

$$SbS_3^{3-} + SbO_3^{3-} + 6H_3O^+ \rightleftharpoons \underline{Sb_2S_3} + 9H_2O \qquad (29\text{-}34)$$

$$SnS_2OH^- + H_3O^+ \rightleftharpoons \underline{SnS_2} + 2H_2O \qquad (29\text{-}35)$$

The mixture of sulfide precipitates is then allowed to react with HCl to redissolve the sulfides as chloro complexes:

$$\underline{Sb_2S_3} + 6H_3O^+ + 8Cl^- \rightleftharpoons 2SbCl_4^- + 3H_2S(g) + 6H_2O \qquad (29\text{-}36)$$

$$\underline{SnS_2} + 4H_3O^+ + 6Cl^- \rightleftharpoons SnCl_6^{2-} + 2H_2S(g) + 4H_2O \qquad (29\text{-}37)$$

The solubility equilibria are shifted to the right by the complexation of Sb(III) and Sn(IV) and by heating the solution to drive off the $H_2S$. Note that the sulfides are here displaying the basic side of their amphoteric behavior.

**Antimony.** The solution containing $SbCl_4^-$ and $SnCl_6^{2-}$ is divided into two parts–one for the antimony test and one for the tin test. The antimony test solution is then allowed to react with oxalic acid ($H_2C_2O_4$); both Sb(III) and Sn(IV) form complex ions under these conditions:

$$SbCl_4^- + 3H_2C_2O_4 + 6H_2O \rightleftharpoons Sb(C_2O_4)_3^{3-} + 4Cl^- + 6H_3O^+ \qquad (29\text{-}38)$$

$$SnCl_6^{2-} + 3H_2C_2O_4 + 6H_2O \rightleftharpoons Sn(C_2O_4)_3^{2-} + 6Cl^- + 6H_3O^+ \qquad (29\text{-}39)$$

The equilibrium constant for Equation (29-39) is so large that no $SnS_2$ will precipitate when ammonium sulfide is added to the solution because of the decreased concentration of $SnCl_6^{2-}$. Only red-orange $Sb_2S_3$ will precipitate, which confirms the presence of antimony:

$$2Sb(C_2O_4)_3^{3-} + 3H_2S \rightarrow \underline{Sb_2S_3} + 6HC_2O_4^- \qquad (29\text{-}40)$$

**Tin.** The Sn(IV) in the second portion on the solution is first reduced to $Sn^{2+}$ by reaction with Al wire:

$$3SnCl_6^{2-} + 2\underline{Al} \rightarrow 3Sn^{2+} + 2Al^{3+} + 18Cl^- \qquad (29\text{-}41)$$

The Sb(III) which is also in solution will be reduced to metallic Sb:

$$SbCl_4^- + \underline{Al} \rightarrow \underline{Sb} + Al^{3+} + 4Cl^- \qquad (29\text{-}42)$$

A side reaction will also reduce some $Sn^{2+}$ to metallic Sn:

$$3Sn^{2+} + 2\underline{Al} \rightarrow 3\underline{Sn} + 2Al^{3+} \qquad (29\text{-}43)$$

However, Sn (but not Sb) can be reoxidized by hot HCl once the Al reduction is complete. This puts all the Sn present in the $Sn^{2+}$ state:

$$\underline{Sn} + 2H_3O^+ \rightarrow Sn^{2+} + H_2(g) + 2H_2O \qquad (29\text{-}44)$$

Finally, tin is confirmed by the familiar reaction with $Hg^{2+}$:

$$Sn^{2+} + 2Hg^{2+} + 8Cl^- \rightarrow SnCl_6^{2-} + \underline{Hg_2Cl_2} \qquad (29\text{-}45)$$

This is the reaction used earlier to confirm Hg. However, since $Sn^{2+}$ is not present in excess this time, $Hg^{2+}$ is reduced only as far as the white precipitate $Hg_2Cl_2$ and not all the way to metallic Hg.

## Chemicals Needed

- Al wire; aluminum
- $Bi^{3+}$ (2 mg/mL)
- $Cu^{2+}$ (2 mg/mL)
- 6M $HC_2H_3O_2$; acetic acid
- $H_2C_2O_4$; oxalic acid
- 2M HCl; hydrochloric acid
- 6M HCl; dil. hydrochloric acid
- 12M HCl; conc. hydrochloric acid
- 6M $HNO_3$; dil. nitric acid
- 16M $HNO_3$: conc. nitric acid
- 3% $H_2O_2$; hydrogen peroxide
- 6M $H_2SO_4$: dil. sulfuric acid
- 18M $H_2SO_4$; conc. sulfuric acid
- $Hg^{2+}$ (2 mg/mL)
- 0.1M $HgCl_2$; mercury(II) chloride
- 0.5M $K_2CrO_4$; potassium chromate
- 0.2M $K_4Fe(CN)_6$; potassium ferrocyanide
- 0.5M KOH; potassium hydroxide
- litmus paper
- 6M $NH_3$; dil. aqueous ammonia
- 15M $NH_3$; conc. aqueous ammonia
- 3M $NH_4C_2H_3O_2$; ammonium acetate
- 6M $NH_4Cl$; ammonium chloride
- 0.2M $NH_4NO_3$; ammonium nitrate
- 22% $(NH_4)_2S$; ammonium sulfide
- 6M NaOH; dil. sodium hydroxide
- pH indicator paper (wide- & short-range)
- $Pb^{2+}$ (2 mg/mL)
- $Sb^{3+}$ (4 mg/mL)
- Sn(IV) (8 mg/mL)
- 0.2M $SnCl_2$; tin(II) chloride

## Procedure

### Analysis of Known and Unknown Samples

1. Use 5 drops of a known solution which contains 2 mg/mL each of $Bi^{3+}$, $Cu^{2+}$, $Hg^{2+}$ and Sb(III), 4 mg/mL of $Pb^{2+}$ and 8 mg/mL of Sn(IV). If the solution is not already acidic to litmus, acidify with 6M HCl.

2. The unknown solution might contain ions of this and later groups. If it is Solution 1.1 from Outline 1 (see Experiment 28), use all of it. If it contains ions of only Group II, treat it as you would the known solution. If the solution is not already acidic to litmus, acidify with 6M HCl.

3. If the unknown is a solid, weigh 75-100 mg of the homogeneous solid into a small casserole or crucible. Dissolve the solid in the smallest possible amount of 6M HCl.

### Preliminary Preparation

Calibration of a Test Tube. Measure out 2.5 mL of water into a clean, dry 4-mL test tube. Mark the position of the meniscus with a strip of label.

Oxidation of Sn(II). Transfer the acidic solution which is to be analyzed to a small casserole or crucible. Add 2 drops of 3% $H_2O_2$ and boil the solution over a microflame until only a few drops remain. Swirl the solution continually in the flame. Be sure to withdraw the dish from the flame just before the volume is as small as you want it to be. The solution will continue to evaporate from the heat of the dish even after it is taken out of the flame. Excessive evaporation will lead to loss of the volatile chlorides of $Hg^{2+}$, Sb(V) and Sn(IV).

Dilute the remaining solution with 3-5 drops of distilled water and transfer it to the calibrated test tube. Wash down the sides and bottom of the dish twice with 3-5 drops of water and transfer these washes to the test tube also. Do not be concerned if the solution looks turbid. This is due to the presence of BiOCl and SbOCl which will both be converted to sulfides in the next step.

### Adjustment of Acidity

Add 6M $NH_3$ dropwise to the solution until it is basic to litmus or wide-range pH indicator paper. If $Cu^{2+}$ is present, you may be able to get an indication from the deep blue color of the ammine complex $Cu(NH_3)_4^{2+}$ in solution. However, this color may be masked by precipitates or complexes of Group III ions if they are present. Add 2M HCl dropwise to the basic solution until the pH is between 2 and 3 as measured with wide-range indicator paper. If the solution becomes too acidic, you can use 6M $NH_3$ to adjust the pH. Then add exactly 0.35 mL of 2M HCl using your calibrated pipet. Check the acidity with short-range indicator paper; it should have a pH of 0.5 corresponding to the $[H_3O^+] = 0.3M$. Add 5 drops of $(NH_4)_2S$ solution and enough water to make the volume exactly 2.5 mL. Mix well. As the sulfides precipitate, note the colors. Use colors in the order of precipitation and $K_{sp}$ values from Table 29.1 to get hints about which ions might be present. To continue the analysis, see Outline 2.

## Outline 2: Precipitation and Subdivision of Group II Cations

Precipitation of the Group. The solution can now be centrifuged and the supernatant liquid withdrawn to another test tube. Check the acidity of the supernatant liquid again using short-range indicator paper. Readjust it with 6M $NH_3$ to pH 0.5, if necessary.[1] Add 3 drops of $(NH_4)_2S$ solution and centrifuge again if more precipitate appears. Repeat this as many times as needed to complete the precipitation. You might need to divide the solution among several test tubes.

Combine your precipitates if you have several. Wash the combined sulfides twice with 20 drops of hot distilled water containing 1 drop of 6M $NH_4Cl$. If your solution contains ions of later groups add the first washing to Solution 2.1.

**Precipitate 2.1. HgS, PbS, $Bi_2S_3$, $Sb_2S_3$, $SnS_2$, S.** Add 10 drops of 0.5M KOH and heat in the water bath to dissolve $Sb_2S_3$ and $SnS_2$. Centrifuge and transfer solution to another test tube. Treat the precipitate with a second portion of KOH and combine the solutions.

**Solution 2.1. Ions of Group III, HCl, $NH_4Cl$, $H_2S$.** If the original sample contained only Group II cations, it can be discarded. If the solution contained ions of Group III, add 1 mL 12M HCl, transfer the solution to a casserole or crucible and evaporate almost to dryness. Then add 5 drops water and 1 drop 6M HCl. Save the solution in a stoppered test tube for the Group III analysis.

**Precipitate 2.2. HgS, PbS, $Bi_2S_3$, CuS, S.** Wash the precipitate twice with 5 drops of 0.2M $NH_4NO_3$.[2] Analyze as in Outline 3.[3]

**Solution 2.2. $SbS_3^{3-}$, $SbO_3^{3-}$, $SnS_3^{2-}$, $SnS_2OH^-$, KOH.** Centrifuge again to remove all traces of Precipitate 2.2. The clear, yellow solution can be transferred to a clean test tube and analyzed according to Outline 4.[4]

### Notes to Outline 2

1. $H_3O^+$ is produced during sulfide precipitation. For example $Cu^{2+} + H_2S + 2H_2O \rightleftharpoons CuS + 2H_3O^+$. This will change the pH.

2. $NH_4NO_3$ helps keep the precipitates from floccutating, that is, forming a colloidal suspension.

3. If you cannot analyze the precipitate immediately, cover the precipitate with water and add a drop of $(NH_4)_2S$ solution. Stopper the test tube. If possible, conduct the analysis as far as the separation of Precipitate 3.1 from Solution 3.1 in Outline 3.

4. If you cannot analyze the solution immediately, add a drop of $(NH_4)_2S$ solution and store it in a stoppered test tube. If the solution stands for more than a day, a precipitate of sulfides sometimes forms. This is not a problem. You can simply continue with the analysis.

## Outline 3. Copper Subgroup Analysis

**Precipitate 2.2 from Outline 2. HgS, PbS, $Bi_2S_3$, CuS, S.** Add 5 drops of water and 5 drops of 6M $HNO_3$ and heat the precipitate in the water bath to dissolve the sulfides of Cu, Bi, and Pb. Wash the remaining precipitate with water and add the washing to Solution 3.1.

**Precipitate 3.1. HgS,** (black or red), **S** (possibly colored by traces of sulfides) or **$2HgS \cdot Hg(NO_3)_2$** (white). Dissolve the precipitate in 4-5 drops of aqua regia[1] heating in the water bath. Place the solution in a crucible and evaporate almost to dryness over a microflame. Add about 1 mL of water and transfer the solution to a test tube. Centrifuge if the solution is not clear and discard any precipitate. Add 2 drops of $SnCl_2$ solution. A white precipitate of $Hg_2Cl_2$ indicates the presence of $Hg^{2+}$. (Sometimes the precipitate might appear gray because some of the $Hg_2Cl_2$ is reduced to metallic Hg.)

**Solution 3.1. $Pb^{2+}$, $Bi^{3+}$, $Cu^{2+}$, $H_3O^+$, $NO_3^-$.** Transfer the solution to a casserole or crucible and add 2 drops of 18M $H_2SO_4$. Evaporate in a microflame under the hood until white dense clouds of $SO_3$ appear.[2] Allow the mixture to cool and carefully add 5 drops of water. With a stirring rod, rub the water over the sides and bottom of the vessel and quickly transfer the solution and suspended precipitate to a test tube. Rinse out the casserole or crucible with two 5-drop portions of water and transfer these to the test tube. Centrifuge and transfer the supernatant liquid to another test tube. Wash the precipitate with 5 drops of water.

**Precipitate 3.2. $PbSO_4$, $[(BiO)_2SO_4]$.**[3] Add 2-3 drops of $NH_4C_2H_3O_2$ solution and warm in a water bath. Centrifuge and discard any precipitate. To the solution add 1 drop of 6M $HC_2H_3O_2$ and 3-4 drops of 0.5M $K_2CrO_4$. A yellow precipitate of $PbCrO_4$ confirms the presence of $Pb^{2+}$.[3]

**Solution 3.2. $Bi^{3+}$, $Cu^{2+}$.** Add 15M $NH_3$ until the solution is basic to litmus. If the solution looks blue, this is an indication of the presence of $Cu(NH_3)_4^{2+}$. Centrifuge even if you do not see a precipitate. Gelatinous $Bi(OH)_3$ is difficult to see. Transfer some of the supernatant liquid to another test tube.

**Precipitate 3.3. $Bi(OH)_3$.** Add 2-3 drops of freshly prepared stannite reagent.[4] An immediate appearance of black metallic Bi confirms the presence of $Bi^{3+}$ in the original solution.

**Solution 3.3.** Acidify with 6M $HC_2H_3O_2$ and add 2 drops of $K_4Fe(CN)_6$ solution. A maroon precipitate of $Cu_2Fe(CN)_6$ confirms the presence of $Cu^{2+}$ in the original solution.

### Notes to Outline 3

1. To prepare aqua regia mix conc. HCl and conc. $HNO_3$ in a 3:1 volume ratio.

2. The appearance of $SO_3$ indicates that all the $HNO_3$ has been evaporated. If you accidently evaporate the solution to dryness, add 2 drops of 6M $H_2SO_4$ and 3 drops of water. Warm in the water bath for several minutes with stirring to help dissolve soluble sulfates.

3. Occasionally $(BiO)_2SO_4$ precipitates here also. The purpose of ammonium acetate and acetic acid is to remove the interference of the Bi compound. $(BiO)_2SO_4$ is less soluble than $PbSO_4$ in $NH_4C_2H_3O_2$ and $(BiO)_2SO_4$ is more soluble than $PbCrO_4$ in $HC_2H_3O_2$.

4. To prepare sodium stannite solution, add 6M NaOH to 1 drop of $SnCl_2$ reagent until the precipitate of $Sn(OH)_2$ redissolves. $Sn^{2+} + 2OH^- \rightarrow \underline{Sn(OH)_2} \xrightarrow{OH^-} Sn(OH)_3^-$.

## Outline 4. Antimony Subgroup

**Solution 2.2 from Outline 2. $SbS_3^{3-}$, $SbO_3^{3-}$, $SnS_3^{2-}$, $Sn_2OH^-$, KOH.** Reprecipitate $Sb_2S_3$ and $SnS_2$ by adding 2 drops $(NH_4)_2S$ solution and then 2M HCl dropwise until the solution is just acid to litmus. Note the colors of any sulfide precipitates as a clue to which cations are present. Centrifuge and discard the supernatant liquid.

**Precipitate 4.1. $Sb_2S_3$** (orange), **$SnS_2$** (yellow), **S.** Dissolve the precipitate by adding 8-10 drops of 12M HCl and heating in the water bath with stirring.[1] Centrifuge and discard any precipitate. Transfer the supernatant liquid to a casserole.[2]

**Solution 4.2. $SbCl_4^-$, $SnCl_6^{2-}$, HCl, $NH_4Cl$, $H_2S$.** Over a microflame evaporate the solution to about half its volume[3] to destroy $H_2S$, and add about 1 mL water. Divide the solution between two test tubes for the Sb and Sn tests.

**Tin Test.** Place a piece of Al wire about 5-6 mm long in the solution and add 10 drops 6M HCl.[4] Warm in the water bath until all the Al has reacted.

**Antimony Test.** Add about half a spatulaful of $H_2C_2O_4$ crystals. Follow this with 3-4 drops of $(NH_4)_2S$ solution. Warm in the water bath if necessary until a dark orange precipitate forms. This precipitate is $Sb_2S_3$ and confirms the presence of Sb(III) in the original solution.

**Precipitate 4.4. Sb** (black). This precipitate provides added confirmation of the presence of Sb.

**Solution 4.4. $Sn^{2+}$, $Al^{3+}$, HCl.** Add an equal volume of water and quickly[6] add 2-3 drops of 0.1M $HgCl_2$. A white to gray precipitate of $Hg_2Cl_2$ and Hg indicates the presence of Sn in the original solution.

**Notes to Outline 4**

1. If the precipitate is black, some HgS might be present, because a little HgS dissolved in the KOH in the step just following precipitation of the Group. This will not interfere with the analysis since the precipitate is discarded.

2. The supernatant liquid should be drawn off with a capillary pipet having a small plug of cotton in the tip.

3. If the solution is evaporated too far, some volatile $SnCl_4$ might be lost.

4. Al can reduce Sn(IV) and $Sn^{2+}$ all the way to metallic Sn. However, Sn will redissolve in HCl while Sb will not.

5. Often the precipitate is initially orange, but darkens because of the slower precipitation of impurities. If Sn is present, a brown precipitate of SnS might form slowly. If the precipitate is very dark, some HgS might be present. If so, decant the solution and dissolve as much of the precipitate as you can in 6M HCl, warming the mixture in the water bath. Centrifuge and discard the HgS, which is not soluble in HCl. Dilute the solution and add $(NH_4)_2S$ to reprecipitate $Sb_2S_3$.

6. If the addition of $HgCl_2$ is delayed too long, $Sn^{2+}$ might be reoxidized to Sn(IV) by air.

## Results

Submit a 3 x 5 card for your unknown giving your name, desk number, ions present, and ions absent. Have your notebook available for your instructor to examine.

## References

1. King, Edward J. "Ionic Reactions and Separations: Experiments in Qualitative Analysis", Harcourt Brace Jovanovich, New York, 1977. pp. 143-162.

2. Sorum, C.H., Lagowski, J.J. "Introduction to Semimicro Qualitative Analysis", Prentice Hall, Englewood Cliffs, 1977. pp. 159-196.

3. Myers, R.J., J. Chem. Educ. 1987, 63, 687.

# EXPERIMENT 30

# QUALITATIVE ANALYSIS OF CATION GROUP III

## Introduction

The cations of group III–$Al^{3+}$, $Cr^{3+}$, $Fe^{3+}$, $Co^{2+}$, $Ni^{2+}$, and $Zn^{2+}$–form hydroxides and sulfides that are insoluble in basic solution. This is the last group of cations we shall deal with. Since there are no other groups in our cation analysis scheme, we shall not precipitate all the ions of Group III together while leaving later cation groups in solution as we did for Groups I and II. Instead, we shall precipitate the insoluble hydroxides first and follow this with the precipitation of the insoluble sulfides. This will enable the separation of Group III into two subgroups at the outset. Except for $Al^{3+}$ and $Zn^{2+}$, all the cations of Group III are transition metals that form many colored complex ions that are easily identified. Hence, the presence of these ions often can be ascertained by observation of their characteristic colors in solution.

## Separation of Group III Cations

As Table 30-1 shows, the ions of Group III form insoluble sulfides and hydroxides.

Table 30-1

$K_{sp}$'s and Colors of Some Sulfides and Hydroxides

| Sulfide | $K_{sp}$ | Color | Hydroxide | $K_{sp}$ | Color |
|---|---|---|---|---|---|
| ZnS | $3.2 \times 10^{-23}$ | white | $Co(OH)_3$ | $3 \times 10^{-45}$ | black-brown |
| CoS | $8 \times 10^{-23}$ | black | $Fe(OH)_3$ | $1.6 \times 10^{-39}$ | red-brown |
| NiS | $8 \times 10^{-19}$ | black | $Al(OH)_3$ | $3 \times 10^{-34}$ | white |
| FeS | $6 \times 10^{-19}$ | black | $Cr(OH)_3$ | $2 \times 10^{-30}$ | green |
| | | | $Zn(OH)_2$ | $3.3 \times 10^{-16}$ | white |
| | | | $Ni(OH)_2$ | $6 \times 10^{-16}$ | green |
| | | | $Fe(OH)_2$ | $7.9 \times 10^{-16}$ | pale green or white |
| | | | $Co(OH)_2$ | $1 \times 10^{-15}$ | rose |

The solution is first buffered to a pH of around 9.3 with ammonia and ammonium chloride. Equation (30-1) shows the equilibrium involving $NH_3$ and $NH_4^+$ and Equation (30-2) give the corresponding equilibrium constant expression:

$$NH_3(aq) + H_2O \rightleftharpoons NH_4^+ + OH^- \quad (30\text{-}1)$$

$$K_b = \frac{[NH_4^+][OH^-]}{[NH_3(aq)]} = 1.8 \times 10^{-5} \quad (30\text{-}2)$$

When the concentrations of $NH_3$ and $NH_4^+$ are adjusted by adding large quantities of $NH_3(aq)$ and $NH_4Cl$, the $[OH^-]$ is fixed in the solution:

$$[OH^-] = K_b\frac{[NH_3]}{[NH_4^+]} = 1.8 \times 10^{-5}\frac{[NH_3]}{[NH_4^+]} \quad (30\text{-}3)$$

If $[NH_3]$ and $[NH_4^+]$ are made roughly equal, the $[OH^-]$ will be numerically equal to $K_b$ which in this case is $1.8 \times 10^{-5}$M. A hydroxide concentration of $1.8 \times 10^{-5}$M corresponds to $[H_3O^+] = 5.6 \times 10^{-10}$ and pH = 9.3.

At this pH, the ion products of the divalent and trivalent hydroxides are:

$$[M^{2+}][OH^-]^2 = (0.01)(1.8 \times 10^{-5})^2 = 3.2 \times 10^{-12} \quad (30\text{-}4)$$

$$[M^{3+}][OH^-]^3 = (0.01)(1.8 \times 10^{-5})^3 = 5.8 \times 10^{-17} \quad (30\text{-}5)$$

These values are larger than the $K_{sp}$'s for all the Group III hydroxides, so the prediction seems to be that all the metals should precipitate as hydroxides. This would not achieve any separation into subgroups. However, $NH_3$ forms complexes with a number of Group III cations. For example,

$$Zn^{2+} + 4NH_3 \rightleftharpoons Zn(NH_3)_4^{2+} \qquad K = 7.8 \times 10^8 \quad (30\text{-}6)$$

This equilibrium constant is so large that effectively all of the $Zn^{2+}(aq)$ is converted to $Zn(NH_3)_4^{2+}$. A calculation shows that $[Zn^{2+}(aq)]$ is around $10^{-10}$M while $[Zn(NH_3)_4]^{2+} = 0.01$M in a buffered solution with a total $Zn^{2+}$ concentration of 0.01M. The $[Zn^{2+}(aq)]$ is so small as a result of complex formation that the ion product no longer exceeds the solubility product value:

$$[Zn^{2+}][OH^-]^2 = 10^{-10}(1.8 \times 10^{-5})^2 = 3.2 \times 10^{-20} < K_{sp} = 3.3 \times 10^{-16} \quad (30\text{-}7)$$

Under these circumstances, no $Zn(OH)_2$ will precipitate. Other ammonia complexes which form and prevent the precipitation of metal hydroxides are blue-purple $Ni(NH_3)_6^{2+}$ and amber $Co(NH_3)_6^{2+}$. The result is that the only hydroxides that precipitate are those of the aluminum subgroup: $Fe(OH)_3$, $Al(OH)_3$ and $Cr(OH)_3$. Another purpose of the buffer is to keep $[OH^-]$ sufficiently small that $Al(OH)_3$ and $Cr(OH)_3$ do not redissolve as the complex ions $Al(OH)_4^-$ and

$Cr(OH)_4^-$. These hydroxides can be centrifuged and the ions remaining in solution, $Zn(NH_3)_4^{2+}$, $Ni(NH_3)_6^{2+}$, and $Co(NH_3)_6^{2+}$, precipitated as sulfides.

At pH 9.3, the predominant sulfur-containing species in solution is $HS^-$ as you can see from the following equations:

$$\frac{[HS^-]}{[H_2S]} = \frac{K_{a1}}{[H_3O^+]} = \frac{1.0 \times 10^{-7}}{5.6 \times 10^{-10}} = 180 \qquad (30\text{-}8)$$

$$\frac{[HS^-]}{[S^=]} = \frac{[H_3O^+]}{K_{a2}} = \frac{5.6 \times 10^{-10}}{10^{-17 \pm 2}} = 6 \times 10^{7 \pm 2} \qquad (30\text{-}9)$$

Thus, the solubility equilibria for MS can be represented as:

$$\underline{MS} + H_2O \rightleftharpoons M^{2+}(aq) + HS^-(aq) + OH^- \qquad (30\text{-}10)$$

This is the equilibrium to which Ksp refers, and we have:

$$K_{sp} = [M^{2+}]\,[HS^-]\,[OH^-] \qquad (30\text{-}11)$$

If we add enough $(NH_4)_2S$ to the buffered solution to make the solution 0.4M in this substance, Equations (30-8) and (30-9) show that all the sulfide ion will be present in the form of $HS^-$. For the case of $Zn^{2+}$ which is mostly present in the form of the $Zn(NH_3)_4^{2+}$ complex:

$$[Zn^{2+}]\,[HS^-]\,[OH^-] = (10^{-10})\,(0.4)\,(1.8 \times 10^{-5}) = 7 \times 10^{-16} > K_{sp} = 3.2 \times 10^{-23} \qquad (30\text{-}12)$$

From this calculation we can conclude that ZnS will precipitate from the buffered solution when $(NH_4)_2S$ solution is added even though the concentration of $Zn^{2+}$ was decreased from 0.01M to $10^{-10}$M by conversion to the ammonia complex. Similar calculations show that NiS and CoS will also precipitate under these conditions separating the cobalt subgroup from the aluminum subgroup.

## Separation and Confirmation of Aluminum Subgroup Ions

**Aluminum and Chromium.** Of the hydroxides of the Al subgroup, only $Al(OH)_3$ is amphoteric. Hence, it can be dissolved in an excess of NaOH:

$$\underline{Al(OH)_3} + OH^- \rightleftharpoons Al(OH)_4^- \qquad (30\text{-}13)$$

During this dissolution process, some $H_2O_2$ is added to the solution; it oxidizes Cr(III) to yellow $CrO_4^{2-}$ and so also dissolves the $Cr(OH)_3$ leaving a precipitate of $Fe(OH)_3$.

$$\underline{2Cr(OH)_3} + 4OH^- + 3H_2O_2 \rightarrow 2CrO_4^{2-} + 8H_2O \qquad (30\text{-}14)$$

The solution containing $Al(OH)_4^-$ is then acidified to convert the hydroxo complex to the $Al^{3+}(aq)$ cation. In the course of the neutralization, you might see the appearance of a precipitate of $Al(OH)_3$ which then reacts further with acid to form $Al^{3+}$:

$$Al(OH)_4^- + H_3O^+ \rightleftharpoons \underline{Al(OH)_3} \xrightarrow{H_3O^+} Al^{3+} + 4H_2O \qquad (30\text{-}15)$$

In the course of addition of acid the yellow chromate ion is converted to orange dichromate:

$$2CrO_4^{2-} + 2H_3O^+ \rightleftharpoons Cr_2O_7^{2-} + 3H_2O \qquad (30\text{-}16)$$

The solution is again buffered to pH 9.3 to reprecipitate $Al(OH)_3$ and the precipitate is again dissolved, this time in an acetic acid/ammonium acetate buffer. Aluminon reagent is added, and $Al^{3+}$ produces a red precipitate of $Al(C_{22}H_{13}O_9)_3$. This gives a specific test for $Al^{3+}$. Formation of the red precipitate is extremely sensitive to pH–hence, the necessity of carefully controlling the pH by repeatedly reprecipitating and redissolving the $Al(OH)_3$ in buffers.

Chromium is confirmed by adjusting the pH and adding lead acetate to precipitate yellow $PbCrO_4$:

$$Cr_2O_7^{2-} + 2OH^- \rightleftharpoons 2CrO_4^{2-} + H_2O \qquad (30\text{-}17)$$

$$CrO_4^{2-} + Pb^{2+} \rightleftharpoons \underline{PbCrO_4} \qquad (30\text{-}18)$$

**Iron.** The precipitate of $Fe(OH)_3$ is redissolved in nitric acid and some KNCS is added to produce the blood-red complex ion $Fe(NCS)^{2+}$, which confirms the presence of $Fe^{3+}$.

$$\underline{Fe(OH)_3} + 3H_3O^+ \rightleftharpoons Fe^{3+} + 6H_2O \qquad (30\text{-}19)$$

$$Fe^{3+} + NCS^- \rightleftharpoons Fe(NCS)^{2+} \qquad (30\text{-}20)$$

## Separation and Confirmation of the Cobalt Subgroup

The precipitate of ZnS, CoS and NiS is then redissolved in acid. You may recall that the sulfides of Group III are soluble in acid solution, and that it was this feature that kept them from precipitating along with Group II cations. The solution is then heated and evaporated almost to dryness to destroy $H_2S$. The residue is dissolved in water and made basic with NaOH.

**Zinc.** The amphoteric property of $Zn(OH)_2$ allows separation of $Zn^{2+}$ from $Co^{2+}$ and $Ni^{2+}$. 6M NaOH precipitates $Co(OH)_2$ and $Ni(OH)_2$ while leaving $Zn(OH)_4^{2-}$ in solution:

$$Zn^{2+} + 2OH^- \rightleftharpoons \underline{Zn(OH)_2} \xrightarrow{2OH^-} Zn(OH)_4^{2-} \qquad (30\text{-}21)$$

$$Co^{2+} + 2OH^- \rightleftharpoons \underline{Co(OH)_2} \qquad (30\text{-}22)$$

$$Ni^{2+} + 2OH^- \rightleftharpoons \underline{Ni(OH)_2} \qquad (30\text{-}23)$$

A final confirmatory test for $Zn^{2+}$ can be made by applying the solution to a piece of paper impregnated with dithizone. The red-purple complex formed between $Zn^{2+}$ and dithizone (Figure 30-1) confirms the presence of $Zn^{2+}$.

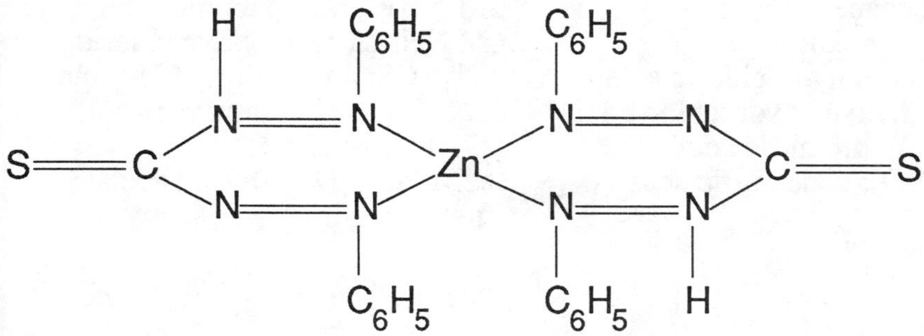

**Figure 30-1**

Zinc complex with dithizone.

**Cobalt and Nickel.** The solution remaining after precipitation of ZnS is divided into two parts–one for the $Co^{2+}$ test and one for the $Ni^{2+}$ test.

Cobalt is detected by the formation of its blue complex with $NCS^-$ ion. The equilibrium constant for the formation of the complex is much higher in acetone than in water, so acetone is added to the solution:

$$Co^{2+} + 4NCS^- \rightleftharpoons \underset{\text{blue in water/acetone}}{Co(NCS)_4^{2-}} \qquad (30\text{-}24)$$

$Ni^{2+}$ is detected by its characteristic red precipitate with dimethylglyoxime (DMG) whose structure appears in Figure 30-2.

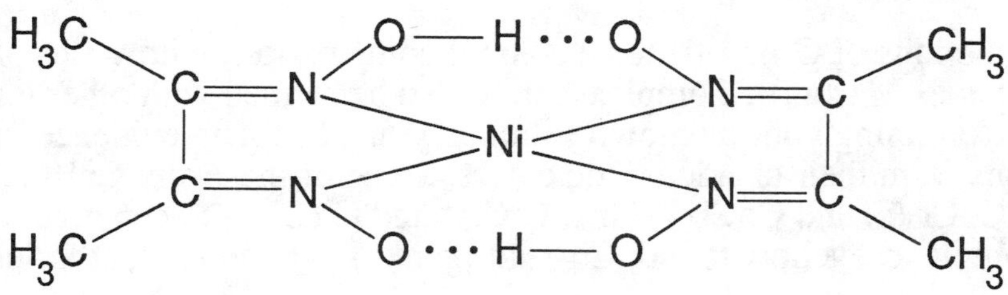

**Figure 30-2**

Nickel dimethylglyoximate.

The solution is made alkaline with $NH_3$ and treated with dimethylglyoxime producing a flocculent red precipitate.

$$Ni^{2+} + 2C_4H_8N_2O_2 + 2NH_3 \rightleftharpoons \underline{Ni(C_4H_7N_2O_2)_2} + 2NH_4^+ \qquad (30\text{-}25)$$

$Co^{2+}$ forms a soluble brown complex with dimethylglyoxime. If this appears, simply add more dimethylglyoxime until enough is present to form the red $Ni(DMG)_2$ precipitate.

## Chemicals Needed

| | |
|---|---|
| acetone | 3% $H_2O_2$; hydrogen peroxide |
| $Al^{3+}$ (2 mg/mL) | 6M $HC_2H_3O_2$; acetic acid |
| aluminon reagent | litmus paper |
| $Co^{2+}$ (2 mg/mL) | 6M NaOH; dil. sodium hydroxide |
| $Cr^{3+}$ (2 mg/mL) | 6M $NH_3$; dil. ammonia |
| 1% dimethylglyoxime (DMG) | 15M $NH_3$; conc. aqueous ammonia |
| dithizone paper | 3M $NH_4C_2H_3O_2$; ammonium acetate |
| $Fe^{3+}$ (2 mg/mL) | 6M $NH_4Cl$; ammonium chloride |
| 6M HCl; dil. hydrochloric acid | $NH_4NCS$; ammonium thiocyanate |
| 12M HCl; conc. hydrochloric acid | 22% $(NH_4)_2S$; ammonium sulfide |
| 6M $HNO_3$; dil. nitric acid | $Ni^{2+}$ (2 mg/mL) |
| 16M $HNO_3$; conc. nitric acid | 0.2M $Pb(C_2H_3O_2)$; lead acetate |
| | $Zn^{2+}$ (4 mg/mL) |

## Procedure

### Analysis of Known and Unknown Samples

1. Use 5 drops of a solution containing 2 mg/mL each of $Al^{3+}$, $Cr^{3+}$, $Ni^{2+}$, and $Co^{2+}$ and 4 mg/mL of $Zn^{2+}$. Note that the known solution contains no iron. Iron is omitted so that you can see the color intensity of the $Fe^{3+}$ test given by impurities in the solution. Since no other ions interfere with the $Fe^{3+}$ test, run the test on a portion of the known solution. Then add 1 drop of the $Fe^{3+}$ test solution and observe the color intensity of a geniune test.

2. If the unknown is Solution 2.1 from Outline 2 (Experiment 29), use all of the solution.

3. If your unknown is a solid, weigh about 75-100 mg of the homogeneous solid into a casserole or crucible. Dissolve the solid, with heating if necessary, in the smallest possible volume of 6M HCl. Transfer the solution to a 4-mL test tube. Cool the solution.

### Preliminary Observations

Some of the ions of Group III form colored complex ions with water of the type $M(H_2O)_m{}^{n+}$ (usually written as $M^{n+}(aq)$). Complexes may also be formed with other ions or molecules in the solution. By examining your unknown solution, you can obtain evidence for the presence of certain of the ions from their characteristic colors. Some of these are: $Cr^{3+}(aq)$, violet; $Ni(NH_3)_6{}^{2+}$, blue-purple; $CoCl_2(s)$ and $CoCl_4{}^{2-}$, blue; $CrCl_2{}^+$ and $Cr(OH)_4{}^-$, deep green; $FeCl^{2+}$, yellow; $Fe^{3+}$ hydroxide complexes, yellow to red; $Co^{2+}(aq)$, pink; $Zn^{2+}$ and $Al^{3+}$, colorless.

## Outline 5. Group III Cation Analysis

Precipitation and Separation of the Group. Dilute the sample with distilled water to a volume of 2 mL. Add 7 drops of 6M $NH_4Cl$ and make the solution alkaline to litmus with 15M $NH_3$; add 1 more drop of $NH_3$. Stir the solution well and centrifuge. Transfer the supernatant liquid to another test tube and test for completeness of precipitation by adding 1 drop of 15M $NH_3$.

**Precipitate 5.1. $Al(OH)_3$, $Cr(OH)_3$, $Fe(OH)_3$.** Wash the Precipitate with 6-7 drops of distilled water to which has been added 1 drop of 6M $NH_3$. Analyze the precipitate according to Outline 6.

**Solution 5.1. $Co(NH_3)_6^{2+}$, $Ni(NH_3)_6^{2+}$, $Zn(NH_3)_4^{2+}$.** Add 6-7 drops of 6M $NH_3$ and 7 drops of $(NH_4)_2S$. Stir the solution well and warm in the water bath. Centrifuge and transfer the supernatant liquid to another test tube. Add 2 drops of $(NH_4)_2S$. If a precipitate forms, save it and combine with the earlier precipitate. If no more precipitate forms, discard the solution.

**Precipitate 5.2. CoS, NiS, ZnS.** Wash the precipitate once with 5-6 drops of water and discard the washings. Analyze this precipitate according to Outline 7.

**Outline 6: Analysis of the Aluminum Subgroup**

**Precipitate 5.1 from Outline 5. $Al(OH)_3$, $Cr(OH)_3$, $Fe(OH)_3$.** Add 6 drops of 6M NaOH and 9 drops of 3% $H_2O_2$. Stir until gas evolution stops. Then add 4 drops more of 3% $H_2O_2$ and heat in the water bath until gas evolution stops. Dilute the solution to 2 mL, centrifuge and transfer the supernatant liquid to another test tube.

**Precipitate 6.1 $Fe(OH)_3$.** Dissolve the precipitate in a minimum amount of 6M HCl. Add several crystals of solid $NH_4NCS$ and stir to dissolve them. A deep red color of the $FeNCS^{2+}$ complex indicates the presence of $Fe^{3+}$. [1]

**Solution 6.1. $Al(OH)_4^-$, $CrO_4^{2-}$.** Acidify the solution to litmus with 6M $HNO_3$. Add 6 drops of 6M $NH_4Cl$ and make the solution just basic to litmus with 15M $NH_3$. Heat the mixture in the water bath for a minute or so to coagulate the $Al(OH)_3$ precipitate.

**Precipitate 6.2. $Al(OH)_3$.** [2] Dissolve the precipitate in 7 drops of $HC_2H_3O_2$. If any solid does not dissolve, centrifuge and discard it. Add 7 drops 3M $NH_4C_2H_3O_2$ and stir. Add 5 drops of aluminon reagent and make the solution barely basic to litmus with 6M $NH_3$ while you stir continually. Centrifuge to settle the red lake whose presence confirms $Al^{3+}$.

**Solution 6.2. $CrO_4^{2-}$.** If the solution is not yellow, $CrO_4^{2-}$ is absent. If $CrO_4^{2-}$ is present, transfer the solution to a casserole or crucible and evaporate it to a volume of 0.5 mL. Acidify with $HNO_3$ and transfer to a test tube. Add 4-5 drops of 0.2M $Pb(C_2H_3O_2)_2$. A yellow precipitate of $PbCrO_4$ confirms the presence of $Cr^{3+}$ in the original solution.

**Notes to Outline 6**

1. If $Co^{2+}$ is present, the solution will be pink. However, this pink color is quite different from the blood red color of $FeNCS^{2+}$. If the pink color is due to $Co^{2+}$, the solution should turn blue if an equal volume of acetone is added, because of the formation of blue $Co(NCS)_4^{2-}$.

2. If the precipitate looks brownish, $Fe(OH)_3$ might be present. Dissolve the $Al(OH)_3$ in 1-2 drops of 6M NaOH and centrifuge to remove brown $Fe(OH)_3$. Neutralize the solution with 6M HCl and reprecipitate the $Al(OH)_3$ with $NH_3$ and $NH_4Cl$ as in the previous step.

## Outline 7. Analysis of the Cobalt Subgroup

**Precipitate 5.2. NiS, CoS, ZnS.** Dissolve the precipitate in 10 drops of 12M HCl.[1] Transfer the solution to a casserole or crucible and evaporate almost to dryness to remove $H_2S$. Add 5 drops of water and transfer the solution to a test tube. Make the solution basic with 6M NaOH and add two more drops.

**Precipitate 7.1. $Ni(OH)_2$, $Co(OH)_2$.** Wash the precipitate with 10 drops of water containing 2 drops of 6M NaOH and discard the washings. Add 3-4 drops of 12M HCl and warm the solution to dissolve the hydroxides. Dilute with 10 drops of water and divide the solution between two test tubes–one for the $Ni^{2+}$ test and one for the $Co^{2+}$ test.

**Solution 7.1. $Zn(OH)_4^{2-}$.** Apply a spot of the solution to dithizone test paper. A purple-red spot indicates the presence of $Zn^{2+}$.

**$Ni^{2+}$ Test.** Add 15M $NH_3$ until the solution is alkaline to litmus. Add several drops of DMG.[2] A deep red precipitate of nickel dimethylglyoximate confirms the presence of $Ni^{2+}$.

**$Co^{2+}$ Test.** Add a spatulaful of $NH_4NCS$ and stir to dissolve. Then add 30-35 drops of acetone. Stir well. The blue color of $Co(NCS)_4^{2-}$ confirms the presence of $Co^{2+}$.

**Notes to Outline 7**

1. If the precipitate has been standing a long time before being analyzed, it might be difficult to dissolve. In this case, add 3-4 drops of 16M $HNO_3$ in addition to the 12M HCl.

2. DMG also reacts with $Co^{2+}$ to give a soluble brown compound. If the solution turns brown, be sure that you add a sufficient excess of DMG to react with $Ni^{2+}$ present.

## Results

Submit a 3 x 5 in. card for your unknown giving your name, desk number, ions present and ions absent. Have your notebook ready for your instructor to examine on request.

## References

1. King, Edward J., "Ionic Reactions and Separations: Experiments in Qualitative Analysis", Harcourt Brace Jovanovich, New York, 1973. pp. 198-209.

2.Sorum, C.H., Lagowski, J. J., "Introduction to Semimicro Qualitative Analysis", 5th ed. Prentice Hall, Englewood Cliffs, 1977. pp. 196-223.

# EXPERIMENT 31

# QUALITATIVE ANALYSIS OF ANIONS

## Introduction

In this experiment, you will test for the presence of seven different anions: $Cl^-$, $Br^-$, $I^-$, $NO_2^-$, $NO_3^-$, $SO_4^=$ and $CO_3^=$. Most of these anions interfere in the tests for other anions. Hence, each test should be done on a fresh portion of the original sample. With each test, procedures are given that will eliminate interfering ions.

You might recall from Experiment 9 that $NO_2^-$, $NO_3^-$, $SO_4^=$ and $CO_3^=$ can be detected by infrared spectroscopy. However, this technique can be applied only to a solid sample. Moreover, if several anions were present in the same sample, overlapping bands would make ions difficult to detect. Hence, we give here the wet chemistry tests for several anions.

## Identification of Anions

**Chloride, bromide and iodide.** A simple preliminary test will show whether any of these ions is present. A solution of $[Ag(NH_3)_2]Cl$ is added. This solution contains silver ion mostly in the form of $Ag(NH_3)_2^+$ and has only a very small concentration of $Ag^+$. However, since all the silver halides have such small $K_{sp}$'s, even this very small $[Ag^+]$ is large enough to give a precipitate of AgCl, AgBr, or AgI if any of these anions is present. At the same time, $[Ag^+]$ is small enough that only the least soluble salts of silver precipitate. In particular, neither $Ag_2SO_4$ nor $Ag_2CO_3$ will precipitate since

$$[Ag^+]^2[SO_4^{2-}] < K_{sp} = 1.5 \times 10^{-5} \quad (31\text{-}1)$$

$$[Ag^+]^2[CO_3^{2-}] < K_{sp} = 8.1 \times 10^{-12} \quad (31\text{-}2)$$

The halide ions are returned to solution by allowing the precipitate to react with Zn metal which reduces $Ag^+$ to Ag metal and forms soluble $Zn^{2+}$ halides. For example,

$$\underline{2AgCl} + Zn \rightarrow \underline{2Ag} + Zn^{2+}(aq) + 2Cl^-(aq) \quad (31\text{-}3)$$

The $Br^-$ and $I^-$ ions in solution are then successively oxidixed to the halogens. Their reduction potentials

$$I_2 + 2e^- \rightarrow 2I^- \qquad E^\circ = +0.54 \text{ volts} \quad (31\text{-}4)$$

$$Br_2 + 2e^- \rightarrow 2Br^- \qquad E^\circ = +1.06 \text{ volts} \quad (31\text{-}5)$$

$$Cl_2 + 2e^- \rightarrow 2Cl^- \qquad E^\circ = +1.36 \text{ volts} \quad (31\text{-}6)$$

show that $I^-$ is most easily oxidized. An oxidizing agent ($NO_2^-$) is chosen that can oxidize $I^-$, but not $Br^-$. The $I_2$ is extracted into $CCl_4$ and identified by its violet color. $MnO_4^-$ is then used to oxidize $Br^-$ to $Br_2$ without oxidizing $Cl^-$. $Br_2$ is also extracted into $CCl_4$ and identified by its brown color. The $Cl^-$ remaining in solution can then be identified by the white precipitate of AgCl which forms on addition of $AgNO_3(aq)$.

**Nitrite.** $NO_2^-$ is identified from the products of its reaction with sulfamic acid:

$$NO_2^- + HSO_3NH_2 \rightarrow N_2(g) + SO_4^= + H_3O^+ \quad (31\text{-}7)$$

The $N_2$ gas can be observed bubbling out of the solution. The $SO_4^=$ can be identified by the white precipitate of $BaSO_4$ which forms on addition of $BaCl_2$ solution.

Obviously, any sulfate already present in solution would interfere by giving a false positive test and so is removed by adding $BaCl_2$ to precipitate it before the nitrite test is run.

**Sulfate.** $SO_4^=$ is easy to detect because it forms a white precipitate of $BaSO_4$ when $BaCl_2$ is added to the solution. None of the other ions in this experiment forms an insoluble $Ba^{2+}$ salt. So none interferes with the sulfate test.

**Nitrate.** $NO_3^-$ is identified by detecting the ammonia produced when Al reduces nitrate in basic solution:

$$3NO_3^- + \underline{8Al} + 5OH^- + 18H_2O \rightarrow 3NH_3(g) + 8Al(OH)_4^- \quad (31\text{-}8)$$

$NH_3$ is detected by allowing it to contact a piece of wet red litmus paper. The red litmus turns blue because of the reaction

$$NH_3 + H_2O \rightleftharpoons NH_4^+ + OH^- \quad (31\text{-}9)$$

The ammonium ion would interfere with this test by acting as a source of $NH_3$ when the solution is made basic via the reverse of Equation (31-8). $NO_2^-$ also interferes because it is reduced to $NH_3$ by Al. Hence, both of these are removed, if present, before the test is begun.

**Carbonate.** When a solution containing $CO_3^=$ is acidifed, $CO_2(g)$ is evolved:

$$CO_3^= + 2H_3O^+ \rightarrow H_2CO_3 \rightarrow CO_2(g) + 3H_2O \quad (31\text{-}10)$$

If the $CO_2$ is allowed to contact a basic solution, it is converted back into the carbonate ion. If the solution also contains $Ba^{2+}$, a precipitate of $BaCO_3$ is produced and serves to identify carbonate:

$$CO_2(aq) + 2OH^- \rightleftharpoons CO_3^{2-} + H_2O \quad (31\text{-}11)$$

$$CO_3^{2-} + Ba^{2+} \rightleftharpoons \underline{BaCO_3} \quad (31\text{-}12)$$

If your unknown is a solid, this test should be made directly on a sample of the solid.

## Chemicals and Equipment Needed

- 0.2M $AgNO_3$; silver nitrate
- 3.3M $AgNO_3/NH_3$; silver nitrate in aq. ammonia
- Al wire
- 0.2M $BaCl_2$; barium chloride
- 0.2M $Ba(OH)_2$; barium hydroxide
- boiling chips
- $CCl_4$; carbon tetrachloride
- corks for test tubes
- 6M $HC_2H_3O_2$; acetic acid
- 6M HCl; dil. hydrochloric acid
- $HNH_2SO_3$; sulfamic acid
- 6M $HNO_3$; nitric acid
- 3% $H_2O_2$; hydrogen peroxide
- 6M $H_2SO_4$; dil. sulfuric acid
- 0.02M $KMnO_4$; potassium permanganate
- 6M $NH_3$; dil. aqueous ammonia
- sat. $(NH_4)_2CO_3/NH_3$; ammonium carbonate in aq. ammonia
- $Na_2CO_3$; sodium carbonate
- $NaNO_2$; sodium nitrite
- 6M NaOH; sodium hydroxide
- 6M $NH_3$; dil. aqueous ammonia
- red litmus paper
- Zn granules; zinc

## Procedure

### Preparing a Solution for Analysis

You will be given solids both as knowns and unknowns. To prepare a solution for analysis, try dissolving about 30 mg of the solid in 30 drops of water. If all the solid does not dissolve, heat in a water bath. If all the sample dissolves, prepare more solution by dissolving 100 mg of the sample in about 5 mL of water.

If all the sample does not dissolve, the solid consists of some insoluble salt(s) or a mixture of soluble salts having anions and cations that form insoluble compounds from solution (for example, a mixture of $Pb(NO_3)_2$ and $ZnCl_2$ would precipitate $PbCl_2$ if you attempted to dissolve it in water). Place 100 mg of a sample that is not completely water soluble in a small flask or beaker along with 5 mL water and 500 mg anhydrous $Na_2CO_3$ and a boiling chip. Boil the contents for 8-10 minutes and divide the contents between two test tube and centriguge. Wash each precipitate with 8 drops of water and add the washings to the combined solutions. The combined solutions are Solution 8.1. Combine the two precipitates.

The addition of $Na_2CO_3$ converts many insoluble compounds to their less soluble carbonates, thus releasing the anions into solution. For example,

$$\underline{2AgCl} + CO_3^{2-} \rightarrow \underline{Ag_2CO_3} + 2Cl^- \qquad (31\text{-}13)$$

Hence, the solution will contain sodium salts of all the anions released in reactions like Equation (31-13).

The precipitate consists of insoluble carbonates and of those compounds whose carbonates are more soluble than the compounds themselves. For example, $K_{sp}$ for $BaCO_3$ is $5.0 \times 10^{-9}$ and for $BaSO_4$, it is $1.1 \times 10^{-10}$. Hence, if $BaSO_4$ is present as one of the insoluble components, it will not be converted to $BaCO_3$ but will simply remain as the sulfate when the sample is treated with $Na_2CO_3$.

The precipitate is then treated with $HC_2H_3O_3$. This dissolves all carbonate precipitates. For example,

$$\underline{Ag_2CO_3} + 2HC_2H_3O_2 \rightarrow 2Ag^+ + CO_2(g) + 2C_2H_3O_2^- + H_2O \qquad (31\text{-}14)$$

Acids convert the carbonate ions to $CO_2$ and $H_2O$. The reaction is complete when the $CO_2$ gas ceases to evolve. The residue will consist of those insoluble salts that were never converted to carbonates: $BaSO_4$. AgCl, AgBr and/or AgI.

Only the first solution (before addition of acetic acid) and the residue need to be analyzed for anions. The acetic acid solution will contain the cations of those salts which were converted to carbonates and can be discarded. Analyze according to Outline 8.

### Outline 8. Anion Analysis

**Residue after $HC_2H_3O_2$ addition. $BaSO_4$, AgCl, AgBr, AgI.** Add 1 mL 6M $HC_2H_3O_2$ and half a spatulaful of Zn granules; heat in the water bath for 10-15 minutes stirring to disperse the precipitate. Add several more drops of acetic acid; centrifuge. A precipitate which remains is $BaSO_4$ (white) which will be contaminated with black Ag metal causing it to look gray. This confirms the presence of $SO_4^{2-}$ in the original sample. The solution containing $Zn^{2+}$, $Cl^-$, $Br^-$ and $I^-$ should be combined with Solution 8.1 for the halide test.

**Halide Test.** The first part of the test should be performed on a 1-mL portion of the original solution. Prepare a solution containing $Ag(NH_3)_2^+$ by taking 3 drops of 3.3M ammoniacal $AgNO_3$ and adding dropwise sufficient ammoniacal $(NH_4)_2CO_3$ so that the precipitate that first forms just dissolves. Add this solution dropwise to the test solution until no more precipitate seems to form. Centrifuge and wash the precipitate with 5 drops of water. Discard the solution and the washings.

**Precipitate 8.1. AgCl, AgBr, AgI.** Add 1 mL 6M $HC_2H_3O_2$ and about half a spatulaful of Zn granules. Heat the solution in the water bath for 10-15 minutes with stirring until the precipitate dissolves and a black precipitate of Ag metal forms. Centrifuge and remove the solution to another test tube.

**Solution 8.1. $Zn^{2+}$, $Cl^-$, $Br^-$, $I^-$.** Combine the solution from treatment of the residue with this solution.

**$I^-$ Test.** Add a quarter spatulaful of $NaNO_2$ and 6 drops of $CCl_4$. Cork the test tube and shake. The pink-purple color of $I_2$ in the $CCl_4$ layer indicates the presence of $I^-$. Discard the $CCl_4$ layer in the waste jar.

**$Br^-$ Test.** Draw off the water layer and acidify it with 6M $HNO_3$. First, all the $I^-$ still left in solution must be destroyed. Add crystals of $KNO_2$ a few at a time with stirring until there is no further increase in the brown color of $I_2$ in aqueous solution. Add 5 drops of $CCl_4$, cork the test tube, and shake to extract the remaining $I_2$ into the $CCl_4$ layer. Draw off the water layer and discard the $CCl_4$ in the waste jar. Repeat this procedure until no more pink color appears on extraction with $CCl_4$. Now that the $I^-$ is destroyed, the $Br^-$ test can be done.

Add 3 drops of $CCl_4$ and 4-5 drops of 0.02M $KMnO_4$. Cork the tube and shake. The brown color of $Br_2$ in the $CCl_4$ layer indicates the presence of $Br^-$ in the original sample. Draw off the water layer and discard the $CCl_4$ layer in the waste jar.

**$Cl^-$ Test.** $Br^-$ must first be destroyed. Boil the solution until all the volatile $Br_2$ is driven off. Keep replenishing the water to maintain the volume. Test for complete disappearance of $Br_2$ by adding 2 drops of $KMnO_4$, 2 drops of $CCl_4$, corking and shaking. Check for the absence of the brown color of $Br_2$ in the $CCl_4$ layer. If $Br_2$ is still present, draw off the water layer and boil again. Repeat the treatment with $KMnO_4$ and $CCl_4$. If any precipitate is present, dissolve it in 1-2 drops of $H_2O_2$.

Acidify the solution with 6M $HNO_3$ and add 0.2M $AgNO_3$ dropwise until no more precipitate forms. The white precipitate of AgCl indicates the presence of $Cl^-$ in the original sample.

**Nitrite Test.** Use 5 drops of the original solution and add 0.2M $BaCl_2$ dropwise until no more white precipitate of $BaSO_4$ seems to form. Centrifuge and discard the precipitate. To the supernatant liquid add 2 drops of $BaCl_2$ solution and half a spatulaful of sulfamic acid crystals. Evolution of $N_2$ gas and formation of a white precipitate of $BaSO_4$ confirms the presence of $NO_2^-$ in the original sample.

**Nitrate Test.** If the test for $NO_2^-$ was positive, add half a spatulaful of sulfamic acid to 7 drops of the original solution and heat in the water bath until $N_2$ evolution ceases. Add a few more crystals to be sure all the $NO_2^-$ has reacted. Make the solution basic to litmus with 6M NaOH and add 5 drops extra.

Transfer the solution to a casserole or crucible and evaporate over a microflame almost to dryness to destroy $NH_4^+$.

Transfer the solution to a test tube and add an equal volume of 6M NaOH being careful not to wet the upper walls of the test tube with the base. Wet a piece of red litmus paper and fold it into a V-shape. Add a couple of pieces of Al wire to the test tube and carefully place the litmus in the mouth of the test tube so that any gas evolving will contact the paper. The tube should stand for several minutes. A blue color of the litmus from evolving $NH_3$ confirms the presence of $NO_3^-$ in the original sample.

**Sulfate Test.** To 6 drops of the original solution, add 0.2M $BaCl_2$ until no more precipitate seems to be forming. The white precipitate of $BaSO_4$ confirms the presence of $SO_4^{2-}$ in the original sample. As an added confirmation, $BaSO_4$ is insoluble in 6M HCl.

**Carbonate Test.** Place a 40-mg sample of the original solid in the bottom of a test tube and add 3-4 drops of 6M $H_2SO_4$. Have ready a medicine dropper into which 0.2M $Ba(OH)_2$ solution has been drawn. Hold this dropper in the test tube so that any evolving gas can enter the bottom of the dropper. A precipitate of $BaCO_3$ (*in the dropper*) produced by evolved $CO_2$ indicates the presence of $CO_3^{2-}$ in the original sample.

## Results

Submit a 3 x 5 in. card for your unknown giving your name, desk number, ions present, and ions absent. Have your notebook available for your instructor to examine on request.

## References


1. King, Edward J., "Ionic Reactions and Separations: Experiments in Qualitative Analysis", Harcourt Brace Jovanovich, New York, 1973. pp. 59-106.

2. Sorum, C.H., Lagowski, J. J., "Introduction to Semimicro Qualitative Analysis", 5th ed. Prentice Hall, Englewood Cliffs, 1977. pp. 243-288.

# APPENDIX I

# SAMPLE LABORATORY REPORT DENSITY DETERMINATION

**Purpose:** To determine by experiment the density of solid sulfur and of an unknown liquid, to gain experience in using a single-pan analytical balance for mass determination, and to practice volume measurement using a graduated cylinder.

**Prodedure:** To determine the density of a solid, a known weight of the substance was placed in a graduated cylinder containing a measured volume of a liquid in which the solid was insoluble. The increase in volume was the volume of the solid. The procedure for the liquid was the same as that described in Experiment 2.

**Data:** A. Density of sulfur

| | |
|---|---|
| Weight of filter paper and sulfur | 25.721 g |
| Weight of filter paper | -1.124 g |
| Weight of sulfur | 24.597 g |
| | |
| Volume of water and sulfur | 42.1 mL |
| Volume of water | -30.1 mL |
| Volume of sulfur | 12.0 mL |

B. Density of unknown liquid Y

| | |
|---|---|
| Weight of beaker and liquid | 50.124 g |
| Weight of beaker | -20.011 |
| Weight of liquid | 30.113 g |
| | |
| Volume of liquid | 30.0 mL |

## Results and Discussion:

The density of a substance is defined as:

density = $\rho$ = weight/volume

for sulfur, $\rho$ = 24.597g/12.0mL = 2.05 g/mL

for liquid Y, $\rho$ = 30.113g/30.0mL = 1.00 g/mL

In weighings made on the analytical balance, the last figure is accurate to ±1 in the last decimal place. This means that the weight is precise to ± 0.001 g or ±0.0001 g, depending on the model of analytical balance used.

The last figure in the volume measurement must be estimated; this estimation is precise to ± 0.1 mL.

The volumes are measured to only three significant figures while the weights are measured to at least five significant figures. The calculated densities can, therefore, be precise to no more than three significant figures. This means that the weighings are really more precise than they need to be. They could have been made on a triple beam balance giving only three significant figures; for these weights, that would mean a pricision of ±0.01g. However, one purpose of the experiment was to gain experience in using the analytical balance. So, the measurements were made using an analytical balance rather than on the triple-beam balance.

Considering the inherent uncertanties in the estimation procedures, the weight of the sulfur sample is actually 24.597 ±0.001 g and the volume is 12.0 ±0.1 mL. Hence, the maximum value which the density could have would result if the true weight were 24.598 g and the actual volume were 11.9 mL. Then $\rho$ = 24.598 g/11.9 mL = 2.07 g/mL. The minumum value for the density would result if the true weight were 24.594 g and the volume were 12.1 mL. Then the calculations as shown above would give a density of 2.03 g. Thus, for sulfur there is an uncertainty of ±0.02 g in the density. The uncertainty is expressed by writing the density as 2.05 ± 0.02 g/mL.

Applying the same treatment to the uncertainty of the liquid density = $\rho$ = 1.00 ± 0.01 g/mL.

According to the C.R.C. Handbook of Chemistry and Physics, the density of sulfur is 2.07 g/mL. Therefore the percent error in the experimental determination is:

$$\frac{\text{(correct value) - (experimental value)}}{\text{Correct value}} \times 100\% =$$

$$\frac{2.07 - 2.05}{2.07} \times 100\% = 1\%$$

**Questions:** none to answer

# APPENDIX II

# SCIENTIFIC MEASUREMENT

Scientists are interested not only in knowing what happens during experiments, but also in making quantitative measurements of materials and occurrences. For example, scientists studying rocks from the moon wanted to know not only what elements these rocks contain, but also the relative amounts of the various elements. When this kind of quantitative information was available, comparison of the moon's composition and Earth's compositon was possible. Most experiments in chemistry laboratories include quantitative measurements, and this Appendix describes how to record and deal with them.

## Precision and Accuracy

*Precision* refers to how well several determinations of one quantity agree with each other. *Accuracy* refers to how a measured value agrees with the true value. Scientists try to make measurements that are both precise and accurate; however, it is possible to be quite precise without being very accurate. For example, if the mass of a beaker were measured several times, the values obtained might be almost the same, for example 84.3 g, 84.1 g, and 84.2 g. But if the balance that was used registers 10.7 g when it is empty, these fairly precise values would not accurately give the mass of the beaker.

Several factors that lessen the accuracy of measurements are said to introduce *systematic* or *determinate errors*, that is, errors within the system that can be determined and eliminated. One source of these errors, inaccurate equipment, was cited above; if you think that equipment is not functioning properly, consult your instructor. Other sources are impure chemicals and inexact procedures; chemicals of the necessary purity will be provided for the work you do, and you can reduce procedural errors by carefully following the instructions given for each experiment.

Both precision and accuracy of measurements are affected by *random* or *indeterminate errors*, errors that are beyond the experiment's control. They include the effects of factors such as small temperature variations during an experiment, absorption of moisture by objects while they are being weighed, and the inherent inexactness of measuring devices. Most indeterminate errors are as likely to affect a measurement in a positive direction as in a negative direction and so a particular measurement might be either slightly too large or slightly too small. Hence, two or more measurements are usually made and their average is used with the expectation that positive and negative random errors will cancel one another. But random errors can never be completely eliminated and, therefore, perfect precision and accuracy are never expected.

## Evaluation of Precision

Two or more measurements of a property are usually made and their average is used as the best estimate of the property being measured, since small random errors tend to cancel one another. We can have confidence in our results if successive determinations of the same quantity deviate only slightly from the average value. For example, three separately made measurements of the mass of an object might be 10.32 g, 10.34 g, and 10.31 g. The average of these three measurements is 10.32 g. But three other people might measure the mass and obtain values of 10.39 g, 10.29 g, and 10.35 g. The average of these values is 10.34 g, very close to the first average, but less precise. An indication of the precision of a set of data is the average amount by which individual measurements deviate from the average value of the measurements; this is called the *average deviation.* The calculation of the average deviation (a.d.) for each of these two sets of data is illustrated on the next page.

| Set I | | Set II | |
|---|---|---|---|
| values | deviations | values | deviations |
| 10.32 | 0.00 | 10.39 | 0.05 |
| 10.34 | 0.02 | 10.29 | 0.05 |
| 10.31 | 0.01 | 10.35 | 0.01 |
| 3 ⌈ 30.97 | 3 ⌈ 0.03 | 3 ⌈ 31.03 | 3 ⌈ 0.11 |
| 10.32 g | 0.01 = a.d. | 10.34 g | 0.04 = a.d. |

The more precise data set, Set I, would be considered more reliable; that is, it is probably more accurate as well as more precise.

## Uncertainty and Significant Figures

Consider the process of making a measurement, using as an example the measurement of the length of a table with a meter stick. If the stick is placed along the table edge, the corner might be between the 96-cm and 97-cm marks, Figure II-1. A closer estimate of the table's length can be obtained by noting that the corner is between 96.8 cm and 96.9 cm. Further refinement of the

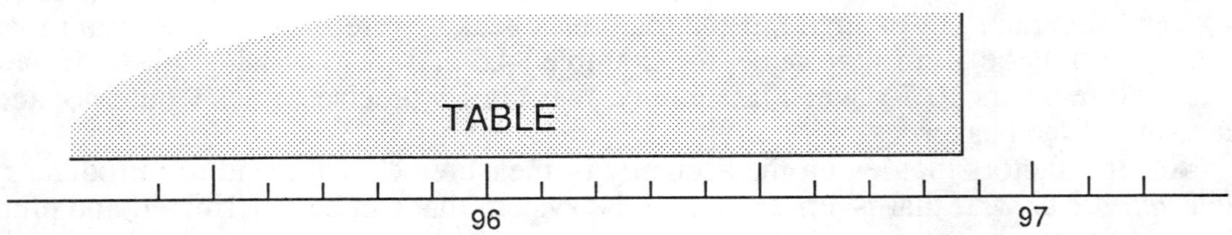


**Figure II-1**

Measuring the length of a table with a meter stick.

measurement can be made by mentally dividing the distance between the 96.8 cm and 96.9 cm marks into 10 divisions and estimating where the corner is. It is about 0.7 of the distance between the two marks, and so 96.87 cm is the best measurement we can make of the table's length.

The last digit in 96.87 cm is uncertain. If the measurement were repeated, the last digit might be estimated to be 5, 6, 8, or 9, rather than 7. This limit on reproducibility exists because it simply is not possible to estimate the measurement any more closely with the unaided eye and a meter stick. The uncertainly, or possible error, in this and any other measurement should be indicated in reporting it. This measurement should be reported as 96.87 ±0.02 cm.

We say that the measurement 96.87 cm has four *significant figures*. Significant figures are all digits in a number that are known without any doubt, plus one digit that is uncertain. In the measurement 96.87 cm, the 9, the 6, and the 8 are certain; the 7 is uncertain because it is estimated. In any measurement that is reported, the last figure can be assumed to be estimated. Just how uncertain this final figure is depends on the measuring process and instruments used. In the absence of any other statement of uncertainty you can assume that the last figure in measurements reported in your textbook and in other scientific books and journals is known to ±1.

### Scientific Notation

We might also express the measured table length of 96.87 cm as 0.9687 meter, 0.0009687 kilometer, or 968,700,000 nanometers. These numbers contain more than four figures but all three numbers are simply different ways of expressing the information obtained from one measurement and not all of the figures are significant. Each of these numbers contains only the three significant figures that were obtained exactly and the fourth significant figure that was estimated in the original measurement; the additional zeros are not significant figures. To write a number in a way that immediately shows how many figures are significant, we write the number in *scientific notation*; that is, we write it as a number between 1 and 10 multiplied by some power of 10; we retain only significant figures. Thus, 0.9687 m would be written $9.687 \times 10^{-1}$ m; 0.0009687 km = $9.687 \times 10^{-4}$ km; and 986,700,000 nm = $9.867 \times 10^{8}$ nm. Even though different units are used, all these numbers are the result of the same measurement and contain only the original four significant figures.

Consider the number 0.000217. The zeros at the beginning of this number have the essential mathematical function of locating the decimal point. However, these zeros do not state a number of units describing the property that was measured; they are not significant in terms of measurement. Only three figures are significant: 217. When this number is written in scientific notation as $2.17 \times 10^{-4}$, it is obvious that it contains only three significant figures. Assuming that the uncertainty in the last figure is ±1, we see that this number has an uncertainty of 1 part in 217, or about 0.5%.

The measurement 1200 g is somewhat more confusing. Does 1200 contain two, three, or four significant figures? The answer depends on the uncertainty in the original measurement. Was the balance precise to ±1 g? If so, all four figures are significant; three figures are exact and the fourth is an estimate; this number is written $1.200 \times 10^{3}$. But if the measurement has an uncertainty of ±10 g, the number is $1.20 \times 10^{3}$. And if the measurement has an uncertainty of ±100 g, the number is $1.2 \times 10^{3}$. Scientific notation removes the ambiguity. You should use it in all your work to communicate information about the uncertainty in measurements just as practicing scientists do.

## Significant Figures in Calculations

The results of many measurements are used in formulas to calculate other quantities. Attention to significant figures is essential in these calculations so that the proper degree of uncertainty is shown. Suppose a square table has a side that is 96.87 cm ±0.02 cm long; this length has an uncertainty of 2 parts in 9687, or about 1 part in 4800, or about 0.02%. To calculate the area of the table, we use the formula $A = s^2 = (96.87\text{ cm})^2 = 93{,}837{,}969\text{ cm}^2$. Using all eight of the figures obtained in the multiplication implies the area has an uncertainty of 1 part in 93,837,969, or about 1 part in 94,000,000, about 0.000001%. But the measurement of the length of the table has an uncertainty of about 1 part in 4800. It is impossible for us to know the calculated quantity with less possible error than we know the measured quantity used in the calculation; simply performing a multiplication cannot decrease the possible error to less than that of the original measurement. Therefore, we cannot determine the area of the table to better than about 1 part in 4800, 9384 $cm^2$ ±2 $cm^2$; reporting this result as 9384 $cm^2$ approximately indicates the uncertainty in this number on the basis of the uncertainty in the original measurement. In any calculation using measurements, *a calculated quantity can be known with no more certainty than the least certain number used in the calculation.*

Here are guidelines for using significant figures in calculations.

1. *For addition and subtraction*, all numbers used in the calculation have the same units. The number that is uncertain by the greatest fraction of a unit determines the uncertainty in the answer. For example,

$$\begin{array}{r} 2.7\text{ g} \\ 0.114\text{ g} \\ \underline{19.77\text{ g}} \\ 22.584\text{ g} \end{array} = 22.6\text{ g}$$

   The first number has the greatest possible error, ±0.1 g, and so the sum of the three numbers can be known only to ±0.1 g.

2. *For multiplication and division,* the units of the numbers used in the calculation and the units of the answer are not the same; a particular fraction of one unit does not have the same meaning as that same fraction of a different unit. To show the proper degree of uncertainty in the answer, the number of significant figures in the answer is the same as the number of significant figures in the number having fewest significant figures. For example

$$\frac{24.6\ \text{m}}{\text{sec}} \times 3.4\ \text{sec} = 83.64\ \text{m} = 84\ \text{m}$$

### Exact Numbers

The measurements discussed so far have all been of properties that can vary continuously and are subject to uncertainty. But there are some numbers that can be known exactly. The counting of discrete objects can be done with complete precision and accuracy. Thus, in using such numbers in calculations the uncertainty of the result is determined by other measurements. For example, if three people weigh 114 lb, 152 lb, and 176 lb, what is the average of their weights?

$$\begin{array}{r l} & 114 \\ & 152 \\ & \underline{176} \\ 3 \,\big) & 442 \\ & 147.33333\ \text{lb} = 147\ \text{lb} = \text{average} \end{array}$$

The average has three significant figures, as does each weight used in the calculation. The counted number 3 is exact and does not limit the number of figures in the result.

Some numbers are exact by definition, for example numerical relationships within one measurement system. One foot is defined as equalling exactly twelve inches. One kilometer is defined as equalling exactly one thousand meters. Therefore, if such relationships are used in calculations they do not limit the number of significant figures in the result. In effect, these relationships contain an unlimited number of significant figures.

# APPENDIX III

# GRAPHING EXPERIMENTAL DATA

Many scientific experiments involve measuring what happens to one property of a system as another property varies. For example, it is possible to measure how the volume of a gas sample changes as pressure on the gas changes.

The following data might be obtained.

| Pressure, atm | Volume, liters |
|---|---|
| 1.0 | 16.0 |
| 2.0 | 8.0 |
| 5.0 | 3.2 |
| 10.0 | 1.6 |

The data clearly show that volume decreases as pressure increases. However, a graph is a more striking presentation of the data; Figure III-1 illustrates how volume drops with increasing pressure.

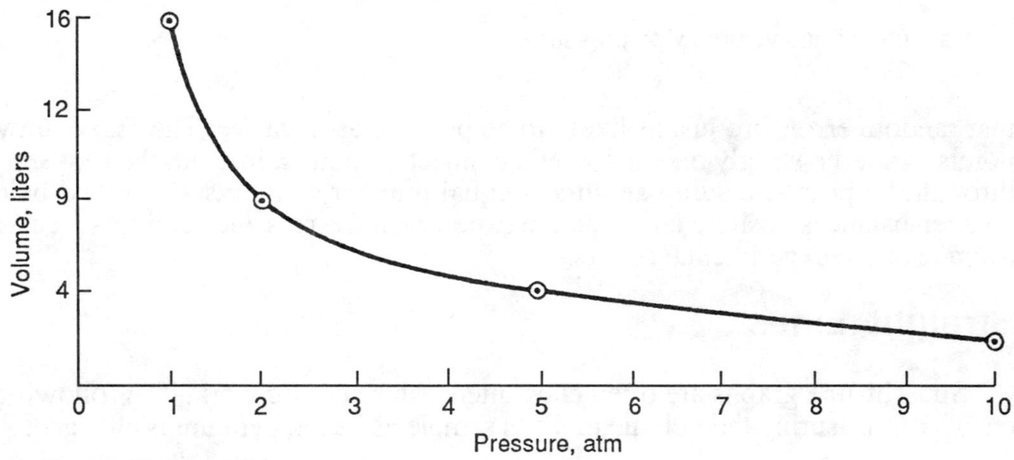


**Figure III-1**

Variation of gas volume with pressure.

In an experiment from which such data would be obtained, the experimenter might control the amount of pressure on the gas; hence, $P$ is an independent variable. Values of the independent variable are usually plotted on the horizontal, or $x$ axis, of a graph (the abscissa). The size of $V$ depends on how large or small the experimenter makes $P$; hence, $V$ is a dependent variable. Values of the dependent variable are usually plotted on the vertical, or $y$ axis, of a graph (the ordinate). Dots are placed at the points corresponding to measured values of $P$ and $V$ and these points are used to draw a smooth curve. We draw a continuous, smooth curve to connect the points (rather than, for example, several straight-line segments) because we recognize that one property varies continuously as some other property on which it is dependent by a specific mathematical relationship varies. Circles can be drawn around experimental points so that the points are easily seen and are not lost on the line or mistaken for normal imperfections in the paper. When two or more different sets of data are plotted on one graph, one set of points can be enclosed in small squares or triangles instead of circles, so that it is obvious which data belong together and were used for a particular line.

The preceding data are more accurate than is usual for data obtained in a laboratory. Actual laboratory measurements are subject to experimental error (see Appendix II on Scientific Measurement). In spite of efforts to be as accurate as possible, random errors occur and lessen the accuracy of measurements. These are unavaoidable errors that remain after accidents, mistakes, and systematic errors have been eliminated. Random errors produce data points that are more scattered on a graph than error-free data points would be. Real data points for an experiment in which *P* and *V* are measured for a gas sample might be similar to those plotted in figure III-2. It is useful to realize

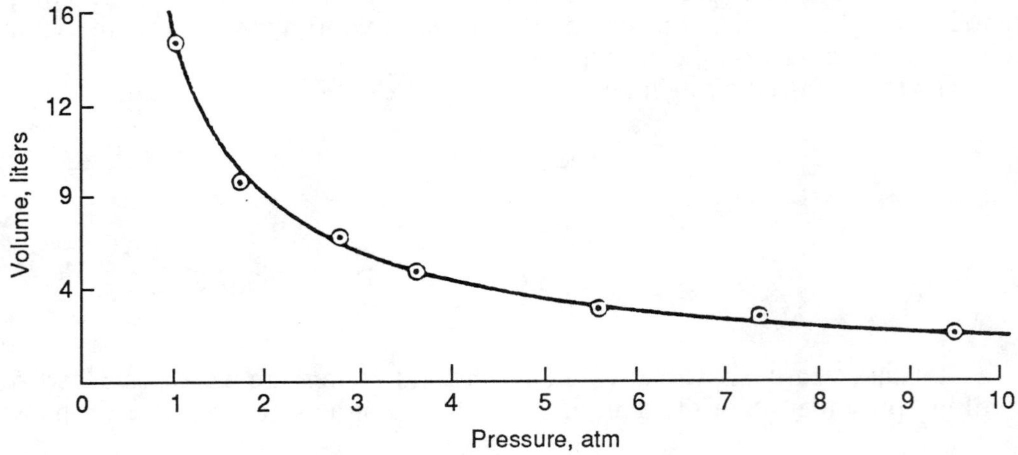


**Figure III-2**

Variation of gas volume with pressure.

that random errors are just as likely to be positive as negative. The curve drawn through the data points is not the wiggly one that would connect all points; instead, the best smooth curve is drawn through the points, leaving an almost equal number of points above and below the curve. This counter-balances positive and negative experimental errors and produces a curve that represents an average of the experimental results.

## Straight Lines

Straight-line graphs are often encountered in scientific work. The following data represent the results of measuring the volume of a gas sample as its temperature is changed.

| Temperature, K | Volume, liters |
|---|---|
| 135 | 11.0 |
| 273 | 21.8 |
| 350 | 28.0 |
| 420 | 33.0 |
| 545 | 43.5 |

A plot of temperature (independent variable) vs. volume (dependent variable) is a straight line (with slight experimental error), as shown in Figure III-3.

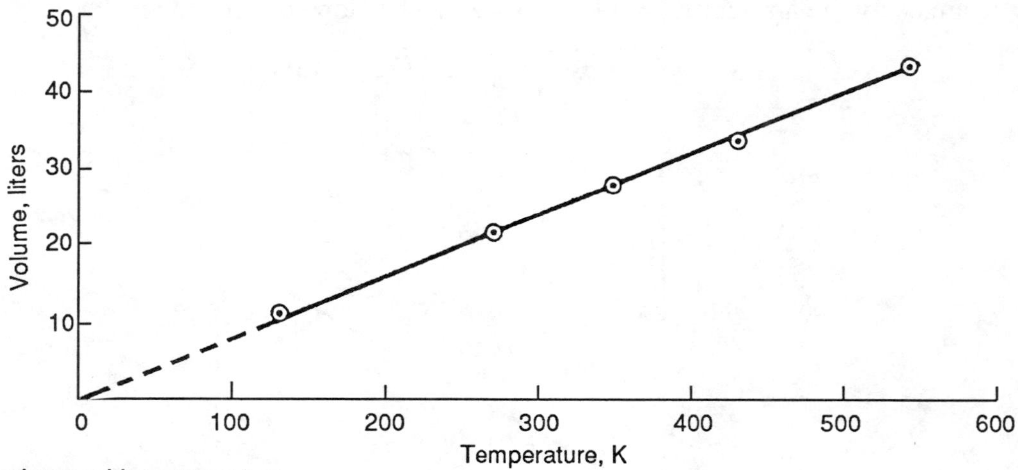


**Figure III-3**

Variation of gas volume with temperature.

The volume is said to vary linearly with temperature. To say that there is a linear variation means that a straight line results when numerical values of one property are plotted along the $y$ axis and numerical values of the other property are plotted along the $x$ axis. Mathematically, a linear plot can be described by the algebraic equation for a straight line:

$$y = mx + b \tag{III-1}$$

The $m$ is constant for a particular line; it is the slope of the line. The $b$ is also constant for a particular line; it is the $y$ intercept.

The slope of a line shows how much $y$ changes when $x$ changes. Consider the straight lines in Figure III-4. For line A, $y$ increases more for a particular change in $x$ than it does for line B; line A has a greater slope. For line C, $y$ decreases as $x$ increases; this line has a negative slope. At the point where $x = 0$ (that is, on the $y$ axis) the general equation for a straight line reduces to

$$y = (m \times 0) + b = b \tag{III-2}$$

The intercept $b$ is the value of $y$ at which the line crosses the $y$ axis.

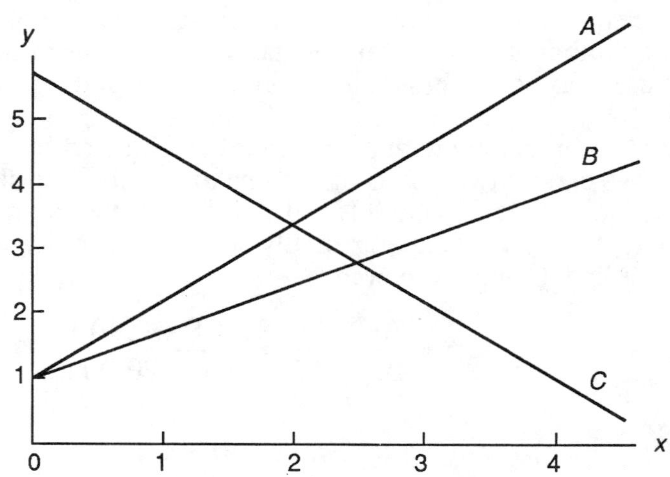


**Figure III-4**

To determine the value of the slope of a line, we choose two points on the line, say points I and II on the line shown in Figure III-5. At point I, $y = 2.1$ and $x = 1.0$; at point II, $y = 4.3$ and $x = 3.0$. Between points I and II, the value of $y$ changes by the amount $\Delta y = 4.3 - 2.1 = 2.2$; $x$ changes by the amount $\Delta x = 3.0 - 1.0 = 2.0$. The slope of this line is

$$m = \frac{\Delta y}{\Delta x} = \frac{2.2}{2.0} = 1.1 \qquad \text{(III-3)}$$

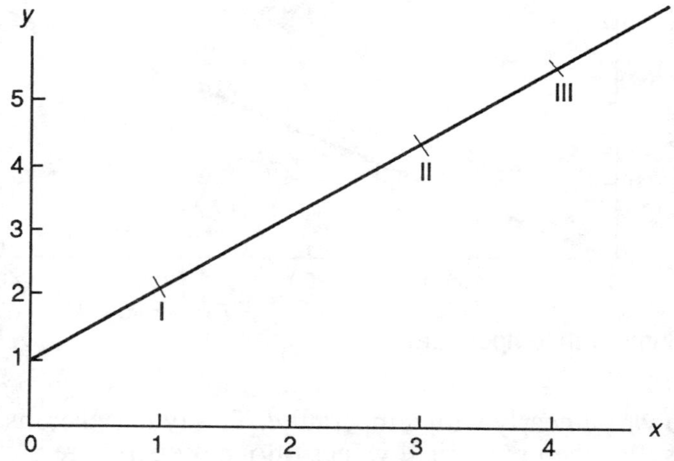


Figure III-5

The value calculated for the slope of the line should be the same regardless of which two points are used. For example, calculate $m$ using point III ($y = 5.4$ and $x = 4.0$) and point I. Occasionally, the use of different sets of points to determine a slope will result in slightly different values; this is because of slight errors in plotting and reading the values for the points and also because of inaccuracies that result from loss of significant figures in the calculation. To minimize the second problem, the two points should be as far apart as possible on the line so that the difference will have as many significant figures as possible. It is also advantageous to choose points on the line that fall exactly at the intersections of lines on the graph paper, so that there is as little doubt as possible in reading the values.

The $y$ intercept can be easily read from a graph. In Figure III-5, the line crosses the $y$ axis (that is, at the point $x = 0$) at the value $y = 1.0$; this is the $y$ intercept.

Thus, the equation for the line in Figure III-5 is

$$y = 1.1x + 1.0 \qquad \text{(III-4)}$$

This is an algebraic equation that enables us to calculate the value of $y$ corresponding to any value of $x$ (or vice versa) without having to make actual measurements of these values. If this equation is recorded for later use, it is not necessary to retain the graph and read it each time such information is required.

Let us now use this procedure with the straight line obtained by plotting $V$ vs. $T$ for a gas sample. (Figure III-3). We choose two points on the line to calculate the slope. Let point I be $V = 16.0$ L and $T = 200$ K. Let point II be $V = 36.0$ L and $T$ 450 K. Note that the points used to calculate $m$ are points on the line; they are not experimentally determined points, which might lie slightly above or below the straight line. Using these points

$$m = \frac{\Delta y}{\Delta x} = \frac{\Delta V}{\Delta T} = \frac{(36.0 - 16.0)\text{ L}}{(450 - 200)\text{ K}} \qquad \text{(III-5)}$$

$$= \frac{20.0\text{ L}}{250\text{ K}} = \frac{0.800\text{ L}}{\text{K}}$$

The slope of the line has units if either or both of the quantities plotted have units. When the line is extrapolated (that is, when it is extended beyond the region for which experimental values are plotted), we see that the point at which the line crosses the y axis is the origin; this is shown by a dashed line in Figure III-3. For this particular line, the y intercept is 0; however, this is not often the case. This value of the intercept means that $V = 0$ when $T = 0$. The equation for this line is

$$V = \frac{0.800\ \text{L}}{\text{K}} T + 0 \qquad \text{(III-6)}$$

Because straight lines are easy to recognize, scientific data are often expressed in forms that produce straight line graphs. For example, the data given at the beginning of this Appendix show that gas volume decreases as pressure increases. Mathematically, the volume and pressure are inversely related and their product is a constant:

$$PV = k \qquad \text{(III-7)}$$

A plot of an equation of the form $xy = k$ is a hyperbola. However, this is often not apparent because the plotted experimental data cover too narrow a range of values or because of the different scales on the two axes. Rearranging Equation (III-7) gives

$$V = \frac{k}{P} \qquad \text{(III-8)}$$

Volume is directly proportional to $\frac{1}{P}$ and a plot of $V$ vs. $\frac{1}{P}$ is a straight line, Figure III-6.

The linearity would be apparent even with only three data points and regardless of the scales used on the two axes.

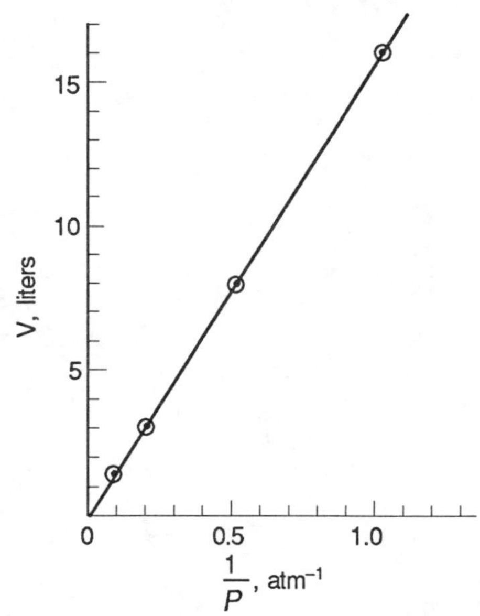


**Figure III-6**

Variation of gas volume with $\frac{1}{P}$

## General Guidelines for Making Graphs.

Keep in mind that a graph is a visual presentation of numerical information; the information should be presented in a manner such that a reader can interpret it quickly, easily, and accurately. Each of the following paragraphs deals with a different aspect of this presentation.

At the top of a graph give it a title that clearly states what information it presents. Note the titles on the graphs in Figures III-1, III-2, III-3, and III-6.

Label each axis with the name or symbol of the variable and the unit in which it is measured. As regular intervals along the axis, give numerical values.

When you first learned to make graphs you probably had one square on the graph paper represent the same number of units along both axes; however, this is not necessary. A graph should fill as much of a sheet of graph paper as is conveniently possible so that the experimental data can be plotted and read with the greatest possible accuracy. To accomplish this, each axis can be scaled in the most convenient manner, perhaps hundreds of units per square along one axis and only a fraction of a unit per square along the other axis. Also, it is not necessary for either scale to start at zero. For example, experimental data involving temperature changes might cover only the temperature range of 72-107°C; if the scale began at 0°C, most of the graph would be blank space containing no experimental points. However, if the $y$ intercept of a straight line plot is to be read from the graph, the $x$ axis must include the point $x = 0$ and the $y$ axis must extend far enough that the $y$ intercept can be read from it. Your decision about the scales will depend on the use to which you intend to put the graph.

For each axis, choose a scale that can be easily used to plot and read values of points on the line. Of course the use of one square to equal one unit (or some decimal multiple or division such as 1000, 10, 0.0001, etc.) gives a scale that is the easiest to interpret. The use of one square = 5, or one square = 2, is also fairly easy to interpret. Other scale graduations, such as one square = 0.25, or one square = 3, or one square = 4, are considerably less convenient to plot and read.

When drawing a line, use a sharp pencil or fine-pointed pen so that you draw a fine line, not a broad line that spreads over a data range. French curves and flexible plastic curves are helpful when drawing curves. Use a ruler to draw a straight line.

If your results require determining the slope of a straight line, mark on the line the points that are used in this calculation.

## SI BASE UNITS

| Physical Quantity | Name of Unit | Symbol |
|---|---|---|
| length | meter | m |
| mass | kilogram | kg |
| time | second | s |
| electric current | ampere | A |
| thermodynamic temperature | kelvin | K |
| amount of substance | mole | mol |

## PREFIXES FOR FRACTIONS AND MULTIPLES OF SI UNITS

| Fraction | Prefix | Symbol | Multiple | Prefix | Symbol |
|---|---|---|---|---|---|
| $10^{-1}$ | deci | d | 10 | deka | da |
| $10^{-2}$ | centi | c | $10^2$ | hecto | h |
| $10^{-3}$ | milli | m | $10^3$ | kilo | k |
| $10^{-6}$ | micro | $\mu$ | $10^6$ | mega | M |
| $10^{-9}$ | nano | n | $10^9$ | giga | G |
| $10^{-12}$ | pico | p | $10^{12}$ | tera | T |
| $10^{-15}$ | femto | f | | | |
| $10^{-18}$ | atto | a | | | |

**Note:** Multiple prefixes, such as $\mu\mu$F for pF, should not be used.

## SPECIAL NAMES AND SYMBOLS FOR CERTAIN SI DERIVED UNITS

| Physical Quantity | Name of SI Unit | Symbol for SI Unit | Definition of Unit |
|---|---|---|---|
| force | newton | N | $kg\ m\ s^{-2}$ |
| pressure | pascal | Pa | $kg\ m^{-1}\ s^{-2}(=N\ m^{-2})$ |
| energy | joule | J | $kg\ m^2\ s^{-2}$ |
| power | watt | W | $kg\ m^2\ s^{-3}(=J\ s^{-1})$ |
| electric charge | coulomb | C | A s |
| electric potential difference | volt | V | $kg\ m^2\ s^{-3}\ A^{-1}(=J\ A^{-1}\ s^{-1})$ |
| frequency | hertz | Hz | $s^{-1}$ (cycle per second) |

## DECIMAL FRACTIONS AND MULTIPLES OF SI UNITS HAVING SPECIAL NAMES

| Physical Quantity | Name of Unit | Symbol for Unit | Definition of Unit |
|---|---|---|---|
| length | ångström | Å | $10^{-10}$ m = $10^{-8}$ cm |
| length | micron* | $\mu$ | $10^{-6}$ m = $\mu$m |
| force | dyne | dyn | $10^{-5}$ N |
| pressure | bar | bar | $10^5$ N $m^{-2}$ |
| energy | erg | erg | $10^{-7}$ J |

*The word micron and the symbol $\mu$ for $\mu$m should be abandoned immediately.



SI (*SYSTEME INTERNATIONALE*) UNITS

## CONVERSION FACTORS FOR COMMONLY USED UNITS

| Length | Å | in. | m | cm |
|---|---|---|---|---|
| 1 ångström (Å) | 1 | $3.94 \times 10^{-9}$ | $10^{-10}$ | $10^{-8}$ |
| 1 inch (in.) | $2.54 \times 10^{8}$ | 1 | $2.54 \times 10^{-2}$ | 2.54 |
| 1 meter (m) | $10^{10}$ | 39.37 | 1 | $10^{2}$ |
| 1 centimeter (cm) | $10^{8}$ | .3937 | $10^{-2}$ | 1 |

| Mass | lb | oz | kg | g |
|---|---|---|---|---|
| 1 pound (lb) | 1 | 32 | .4536 | 453.6 |
| 1 ounce (oz) | .0625 | 1 | $2.836 \times 10^{-2}$ | 28.36 |
| 1 kilogram (kg) | 2.204 | 35.3 | 1 | 1000 |
| 1 gram (g) | $2.204 \times 10^{-3}$ | .0353 | .001 | 1 |

| Pressure | atm | Pa | mm Hg | dyn cm$^{-2}$ | lb in.$^{-2}$ |
|---|---|---|---|---|---|
| 1 atmosphere (atm) | 1 | $1.013 \times 10^{5}$ | 760 | 1.013 | 14.70 |
| 1 pascal (Pa) | $9.872 \times 10^{-6}$ | 1 | $7.502 \times 10^{-3}$ | $10^{-5}$ | $1.451 \times 10^{-4}$ |
| 1 torr (mm Hg) | $1.32 \times 10^{-3}$ | $1.333 \times 10^{2}$ | 1 | $1.33 \times 10^{-5}$ | $1.93 \times 10^{-2}$ |
| 1 bar (dyn cm$^{-2}$) | $9.872 \times 10^{-1}$ | $10^{5}$ | $7.502 \times 10^{2}$ | 1 | $1.451 \times 10^{1}$ |
| 1 pound per square inch (lb in.$^{-2}$) | $6.803 \times 10^{-2}$ | $6.891 \times 10^{3}$ | $5.17 \times 10^{1}$ | $6.891 \times 10^{-2}$ | 1 |

| Temperature | K | °F | °C |
|---|---|---|---|
| 1 degree Kelvin (K) | 1 | $^9/_5$ (K) – 459.7 | K + 273.16* |
| 1 degree Fahrenheit (°F) | $^5/_9$(°F) + 255.4 | 1 | $^5/_9$(°F – 32) |
| 1 degree Centigrade (°C) | °C – 273 | $^9/_5$(°C) + 32 | 1 |

| Energy | J | erg | cal | kWh |
|---|---|---|---|---|
| 1 joule (J) | 1 | $10^{7}$ | .2390 | $2.8 \times 10^{-7}$ |
| 1 erg (erg) | $10^{-7}$ | 1 | $2.390 \times 10^{-8}$ | $2.8 \times 10^{-14}$ |
| 1 thermochemical calorie (cal) | 4.184 | $4.184 \times 10^{7}$ | 1 | $1.162 \times 10^{-6}$ |
| 1 kilowatt-hour (kWh) | $3.6 \times 10^{6}$ | $3.6 \times 10^{13}$ | $8.604 \times 10^{5}$ | 1 |

| Volume | ml | cm$^3$ | qt | oz |
|---|---|---|---|---|
| 1 milliliter (ml) | 1 | 1 | $1.06 \times 10^{-3}$ | $3.392 \times 10^{-2}$ |
| 1 cubic centimeter (cm$^3$) | 1 | 1 | $1.06 \times 10^{-3}$ | $3.392 \times 10^{-2}$ |
| 1 quart (qt) | 943 | 943 | 1 | 32 |
| 1 fluid ounce (oz) | 29.5 | 29.5 | $3.125 \times 10^{-2}$ | 1 |

*Absolute zero (K) = –273.16°C



CONVERSION FACTORS FOR COMMONLY USED METRIC AND ENGLISH UNITS

# PERIODIC TABLES OF THE ELEMENTS
## with Element Names

KEY

| | |
|---|---|
| 6 | ← Atomic Number |
| **C** | ← Symbol |
| 12.01 | ← Atomic Weight |
| Carbon | ← Name |

| | | | | | | | | | | | | | | | | | |
|---|---|---|---|---|---|---|---|---|---|---|---|---|---|---|---|---|---|
| 1<br>**H**<br>1.0079<br>Hydrogen | | | | | | | | | | | | | | | | | 2<br>**He**<br>4.003<br>Helium |
| 3<br>**Li**<br>6.941<br>Lithium | 4<br>**Be**<br>9.012<br>Beryllium | | | | | | | | | | | 5<br>**B**<br>10.81<br>Boron | 6<br>**C**<br>12.011<br>Carbon | 7<br>**N**<br>14.007<br>Nitrogen | 8<br>**O**<br>15.999<br>Oxygen | 9<br>**F**<br>19.00<br>Fluorine | 10<br>**Ne**<br>20.18<br>Neon |
| 11<br>**Na**<br>22.99<br>Sodium | 12<br>**Mg**<br>24.31<br>Magnesium | | | | | | | | | | | 13<br>**Al**<br>26.98<br>Aluminum | 14<br>**Si**<br>28.09<br>Silicon | 15<br>**P**<br>30.97<br>Phosphorus | 16<br>**S**<br>32.066<br>Sulfur | 17<br>**Cl**<br>35.453<br>Chlorine | 18<br>**Ar**<br>39.95<br>Argon |
| 19<br>**K**<br>39.10<br>Potassium | 20<br>**Ca**<br>40.08<br>Calcium | 21<br>**Sc**<br>44.96<br>Scandium | 22<br>**Ti**<br>47.88<br>Titanium | 23<br>**V**<br>50.94<br>Vanadium | 24<br>**Cr**<br>52.00<br>Chromium | 25<br>**Mn**<br>54.94<br>Manganese | 26<br>**Fe**<br>55.85<br>Iron | 27<br>**Co**<br>58.93<br>Cobalt | 28<br>**Ni**<br>58.71<br>Nickel | 29<br>**Cu**<br>63.54<br>Copper | 30<br>**Zn**<br>65.38<br>Zinc | 31<br>**Ga**<br>69.72<br>Gallium | 32<br>**Ge**<br>72.59<br>Germanium | 33<br>**As**<br>74.92<br>Arsenic | 34<br>**Se**<br>78.96<br>Selenium | 35<br>**Br**<br>79.90<br>Bromine | 36<br>**Kr**<br>83.80<br>Krypton |
| 37<br>**Rb**<br>85.47<br>Rubidium | 38<br>**Sr**<br>87.62<br>Strontium | 39<br>**Y**<br>88.91<br>Yttrium | 40<br>**Zr**<br>91.22<br>Zirconium | 41<br>**Nb**<br>92.91<br>Niobium | 42<br>**Mo**<br>95.94<br>Molybdenum | 43<br>**Tc**<br>(99)<br>Technetium | 44<br>**Ru**<br>101.01<br>Ruthenium | 45<br>**Rh**<br>102.91<br>Rhodium | 46<br>**Pd**<br>106.4<br>Palladium | 47<br>**Ag**<br>107.87<br>Silver | 48<br>**Cd**<br>112.41<br>Cadmium | 49<br>**In**<br>114.82<br>Indium | 50<br>**Sn**<br>118.69<br>Tin | 51<br>**Sb**<br>121.75<br>Antimony | 52<br>**Te**<br>127.60<br>Tellurium | 53<br>**I**<br>126.90<br>Iodine | 54<br>**Xe**<br>131.30<br>Xenon |
| 55<br>**Cs**<br>132.91<br>Cesium | 56<br>**Ba**<br>137.3<br>Barium | 57<br>**La**<br>138.91<br>Lanthanum | 72<br>**Hf**<br>178.49<br>Hafnium | 73<br>**Ta**<br>180.95<br>Tantalum | 74<br>**W**<br>183.85<br>Wolfram | 75<br>**Re**<br>186.2<br>Rhenium | 76<br>**Os**<br>190.2<br>Osmium | 77<br>**Ir**<br>192.2<br>Iridium | 78<br>**Pt**<br>195.09<br>Platinum | 79<br>**Au**<br>197.0<br>Gold | 80<br>**Hg**<br>200.59<br>Mercury | 81<br>**Tl**<br>204.38<br>Thallium | 82<br>**Pb**<br>207.2<br>Lead | 83<br>**Bi**<br>208.98<br>Bismuth | 84<br>**Po**<br>(210)<br>Polonium | 85<br>**At**<br>(210)<br>Astatine | 86<br>**Rn**<br>(222)<br>Radon |
| 87<br>**Fr**<br>(223)<br>Francium | 88<br>**Ra**<br>(226)<br>Radium | 89<br>**Ac**<br>(227)<br>Actinium | 104<br>**Unq**<br>(261) | 105<br>**Unp**<br>(262) | 106<br>**Unh**<br>(263) | 107<br>**Uns**<br>(262) | 108<br>**Uno**<br>(265) | 109<br>**Une**<br>(267) | | | | | | | | | |

| | | | | | | | | | | | | | |
|---|---|---|---|---|---|---|---|---|---|---|---|---|---|
| 58<br>**Ce**<br>140.13<br>Cerium | 59<br>**Pr**<br>140.92<br>Praseodymium | 60<br>**Nd**<br>144.27<br>Neodymium | 61<br>**Pm**<br>(147)<br>Promethium | 62<br>**Sm**<br>150.35<br>Samarium | 63<br>**Eu**<br>152.0<br>Europium | 64<br>**Gd**<br>157.26<br>Gadolinium | 65<br>**Tb**<br>158.93<br>Terbium | 66<br>**Dy**<br>162.51<br>Dysprosium | 67<br>**Ho**<br>164.94<br>Gadolinium | 68<br>**Er**<br>167.27<br>Erbium | 69<br>**Tm**<br>168.94<br>Thulium | 70<br>**Yb**<br>173.04<br>Ytterbium | 71<br>**Lu**<br>174.99<br>Lutetium |
| 90<br>**Th**<br>(232)<br>Thorium | 91<br>**Pa**<br>(231)<br>Protactinium | 92<br>**U**<br>238.07<br>Uranium | 93<br>**Np**<br>(237)<br>Neptunium | 94<br>**Pu**<br>(242)<br>Plutonium | 95<br>**Am**<br>(243)<br>Americium | 96<br>**Cm**<br>(247)<br>Curium | 97<br>**Bk**<br>(249)<br>Berkelium | 98<br>**Cf**<br>(251)<br>Californium | 99<br>**Es**<br>(254)<br>Einsteinium | 100<br>**Fm**<br>(253)<br>Fermium | 101<br>**Md**<br>(256)<br>Mendelevium | 102<br>**No**<br>(253)<br>Nobelium | 103<br>**Lr**<br>257<br>Lawrencium |